The Joan Palevsky  Imprint in Classical Literature

In honor of beloved Virgil—

"O degli altri poeti onore e lume . . ."

—Dante, *Inferno*

The publisher gratefully acknowledges the generous contribution to this book provided by the Classical Literature Endowment of the University of California Press Associates, which is supported by a major gift from Joan Palevsky.

Cleomedes' Lectures on Astronomy

HELLENISTIC CULTURE AND SOCIETY

General Editors: Anthony W. Bulloch, Erich S. Gruen, A. A. Long, and Andrew F. Stewart

I. *Alexander to Actium: The Historical Evolution of the Hellenistic Age*, by Peter Green
II. *Hellenism in the East: the Interaction of Greek and Non-Greek Civilizations from Syria to Central Asia after Alexander*, edited by Amélie Kuhrt and Susan Sherwin-White
III. *The Question of "Eclecticism": Studies in Later Greek Philosophy*, edited by J. M. Dillon and A. A. Long
IV. *Antigonos the One-Eyed and the Creation of the Hellenistic State*, by Richard A. Billows
V. *A History of Macedonia*, by R. Malcolm Errington, translated by Catherine Errington
VI. *Attic Letter-Cutters of 229 to 86 B.C.*, by Stephen V. Tracy
VII. *The Vanished Library: A Wonder of the Ancient World*, by Luciano Canfora
VIII. *Hellenistic Philosophy of Mind*, by Julia Annas
IX. *Hellenistic History and Culture*, edited by Peter Green
X. *The Best of the Argonauts: The Redefinition of the Epic Hero in Book One of Apollonius'* Argonautica, by James J. Clauss
XI. *Faces of Power: Alexander's Image and Hellenistic Politics*, by Andrew Stewart
XII. *Images and Ideologies: Self-definition in the Hellenistic World*, edited by A. W. Bulloch, E. S. Gruen, A. A. Long, and A. Stewart
XIII. *From Samarkhand to Sardis: A New Approach to the Seleucid Empire*, by Susan Sherwin-White and Amélie Kuhrt
XIV. *Regionalism and Change in the Economy of Independent Delos, 314-167 B.C.*, by Gary Reger
XV. *Hegemony to Empire: The Development of the Roman Imperium in the East from 148 to 62 B.C.*, by Robert Kallet-Marx
XVI. *Moral Vision in* The Histories *of Polybius*, by Arthur M. Eckstein
XVII. *The Hellenistic Settlements in Europe, the Islands, and Asia Minor*, by Getzel M. Cohen
XVIII. *Interstate Arbitrations in the Greek World, 337–90 B.C.*, by Sheila L. Ager
XIX. *Theocritus's Urban Mimes: Mobility, Gender, and Patronage*, by Joan B. Burton
XX. *Athenian Democracy in Transition: Attic Letter-Cutters of 340 to 290 B.C.*, by Stephen V. Tracy
XXI. *Pseudo-Hecataeus, "On the Jews": Legitimizing the Jewish Diaspora*, by Bezalel Bar-Kochva

XXII. *Asylia: Territorial Inviolability in the Hellenistic World*, by Kent J. Rigsby

XXIII. *The Cynics: The Cynic Movement in Antiquity and Its Legacy*, edited by R. Bracht Branham and Marie-Odile Goulet-Cazé

XXIV. *The Politics of Plunder: Aitolians and Their* Koinon *in the Early Hellenistic Era, 279–217 B.C.*, by Joseph B. Scholten

XXV. *The Argonautika*, by Apollonios Rhodios, translated, with introduction, commentary, and glossary, by Peter Green

XXVI. *Hellenistic Constructs: Essays in Culture, History, and Historiography*, edited by Paul Cartledge, Peter Garnsey, and Erich Gruen

XXVII. *Josephus's Interpretation of the Bible*, by Louis H. Feldman

XXVIII. *Poetic Garlands: Hellenistic Epigrams in Context*, by Kathryn J. Gutzwiller

XXIX. *Religion in Hellenistic Athens*, by Jon D. Mikalson

XXX. *Heritage and Hellenism: The Reinvention of Jewish Tradition*, by Erich S. Gruen

XXXI. *The Beginnings of Jewishness*, by Shaye D. Cohen

XXXII. *Thundering Zeus: The Making of Hellenistic Bactria*, by Frank L. Holt

XXXIII. *Jews in the Mediterranean Diaspora from Alexander to Trajan (323 BCE–117 CE)*, by John M. G. Barclay

XXXIV. *From Pergamon to Sperlonga: Sculpture and Context*, edited by Nancy T. de Grummond and Brunilde Sismondo Ridgway

XXXV. *Polyeideia: The Iambi of Callimachus and the Archaic Iambic Tradition*, by Benjamin Acosta-Hughes

XXXVI. *Stoic Studies*, by A. A. Long

XXXVII. *Seeing Double: Intercultural Poetics in Ptolemaic Alexandria*, by Susan A. Stephens

XXXVIII. *Athens and Macedon: Attic Letter-Cutters of 300 to 229 B.C.*, by Stephen V. Tracy

XXXIX. *Encomium of Ptolemy Philadelphus*, by Theocritus, translated, with an introduction and commentary, by Richard Hunter

XL. *The Making of Fornication: Eros, Ethics, and Political Reform in Greek Philosophy and Early Christianity*, by Kathy L. Gaca

XLI. *Cultural Politics in Polybius'* Histories, by Craige Champion

XLII. *Cleomedes' Lectures on Astronomy: A Translation of* The Heavens, with an introduction and commentary, by Alan C. Bowen and Robert B. Todd

XLIII. *History as It Should Have Been: Third Maccabees, Historical Fictions and Jewish Self-Fashioning in the Hellenistic Period*, by Sara Raup Johnson

XLIV. *Alexander the Great and the Mystery of the Elephant Medallions*, by Frank L. Holt

Cleomedes' Lectures on Astronomy

A Translation of The Heavens
With an Introduction and Commentary by

Alan C. Bowen and Robert B. Todd

UNIVERSITY OF CALIFORNIA PRESS

Berkeley   Los Angeles   London

University of California Press
Berkeley and Los Angeles, California

University of California Press, Ltd.
London, England

Library of Congress Cataloging-in-Publication Data

Cleomedes.
[Elementary theory of the heavens]
Cleomedes' lectures on astronomy : a translation of the heavens with an introduction and commentary / by Alan C. Bowen and Robert B. Todd.
p. cm.—(Hellenistic culture and society ; 42)
Includes bibliographical references and index.
ISBN 0-520-23325-5 (cloth : alk. paper)
1. Astronomy, Greek. 2. Astronomy—Early works to 1800. I. Title: Lectures on astronomy. II. Bowen, Alan C. III. Todd, Robert B. IV. Title. V. Series.

QB21.C53 2004
520'.938—dc21 2003055222

Manufactured in the United States of America
13 12 11 10 09 08 07 06 05 04
10 9 8 7 6 5 4 3 2 1

The paper used in this publication is both acid-free and totally chlorine-free (TCF). It meets the minimum requirements of ANSI/NISO Z39.48-1992 (R 1997) (*Permanence of Paper*). ♾

To David Furley, in gratitude

CONTENTS

PREFACE

The sole surviving treatise by the Stoic Cleomedes may belong chronologically to some time around A.D. 200, but philosophically it is rooted in the Hellenistic period: in the third century B.C. when Stoicism was first established, and in the first century B.C. when that school underwent a renaissance at the hands of Posidonius of Apamea. The treatise itself, a digression on astronomy and some aspects of cosmology, was prepared for pedagogical purposes as part of a larger survey of Stoicism. Had the works of the major Stoics survived, Cleomedes would be an insignificant footnote in the history of this school, marginalized as the minor Platonists of his era are by the survival of the Platonic corpus. But since the foundational works of Stoicism are lost, his treatise takes on a significance that exceeds its merits, but reflects its uniqueness. As the only work of "school" Stoicism from its period to survive intact, it fully warrants the close scrutiny that it receives in the present study.

Cleomedes maintained the doctrines of the early Stoa of Chrysippus, but, as we argue in the Introduction, was also influenced to some degree by Posidonius' important interventions in Stoic philosophy. This major thinker had in particular sought to redefine Stoic physical theory in relation to the science of astronomy, which had made such spectacular advances during his own lifetime. While Cleomedes' own account of astronomy is relatively unsophisticated, his inclusion of this science in an outline of Stoic philosophy at all is testimony to that earlier encounter between science and philosophy, and proof that the Stoa of the first two centuries of the Roman Empire did not, as is often assumed, restrict itself to moral philosophy.

We have anticipated a varied readership for this work, most of whom will not know the ancient languages. We thus offer the first English translation of Cleomedes (the sixth into all languages since the late fifteenth century), and the first based on a critical edition of the Greek text. We have addressed the interests of students of Stoic philosophy (in particular of physical theory and epistemology, the two theoretical components that dominate Cleomedes' treatise), though we have frequently referred them to the rich body of recent scholarship on Stoicism for further discussion of issues that could not be pursued in detail in relation to the Cleomedean evidence.

We have also aimed to reach students of ancient mathematical astronomy, and of the general history of astronomy, though, given the wide range of our intended readership, we have not restructured Cleomedes' text in order to correlate it closely with the technical figures appended to the translation. Cleomedes earned the right to an eponymous lunar crater primarily because he provided the fullest surviving account of the two major ancient attempts to evaluate the circumference of the Earth—projects that retain a perennial interest, and that are summarized in most introductions to astronomy. We make these texts, like the other astronomical material in the treatise, available within a complete presentation of the work from which they are often excerpted, since whatever Cleomedes has to offer contemporary readers is surely best assimilated in the full context in which it was originally composed.

Robert Todd acknowledges the considerable support of the Social Sciences and Humanities Research Council of Canada for his Cleomedean researches over several years. We both thank Henry Mendell and a second reader for the University of California Press for their helpful criticism and suggestions, as well as Stephen Menn for his comments. We are grateful to Tony Long for his generous support and assistance and indebted to Heinrich von Staden for arranging for us to use the facilities of the Institute for Advanced Study at Princeton in preparing the final version of the manuscript.

Two scholars in particular have in differing ways influenced our study. First, like all students of Posidonius, we are indebted to the monumental labors of Ian Kidd. If we take issue with some of his interpretations, we do so out of the deepest professional respect toward a scholar who has laid such solid foundations for future work. Second, we dedicate our book to David Furley. His writings have long been an inspiration to all students of ancient cosmology, and he has over several decades been to both of us a friend, teacher, and mentor.

Princeton, N.J.
Vancouver, B.C.
February 2003

ABBREVIATIONS

For other abbreviated references to both primary and secondary sources see the Bibliography and the Index Locorum.

Caelestia — R. B. Todd ed., *Cleomedis Caelestia* (*ΜΕΤΕΩΡΑ*) (Teubner: Leipzig, 1990). Since the chapters in this edition have separately numbered lines, references are tripartite, with a *Roman* numeral employed for the two books: e.g., I.5.20. The page numbers of H. Ziegler's earlier edition, *Cleomedis de motu circulari corporum caelestium libri duo* (Leipzig, 1891) are cited in the margins of Todd's edition.

OTHER WORKS

Ach. *Isag.* — Achilles, *Eisagōgē (Introductio in Aratum)*. Ed. E. Maass, *Commentariorum in Aratum Reliquiae* (Berlin, 1898), 25–75. On this author, often mistakenly called "Achilles Tatius," see Mansfeld and Runia (1997) 299–305.

DK	H. Diels and W. Kranz, *Die Fragmente der Vorsokratiker.* 6th ed. (Berlin, 1952).
EK	L. Edelstein and I. G. Kidd eds., *Posidonius: Volume I. The Fragments.* Cambridge Classical Texts and Commentaries 13, 2nd ed. (Cambridge, 1989); with I. G. Kidd, *Posidonius: Volume III. The Translation of the Fragments.* Cambridge Classical Texts and Commentaries 36 (Cambridge, 1999). Posidonius is usually cited from this collection, which includes only fragments that mention him by name.
Gem. *Isag.*	Geminus, *Eisagōgē eis ta phainomena.* Ed. G. Aujac (Paris, 1975).
Goulet	R. Goulet, *Cléomède: Théorie Élémentaire. Texte présenté, traduit et commenté* (Paris, 1980).
Kidd, *Comm.*	I. G. Kidd, *Posidonius: Volume II: The Commentary.* Cambridge Classical Texts and Commentaries 14A and 14B (Cambridge, 1988). The two volumes are paginated continuously and are cited only by page number.
LS	A. A. Long and D. N. Sedley, *The Hellenistic Philosophers.* 2 vols. (Cambridge, 1987).
Ptol. *Alm.*	Ptolemy, *Almagest.* Ed. J. L. Heiberg. 2 vols. (Leipzig, 1898, 1903).
SVF	*Stoicorum Veterum Fragmenta.* Ed. H. von Arnim. 3 vols. (Leipzig, 1903, 1905).
Theiler	W. Theiler ed., *Poseidonios: Die Fragmente. I. Texte; II. Erläuterungen* (Berlin, 1982).
Theon *Expos.*	Theon of Smyrna, *Expositio rerum mathematicarum.* Ed. E. Hiller (Leipzig, 1878).
Usener *Epicur.*	H. Usener ed., *Epicurea* (Leipzig, 1887).

INTRODUCTION

The Heavens (in Latin *Caelestia;* in Greek *Meteōra;* literally "Things in the Heavens")[1] is the only surviving work by the Stoic philosopher Cleomedes. In the absence of any external biographical information on him,[2] his *floruit* has to be inferred from the probable date of his treatise.

1. On the title of the treatise see Goulet 35 n. 1, Todd (1985), *Caelestia* ed. Todd at *Praefatio* xx–xxi and Goulet (1994). The original title, assuming that there was one, is unknown, and we depend entirely on what is preserved in the Greek manuscript tradition. The use of *Meteōra* in the present study reflects a choice made within that tradition (see Todd [1985]) over another title, favored by Goulet 35 (see Todd [1985] 259–260 and Goulet [1994] 438), *Kuklikē Theōria* ("Elementary Theory," with *kuklikos* taken in a pedagogical sense). The title used in editions of the Renaissance, *De motu circulari corporum caelestium,* and retained in H. Ziegler's edition of 1891, is a loose translation of a later synthesis in the Greek manuscript tradition of these two titles: *Kuklikē Theōria Meteōrōn* ("The theory of heavenly bodies as it pertains to motion").

2. For an introductory survey on Cleomedes see Goulet (1994). On the secondary literature see Todd (1992) and (2004). The suggestion at Neugebauer (1941) (= Neugebauer [1975] 960–961) that Cleomedes was from Lysimachia is unduly speculative given that this location is introduced in a passage that is an elaborate reductio ad absurdum of the theory that the Earth is flat; see further I.5 n. 16.

CLEOMEDES' DATE

Clearly this work cannot have been composed earlier than the time of the latest historical figure mentioned in it, Posidonius of Apamea (ca. 135–ca. 51 B.C.).[3] It is also unlikely to have been composed much after A.D. 200 on the general grounds that Cleomedes' polemics against Peripatetics (followers of Aristotle) and Epicureans are typical of debates between the Stoics and their philosophical opponents during the first and second centuries A.D., and are unparalleled after the early third century A.D., as are the pedagogical presentations of Stoic philosophy of which the *Caelestia* is clearly typical.[4] Cleomedes, that is, was a professional teacher who, as internal evidence shows, offered, presumably in the form of oral teaching, accounts of the main topics in Stoic philosophy.[5] Thus, while in the *Caelestia* he deals primarily with elementary astronomy and some aspects of cosmology, he also anticipates and assumes instruction in Stoic physical theory,[6] and cites without explana-

3. This earlier Stoic is named in connection with terrestrial zones (I.4.94, 110, 113, 115), measurements of the Earth and the Sun (I.7.2, 4, 18, 27, 48; II.1.273), refraction (II.1.54), lunar density (II.4.98), and as a general source for II.1 (I.8.162) as well as for "most" of the treatise (II.7.14).

4. On Stoic scholasticism in the first two centuries A.D. see Todd (1989), and in the *Caelestia* see especially I.1.81–111 and II.1 *passim*, and also I.1 n. 31 on parallels in the works of Alexander of Aphrodisias (*fl.* A.D. 200) to the Peripatetic arguments attacked by Cleomedes. As Sharples (1990) 111 notes, Alexander also marked the end of the Peripatetic tradition. See also Algra (2000) 172–173 for the same general view.

5. The origins of this work in oral presentations are suggested by its stereotyped and elementary argumentation, its frequent explications of terminology, and particularly by self-descriptive language such as "introduction" (*eisagōgē*, I.8.60), "lecture courses" (*skholika*, II.2.7; *skholai*, II.7.12; see II.7 n. 7), "instruction" (*didaskalia*, I.1.191), and "learning" (*katamanthanein;* e.g., I.1.114, 189; I.5.8, 30).

6. For an account of the centripetal motion of the elements as the explanation of the stability and sphericity of the cosmos designed for inclusion in a later, but now lost, exposition see I.1.94–95, 173–174, 191–192. For other central

tion doctrines from Stoic metaphysics, epistemology, semantics, and logic.[7] Far from being a "sample of scientific literature in late antiquity,"[8] his treatise is therefore best characterized as the presentation of ancillary material within a larger exposition of Stoicism.[9] A work of this kind is unlikely to have been composed much after A.D. 200.[10]

If there is any other evidence for dating the treatise more precisely between ca. 50 B.C. and, say, ca. A.D. 250, it can be found only in the as-

---

doctrines of Stoic physics see I.1.7 and 126–127 (the finitude of the cosmos); I.1.43–49 and 101–103 (the conflagration); I.1.72–73 and 98–99, and I.5.131–132 (pneuma, tension, and *hexis*); and I.1.116–119 (the elements). For the teleological view of nature see I.1.11–16 and 268–269, I.2.2–3, I.4.40–43, I.8.98–99, and II.1.396–399, and for the physical theology that makes planets deities see II.4.129–131, and II.5.5, 75, 92, 100.

7. These include the incorporeals (I.1.141–142 and II.5.100–101), the general definition of body (I.1.66–67), the criterion of truth (I.5.1–6, and II.1.2–5, 140–142), and elements of Stoic logic (I.5.20–22, I.6.2–4 and II.5.92–101; cf. I.1 n. 67). Reports of Stoic arrangements of the parts of their system are so programmatic (see Ierodiakonou [1993]) that we cannot be sure of the order in which Cleomedes introduced the earlier material noted here and in the preceding note. Cf. also II.1 n. 95.

8. Neugebauer (1975) 959.

9. Goulet 15–21 argues that the treatise may have been designed as an introduction to the astronomical poem, the *Phaenomena*, by Aratus (*ca.* 315–240/39 B.C.), and Mansfeld (1994) 197–198 notes the relevance of some ancient scholia on Aratus (cf. II.1 nn. 11, 24 and 31). Yet any such relation to Aratus can only be secondary to the role of the *Caelestia* within a wider exposition of Stoic thought, and is no more significant than Cleomedes' use of Homeric quotations (notably at II.1.470–492; cf. also I.8.60–61 [= II.6.22–23], and II.1.470–471; on this Stoic practice in general see Long [1992]).

10. In addition to the evidence in Todd (1989), the epideictic oratory in the anti-Epicurean polemic (II.1.467–524) links the *Caelestia* with the culture of the "Second Sophistic." Such evidence complements the conclusions of a study of Cleomedes' language and style by Schumacher (1975) which suggests from the evidence gathered of Atticizing an approximate *terminus post quem* of the end of the first century B.C.

tronomical data and theories that Cleomedes presents. One of the observations that he records (I.8.46–56) was used by Otto Neugebauer to assign the work to the fourth century A.D., but, in our view (see I.8 n. 16), the details of the relevant text do not justify such an inference. *Caelestia* I.4.72–89, on the other hand, has been thought to allude to the equation of time. Since this is a subtle concept that Ptolemy (*Alm.* 3.9) was the first to articulate, this passage would justify assigning the *Caelestia* a date no earlier than the middle of the second century A.D. (a *terminus post quem* consistent, as we have seen, with his philosophical culture).[11] However, the Cleomedean text in question just rehearses a point identical to that made by Geminus (first cent. B.C.) in his *Eisagōgē* 6.1–4 regarding the length of the day, with both texts providing an account of the variation in its length that is pre-Ptolemaic.[12] Thus the date of the *Caelestia* cannot be determined by any line of dependence from Ptolemy, and if it is post-Ptolemaic (as cultural considerations would allow; see n. 4 above), then this would mean that Cleomedes was either unaware of or indifferent to the *Almagest*.[13]

11. For this argument see Bowen and Goldstein (1996) 171 n. 27. Ptolemy's *floruit* was ca. A.D. 130–ca. A.D. 170. The *Almagest* itself cannot be dated precisely.

12. Both Geminus and Cleomedes define the day as the interval from one sunrise to the next, and bring into consideration the rising time of the arc which the Sun travels along the zodiacal circle during the course of one revolution of the cosmos. (For a translation of the passage from Geminus see I.4, n. 16.) Ptolemy, by contrast, defines the day as the interval from one meridian crossing by the Sun to the next, and focuses on the time it takes the arc that the Sun travels during a day to cross the meridian circle. In effect, Ptolemy isolates the trigonometric contribution of the Sun's motion to the variation in the length of the day (that is, to the equation of time) by subtracting the effect of latitude. So *Caelestia* 1.4.72–89, far from showing Cleomedes' reliance on Ptolemy, presents an account of the length of the day that Ptolemy clarified and made precise.

13. See Algra (2000) 168 with n. 16 for this general conclusion.

CLEOMEDES AND POSIDONIUS
Physics and Astronomy

The *Caelestia* was read almost exclusively as an astronomical handbook until around 1900,[14] when German classical philologists began to define the historical context of the treatise through *Quellenforschung*, the procedure of "source hunting" by which the ideas in lost works were reconstructed from purportedly later residues.[15] Since Cleomedes admitted using material from Posidonius (n. 3 above), his work was sometimes seen simply as a repository of Posidonian doctrines and treatises, or, in more modified claims, as an amalgam of earlier Stoic literature and Posidonian components.[16] But since almost every Cleomedean sentence could be linked with some Stoic or Posidonian fragment, as well as with passages in manuals of elementary astronomy,[17] interpretations based on the parallels gathered from such sources inevitably left its author identified generically as a post-Posidonian Stoic scholastic with an interest in astronomy. They did not succeed in placing him in any specific historical and intellectual context.

In the early 1920s Karl Reinhardt adopted a more promising approach.[18] He identified what he called the "inner form" of Cleomedes'

14. See Todd (1992) 2–5. Neugebauer (1975) 959–965 continues this tradition. The astronomy in the *Caelestia* did, however, ensure its survival in late antiquity when most other Stoic scholastic material was lost.

15. On this older literature see Todd (2004).

16. Cleomedes' Stoicism generally conforms to the body of doctrine associated with this school's major figure in the Hellenistic period, Chrysippus (280/76–208/4 B.C.); see Algra (1995) 268–270, and the survey at Goulet 9–11. Apart from the theory of lunar illumination in II.4 (see II.4, nn. 1, 8 and 19), Posidonius is not identified as the exclusive source of any of the fundamental theories mentioned in the *Caelestia*, although, as we argue below, he exercised a strong influence on its general principles and methodology. Cf. also n. 52 below.

17. See the parallels and *similia* in Todd's edition of the *Caelestia*, and at Goulet 11–15.

18. See Reinhardt (1921) 183–207, and the recapitulations at Reinhardt (1926) 124, with n. 2, and (1953) 685–686. For some discussion of his views see Todd (2004).

treatise by drawing on a Posidonian fragment (F18EK; translated and discussed in our Appendix), which, along with some associated texts,[19] posits a hierarchical relation between the two major components of the *Caelestia*, physical theory and astronomy.[20]

Physical theory, according to this fragment, deals with matter, causal relations, and teleological explanation, while astronomy is defined as an activity that uses geometry and mathematics to analyze the shape, size, motions, and interactions of the principal heavenly bodies.[21] The two disciplines might address identical topics (for example, the sphericity of the Earth, the size of the Sun, or solar and lunar eclipses), but will conduct their demonstrations in systematically different ways. For while astronomy will be based on observations, it will acknowledge physical theory as foundational. Thus in the all-important explanation of the motion of the heavenly bodies, physical theory supplies the "first principles" *(arkhai)* that astronomy has to adopt and follow.[22]

The distinction, and hierarchical relation, posited between physics and astronomy in F18EK is, as Reinhardt saw, respected throughout the *Caelestia*, where physical theory is acknowledged as defining the cosmology presupposed in astronomical observations and calculations. The treatise in fact opens (I.1.20–149) with a lengthy demonstration that the cosmos

19. On this material see I. G. Kidd (1978a) and *Comm.* 134–136. The texts in question are Diog. Laert. 7.132–133 (= F254 Theiler), Sen. *Ep.* 88.25–28 (F90EK; F447 Theiler), and Strabo 1.1.20 and 2.5.2 (F3b and 3c Theiler). Cf. also Theon *Expos.* 166.4–10 and 188.15–24 (on which see Appendix n. 30) and 199.9–200.12.

20. The reader may now wish to consult the Appendix. The relevance of F18 to the *Caelestia* was also noted at Todd (1989) 1368–1369, and has recently been briefly discussed by Algra (2000) 175–177.

21. Posid. F18.5–18 EK; cf. Sen. *Ep.* 88.26.

22. F18.46–50 EK; see Appendix n. 8, and also Appendix n. 2 for Geminus' concern with the same issue. For the same autonomy of physical theory see Strabo 2.5.2, and Sen. *Ep.* 88.28. F18.46–50 EK and Strabo 2.5.2 also refer to the theory of planetary motion in eccentric circuits found at *Caelestia* II.5.102–141.

(the realm of astronomical observation) is a finite material continuum surrounded by unlimited void. In line with F18EK, its analysis and arguments focus on matter, causal relations, and teleology. Also in that opening chapter, the stability and sphericity of the cosmos are said to be dependent on the centripetal motion of the elements, and while this particular physical theory goes undemonstrated (cf. n. 6 above), its foundational status is obvious, and it is later cited as the basis for the Earth's stability at the center of the cosmos (I.6.41–43).[23] Then at I.5.126–138 the properties of the lighter elements, air and aether, are identified as the cause of the sphericity of the cosmos, while elsewhere the differing density of the elements serves to demonstrate both how the Earth and the heavenly bodies are organically sustained by a mutual interchange of matter (I.8.79–95), and why different heavenly bodies move at different velocities (II.1.334–338). Finally, the lunar theory of II.4–6 consists of an opening chapter on the physical theory of lunar illumination as a prelude to an analysis of lunar phases and eclipses. In all these cases physical causality, as Posidonius claims in general terms in F18EK, serves to define the cosmological structure within which astronomical observations are made and utilized.[24] But the most explicit, and philosophically most interesting, links between the Posidonian program in F18EK and the *Caelestia* are to be found not in physical theory so much as in epistemology and methodology.

23. In fact, this physical assumption determines the way that observations are described throughout the treatise. Without it, the same phenomena (such as the risings and settings of the Sun, as well as the lengths of daytimes and nighttimes) that in I.5–6 are said to entail the sphericity and stable centrality of the Earth could be explained by the rotation of the Earth and the stability of the Sun, as Posidonius recognized in his critique of Heraclides of Pontus at F18.39–42 EK (see Appendix and cf. I.6 n. 14).

24. At the same time, physical causality is exemplified at *Caelestia* I.2–4 through the motion of the Sun, and the diurnal and seasonal changes that it causes. See further Appendix n. 23.

Epistemology and Scientific Method

Physical theory, F18EK argues (lines 30–32 and 42–46), cannot be directly established from "the phenomena" (that is, from the evidence of primarily celestial observations), since, in a famous phrase, these can equally well be "saved" by other "hypotheses." This undesirable possibility is exemplified (F18.39–42 EK) by the cosmology of a mobile Earth and immobile Sun that had been hypothesized on the basis of the phenomena by Heraclides of Pontus (a pupil of Plato). Posidonius, to borrow language from W. V. O. Quine,[25] can thus be said to have regarded the phenomena as "underdetermining" the theory of the basic structure of the cosmos in that they can support diverse and even conflicting accounts. For Posidonius, "hypothesis" is thus a derogatory term for any theory that, while in principle foundational to cosmology, is mistakenly formulated only on the basis of observations. Posidonius' aim was to eliminate hypotheses of this type entirely from scientific reasoning.[26]

Cleomedes' treatment of these central themes in the philosophy of science is, by contrast, indirect and unreflective. He never overtly addresses the status and limitations of astronomy as an observational science, apart from acknowledging at one place (I.1.172–175) that a spherical and geostatic cosmology cannot be established conclusively "from the phenomena" but only from the theory of the centripetal motion of the elements.[27] He is, however, concerned with the way that observations of the heavens, if used uncritically, can lead to false theories, and in one place (I.5.1–6) articulates this concern with explicit reference to the Stoic concept of the criterion of truth.

In Stoicism this criterion is defined as the "cognitive presentation"

25. See Quine (1975).

26. On the special role of F18.32–39 EK in establishing this position see the Appendix 195–199.

27. Cf. n. 23 above, and for the significance of density for the Earth's sphericity and stability see, respectively, I.5.6–7 and I.6.41–43 (with I.6 n. 14), brief references in chapters that otherwise present arguments based on observations.

(phantasia katalēptikē), which, whether perceptual or based on reasoning,[28] guarantees its own veridicality, or, as it is sometimes described, is "self-certifiable." In effect, it posits a fit between human reason and the content with which it was presented—one that is proof against misleading (or noncognitive) presentations.[29] At an earlier point in his exposition of Stoic philosophy Cleomedes must have introduced this doctrine, probably in an account of the logical theory of which he also presupposes knowledge on the part of his hearers.[30] However, he does not draw close links between general Stoic epistemology and the numerous arguments by which he demonstrates the presuppositions of spherical astronomy (I.5–6 and 8), estimates and calculates the sizes of heavenly bodies (II.1–3), and establishes the central features of lunar theory (II.4–6). But there is sufficient evidence to consider the influence that the general epistemological principles of his philosophical system exercised on this specialized treatise.

As far as his use of the key term *kritērion* is concerned, it serves in four places (I.5.2; II.1.4, 103, 141) to strengthen an objection to the use of sense perceptions to draw immediate and uncritical inferences. Thus in the first instance "sight itself" *(autē hē opsis)* is said to be an invalid criterion on the principle that "it is not the case that everything in fact usually appears to us as it [really] is" (I.5.2–3). So even where sight correctly "suggests," for example, the sphericity of the cosmos (I.5.2), it is still not the criterion of its shape, and obviously not so when someone incorrectly "follows"[31] sight

28. For the distinction between "perceptual" *(aisthētikai)* and "rational" *(logikai)* presentations see Diog. Laert. 7.51 (at *SVF* 2.61). The second category must imply that the presentations are both non-perceptual and based on reasoning.

29. See LS 1.250–251 for this widely accepted interpretation, the use of "self-certifiability" of this presentation, and the coordination between human reasoning and the input it receives.

30. See I.5.1–4 with I.5 n. 3, and II.1.408 with II.1 n. 95.

31. The compound verb for "follow" *(katakolouthein)* is specifically used at I.5.12 and II.1.4 and 140 to describe this mental error of uncritically accepting the apparent properties of an observation. The simple form *akolouthein* is used, at least in this treatise, to describe inferences drawn from observations or theoretical propositions (e.g., I.1.106; cf. *akolouthia* in the present context at I.5.5).

to infer that the Earth is flat (because the horizon appears as a plane) (I.5.11–13), or to claim, as Cleomedes alleges the Epicureans did, that the Sun is the minuscule size it appears to be (II.1.2–5; cf. 102–3 and 140–142). The lengthy polemic over the size of the Sun in II.1 probably reflects a general attack on the Epicurean claim that all appearances are true, which, however charitably interpreted (cf. II.1 n. 3), left its proponents open to the charge of treating celestial observations uncritically. But in the *Caelestia* it was subsumed under a wider program of rejecting "mere sight" as a criterion simply because distances and distorting factors made *all* astronomical observations misleading if taken at face value.[32]

So "sight itself" must be replaced (we learn at I.5.3–6) by demonstrations offered *(a)* "on the basis of what is 'clearer' (that is, of what is presented cognitively *(katalēptikos)* to us" and *(b)* "in accordance with what is patently implied." Now *(a)* points directly to the Stoic criterion of truth, the "cognitive presentation" (although this is the only place where the key term *katalēptikōs* appears in this treatise), while *(b)* defines, at least, programmatically a system of inference (*akolouthia*; cf. n. 30) distinct from, and superior to, an immediate and uncritical response to "sight itself." The argumentation identified in *(b)*, though described somewhat vaguely as "patent" *(phainomenē)*, indicates that conclusions about "things that are not in and of themselves fully displayed" (*ta mē autothen ekphanē*; I.5.4–5) are necessitated just because they are reached through cognitive presentations that are either perceptual or based on reasoning.[33]

Now unfortunately this morsel of evidence, which is offered as a brief prelude to the arguments for the sphericity of the Earth in I.5, represents the sum total of Cleomedes' epistemological and methodological

32. Note II.1.142–143 where he is passionate about "taking to heart" *(enthumeisthai)* the significant damage usually caused when we are misled by superficial sense presentations.

33. See I.5 n. 4 on the context of *(b)*, which accords with Diog. Laert. 7.45 (*SVF* 2.35) where a demonstration is defined as an argument that reaches a conclusion that is relatively less accessible (the term is *katalambanomenon*, from the same root as *katalēptikos*, "cognitive") than its premises.

reflection. But it does offer a basis for judging how far the other arguments in the *Caelestia* for astronomical and cosmological theses extend and elaborate Stoic epistemology, and, in particular, how the Stoic criterion should be interpreted in this context of constructive argumentation.

The Criterion and Demonstrative Procedures

Cleomedes calls his various proofs and counter-arguments "demonstrations" *(apodeixeis)* and "procedures" *(ephodoi)*.[34] The terms are virtually equivalent,[35] though *ephodos* may be a technical term that postdates the early Stoa. Aristotle uses it to describe a process of systematic reasoning,[36] and, along with cognate verbal forms *(ephodeuesthai, ephodiazesthai)*, it carries this general sense in various authors of the first and second centuries A.D.[37] The hallmark of the "procedures" in the *Caelestia* is the presence, whether explicit or implicit, of an independently identifiable truth or principle that is the foundation, and so, in effect, the axiom, of the argument.[38] Arguments themselves employ both observational and non-observational, or theoretical, premises, and, in keeping with the pro-

34. The underlying metaphor (elsewhere the literal sense) is that of being supplied for a journey (cf. 202 n. 30). In its logical use the term therefore describes how a journey from premises to a conclusion starts out and is equipped. Cf. the use of the verb *proienai* ("go forward") at I.5.25 (cf. II.1.226), and *hormasthai* ("start out") at I.5.105, I.6.3, II.1.275 and II.2.6.

35. At I.7.69 an *ephodos* is said to "demonstrate" *(apodeiknunein)* a conclusion.

36. See Arist. *Top.* 105a13–14 where it refers to an "induction" *(epagōgē)* from particulars to universals.

37. See 202 n. 30 for one such example that may be linked to Posidonius. Among authors of the second century A.D., Ptolemy and Sextus Empiricus, for example, occasionally employ this language.

38. For *ephodoi* explicitly identified see I.5.20–29, I.6.1–8, I.7 *passim*, II.1.156, 225–333. In several cases the underlying structure of the reasoning is implicit; see I.5 n. 10, I.7 nn. 9 and 21, II.1 nn. 50, 58, and 73, II.2 n. 9, II.3 n. 11, II.4 nn. 31 and 32, II.5 n. 8, and II.6 n. 7. Cleomedes uses the term *methodos* for procedures once, at II.1.343.

grammatic rejection of "sight itself" as a criterion, lead to conclusions that (as the general Stoic definition of a demonstration prescribes) go beyond what is directly accessible by sense perception.[39]

Procedures vary considerably in their complexity and in the detail with which they are articulated. Some involve direct inferences from the phenomena,[40] others implicit axioms.[41] The exclusionary proofs for the Earth's sphericity and centrality depend on overtly identified principles from Stoic logic (I.5.24–25; I.6.2–4), while others that deal with solar and lunar eclipses rely on optical principles (II.4.119–121; II.5.51–54).[42] The elaborate calculations of the size of the Earth and Sun in I.7 and II.1 ultimately depend on implicit high-level axioms, and also need a greater

39. See n. 33 above. At I.7.5 Cleomedes uses *apodeixis* ("demonstration") with particular reference to the conclusions of procedures that he is about to describe. Arguments can, of course, proceed directly to their conclusions where the phenomena are reliable.

40. See I.1.68–80 (the absence of intracosmic void), I.1.172–191 (the sphericity of the cosmos), I.5.104–113 (the sphericity of the Earth, explicitly contrasted with a formal procedure at I.5.102–105), II.1.225–239 (the Sun's being of significant size; again a contrast with a more elaborate *ephodos* at II.1.240–244), II.2.4–18 (the Sun's being larger than the Earth), II.3.84–99 (the Moon's being the closest planet to the Earth), and II.6.35–78 (the Moon's being eclipsed by falling into the Earth's shadow). But in most of these cases more rigorous demonstrations are, or can be, supplied: e.g., from physical theory in the case of the void and the sphericity of the cosmos (cf. n. 6 above), or in the form of elaborate *ephodoi* in the case of the size of the Sun (at II.1.269–286 and 287–333).

41. Thus the arguments at II.1.144–268 that prove that the Sun is larger than it appears are provisional *ephodoi*, with the underlying axioms omitted as self-evident (see II.1 n. 38). They may have been used as a way of easing students into more elaborate procedures for estimating the size of the Sun (II.1.269–333). At II.1.269, the phrase "the following kind of procedure" *(hē toiautē ephodos)* suggests a deliberate contrast with the more informal kind of demonstration that has preceded.

42. In the *ephodos* at II.4.118–126 that demonstrates the cause of a solar eclipse the axiomatic optical principle (see II.4 nn. 31 and 32) is directly applied to celestial observations (119–121) before being generalized (121–22), rationalized in terms of familiar observations (123–124), and further clarified (124–126).

number of supplementary "assumptions" *(hupotheses)* in the form of definitions, and of numerical and geographical data.[43] In some cases reasoning is even facilitated by purely stipulative hypotheses.[44] The detailed analysis of these arguments can be left to the commentary. Here we shall consider only their relevance to Stoic epistemology, and indicate their probable origins.

Our thesis is that the concept of the criterion of truth is extended in the *Caelestia* from the domain of cognitive, or self-certifiable, presentation to structures of argument within which such presentations are included as premises. Good evidence for this development is that in II.1 Cleomedes winds up a litany of the consequences resulting from uncritically "following" visual sense presentations (as the Epicureans allegedly do when arguing for the minuscule size of the Sun) by saying (II.1.141–142) that "for [bodies] of such size some other criterion must be established." By "criterion" here he cannot mean another *type* of sense presentation (since, as we have seen, he does not regard any visual sense presentation of the heavens as a plausible criterion), but *another way of using the same observational evidence.* This implication is borne out when he immediately introduces a series of procedures in which observations are used, in conjunction with other premises, to reach at first estimates (II.1.144–268), and later specific values (II.1.269–352) for the size of the Sun.

More indirect, but equally compelling, evidence is found in I.7 and II.1, where Cleomedes presents two pairs of procedures that conclude with numerical values (for the sizes of the Earth and Sun respectively), rather than, as is usually the case, demonstrate qualitative or causal the-

43. See I.7 nn. 10 and 20, and II.1 nn. 58 and 73. Bowen (2002) discusses further the two procedures used in I.7.

44. On hypotheses see I.7 nn. 4 and 11, and II.1 n. 57. Hypotheses, or, on occasion, thought experiments, furnish the observational premises used within *ephodoi* at II.1.144–154, 166–170, 255–256 and 304–305, and at II.1.155–166 an observation is derived from a historical report. Whatever its form, an observational statement of some form is always a premise in an *ephodos.*

ses.[45] In I.7 he describes one of them (that of Posidonius) as "less complicated" (I.7.4), and favors the other (that of Eratosthenes) to the extent that he dismisses the view that it is "somewhat more obscure" (I.7.49–50). Now since both Eratosthenes' and Posidonius' procedures depend on *valid* and similar axiomatic foundations (see n. 50), if one of them can be considered superior to the other (as implied by Cleomedes' use of Eratosthenes' value elsewhere, at I.5.272 and II.1.294–295), then, in effect, a criterion of *truth* is applied to their arguments. That seems to be the theoretical background to a qualified scepticism conveyed (at I.7.46–47) about a crucial premise (a terrestrial distance) in Posidonius' calculation.[46]

A similar concern with the truth of the premises in a procedure is evident in Cleomedes' remark (II.1.286) that one of the two calculations he records of the size of the Sun "is considered to carry a greater degree of cognitive reliability" *(enargesterou tinos mallon ekhesthai)* than the first (that of Posidonius) because the latter contains as one of its premises an arbitrary stipulative hypothesis (cf. II.1.282–285)—a defect that makes it "*less* cognitively reliable." Here we can see an extension of the evaluative term *enargēs* (which we have here translated "cognitively reliable," as at II.1.114 and II.6.195) from its application to credible presentations (cf. I.5.4, where it is used interchangeably with *katalēptikos*, the usual term for "cognitive") to the role of characterizing the probative value of a whole argument. Yet the link with general Stoic epistemology can be maintained,

45. It would be implausible for a Stoic to entertain competing demonstrations of theses such as the Earth's sphericity or its centrality in the cosmos, or of the causes of lunar and solar eclipses. As we have seen, such demonstrations can be offered both from observations and from physical theory, though the latter type is considered superior and authoritative (see n. 40). Moreover, within either category of qualitative demonstration, there can be only one way of reasoning correctly. In quantitative demonstrations, or calculations, by contrast, there may be different ways of determining measurements, and an inferior procedure can be rejected without compromising any fundamental physical theory.

46. See I.7 nn. 4 and 11, and Bowen (2002).

since individual premises, whether observational or based on reasoning, must still be "cognitive"[47] if *collectively* they are to entail a true conclusion. But, as we have seen, when doubt is raised about the quality of a given premise, one procedure may be considered superior (namely, closer to the truth) relative to another.

And so, despite the paucity of direct evidence, and the related need for caution in speculative reconstruction, we conclude that the Stoic criterion is adapted in the *Caelestia* to a program of establishing knowledge of astronomical and cosmological matters. This is what we would expect of a work that is a part of a comprehensive and ongoing survey of Stoicism (see nn. 6 and 7), though, as we shall argue next, it also reflects earlier work by Cleomedes' primary source, Posidonius.

Posidonius' Legacy

There are indirect links with what we know of Posidonius that make him a likely source of the procedures found in the *Caelestia*.[48] First, as we have seen, they all reflect Posidonius' general prescription for astronomy in F18EK, in that they presuppose an independently established cosmic structure, and accommodate observations within it, often by using arithmetic and geometry in the manner prescribed for astronomy at F18.15–16 EK. This firm theoretical context guarantees that even if (as in calculations of the sizes of the Earth or heavenly bodies) the procedures may not necessarily yield the truth, they are at least not in conflict with phys-

47. Such premises can be called "presentations" *(phantasiai)*, since in general Stoic epistemology these can be "perceptual" or "rational" *(logikai)*; see n. 28 above.

48. Diog. Laert. 7.54 (*SVF* 1.631; F42EK; LS sect. 40A) reports that Posidonius identified "right reasoning" *(orthos logos)* as a criterion of truth. I. G. Kidd (1978b) 275–276 and 282, and, more fully at (1989), especially 148, links this report with the philosophy of science in F18EK. It could equally well serve as a programmatic rationale for *ephodoi*. LS 2.243, however, legitimately question the reliability of Diogenes' doxography.

ical first principles. Second, since Posidonius had a general interest in logic and the foundations of mathematics,[49] we can assume that it extended to the demonstrations in physical theory and astronomy that he mentions in F18EK. Certainly, at F18.21 EK *hodos*, the root term in *ephodos*, refers to the inferential procedures that are used to demonstrate two theses that are also demonstrated in the *Caelestia:* the sphericity of the Earth (I.5) and the magnitude of the Sun (II.1).

Finally, on a point of detail, in the two pairs of calculations of the size of the Earth and Sun at, respectively, *Caelestia* I.7 and II.1, one member of each pair is attributed to Posidonius (I.7.8–47; II.1.269–285), and all four are founded on the implicit principle that, when two quantities are in a ratio, equimultiples of these quantities taken in corresponding order are in the same ratio. Since, as we learn from Galen, Posidonius studied relational syllogisms that rest on the particularizations of ratios among pairs of quanta in proportion, it is plausible that all of Cleomedes' reports in these particular texts are ultimately derived from Posidonius.[50] This is quite consistent with Cleomedes' preferring, as we have seen, the non-Posidonian alternative in these paired calculations, since Posidonius might have reported other philosophers' views as embodying the axiomatic principle in question.

In conclusion, if *any* source is to be assigned to the conjunction of rigorous reasoning, observations, and physical theory that is so pervasive in the *Caelestia*, the only possible candidate is Posidonius, even if Cleomedean demonstrative procedures are not regarded as Posidonian in every detail.[51] What we can identify is a general Posidonian provenance. As F18EK shows, Posidonius was the only major earlier Stoic who was engaged with the science of astronomy, and who took Stoic epis-

49. See I. G. Kidd (1978b).

50. See Galen *Inst. log.* 18 (F191EK), I.7 nn. 9 and 21, and II.1 nn. 58 and 73. Cf. also I.5 n. 22 and II.3 n. 11.

51. See especially I. 7 nn. 9 and 11 below.

temology into the realms of the philosophy of science. Cleomedes is unquestionably maintaining that general innovation, even if in a less sophisticated and self-conscious manner. *Quellenforschung* may well be a discredited methodology, but in the present case it allows us to identify Cleomedes' *Caelestia* as a remote tribute by a minor Stoic to the ideas of a major predecessor.[52] In fact, without Posidonius, astronomy would probably never have been included in Cleomedes' program of Stoic teaching.

TEXT AND TRANSLATION

Our translation is based on the text in Todd's edition, except for a few changes.[53] Since Cleomedes' scholastic prose is often elliptical (probably reflecting its origins in oral teaching; see n. 5 above), we have introduced a number of supplements, indicated by square brackets. (Angle brackets identify supplements introduced by emendation into the Greek text.) In matters of terminology we have tried to be consistent, but in some cases have had to be flexible.

We have used Arabic numerals for all whole numbers (cardinal and ordinal) above nine, as well as for all numbers (whether whole or fractional) in passages involving calculations or the presentation of measurements and quantified data. In other passages we have followed standard usage and written out common fractions except where they are accompanied by a whole number. For consistency throughout, however,

52. Cleomedean material can therefore be identified as Posidonian in less restrictive terms than are adopted in EK, where a "fragment" has to include Posidonius' name (see most recently I. G. Kidd [1997]). But specific cases must always be judged on their merits; see I.4 n. 33, I.5 n. 22, I.6 n. 10, I.8 n. 39, and II.4 nn. 8 and 19. W. Theiler's collection of Posidonian evidence from Cleomedes is based on Karl Reinhardt's general approach (cf. n. 18 above), and so includes more extensive quotations from the *Caelestia*.

53. See I.4 nn. 8 and 42, II.4 n. 3 and II.6 n. 27.

we have used '1 foot wide' for the term *podiaios* in all cases in which it is applied to the heavenly bodies.

Finally, we have used transliteration to make the original terminology as accessible as possible, and Greek where textual matters are involved, as well as where some nouns appear in oblique cases and verbs in forms other than the infinitive.

Book One of Cleomedes' The Heavens

OUTLINE*

Book One

Chapter 1. The cosmos is a finite and stable structure surrounded by infinite void; the Earth and the heavens are homocentric spherical bodies with corresponding zones of latitude.

Chapter 2. The fixed stars are different from the planets which move within the zodiacal band in a direction opposite to the daily rotation.

Chapter 3. The latitudes of the Earth have differing seasons caused by the motion of the Sun in the zodiacal circle.

Chapter 4. The lengths of daytimes and nighttimes differ at differing latitudes because of the motion of the Sun in the zodiacal circle. (*Digression:* The torrid zone of the Earth is uninhabitable.)

Chapter 5. The shape of the Earth is spherical.

Chapter 6. The Earth is at the center of the cosmos.

Chapter 7. *Digression:* Posidonius' and Eratosthenes' measurements of the circumference of the Earth.

Chapter 8. The size of the Earth is discountable in observations of all heavenly bodies except the Moon.

* The outlines here and at the beginning of Book Two (page 98) are included for the sake of the reader and do not belong to the translation proper. The division of the *Caelestia* into books reflects its original structure as two lecture courses (see II.7 n. 7), but the division into chapters dates only from the editions of the Renaissance, and, while, generally logical, and supported by marginalia in manuscript sources, it has been revised for I.1–4; see Todd ed. *Caelestia Praef.* xx, and also I.1 n. 58, I.2 n. 15, and I.4 n. 1. Cleomedes himself refers to *logoi* ("discussions," perhaps single lectures), which in some cases correspond to our existing chapters; see II.2.29–30 (identifying II.6 as part of II.4–6), II.4.136–137 (identifying II.5), and II.5.150 (identifying II.6).

CHAPTER ONE[1]

3: "Cosmos" is used in many senses, but our present discussion[2] concerns it with reference to its final arrangement,[3] which is defined as follows: a cosmos is a construct formed from the heavens, the Earth, and the natural substances within them.[4] This [cosmos] encompasses all bodies, since, as is demonstrated elsewhere, there is, without qualification, no body existing outside the cosmos.[5] Yet the cosmos is not unlimited, but is limited, as is clear from its being administered throughout by Nature. For it is impossible for Nature to belong to anything unlimited, since Nature must control what it belongs to.

1. On the title of the treatise see Introduction n. 1 above.

2. I.e., the present chapter, which is a preliminary overview of the structure of the cosmos; see further n. 58 below.

3. *diakosmēsis;* cf. *SVF* 2.526–527 and 2.558 where it identifies the distribution of the elements in a fully established cosmos. Cf. also n. 6 below.

4. This is a standard definition; e.g., *SVF* 2.638 (192.35–36), Posid. F14EK, Ps.-Arist. *De mundo* 391b9–10, and *Aratea* 127.14–15.

5. Cleomedes must already have demonstrated that the cosmos is a finite continuum; cf. also lines 126–127 below. At lines 69–73 below, and at I.5.128–134 he alludes to the conditions that maintain this continuum.

11: And that the cosmos has Nature as that which administers it is evident from the following: the ordering of the parts within it;[6] the orderly succession of what comes into existence;[7] the sympathy of the parts in it for one another;[8] the fact that all individual entities are created in relation to something else; and, finally, the fact that everything in the cosmos renders very beneficial services.[9] (These are also properties of individual natural substances.) So since the cosmos has Nature administering <it> throughout, it is itself necessarily limited, whereas what is outside it is a void that extends without limit in every direction. Of this [void] the [part] that is occupied by body is called "place," while that which is not occupied will be void.[10]

20: We shall now briefly summarize [the argument] that there is a void: Every body is necessarily present in something; but the thing that a body is present in, given that it is incorporeal and as such without physical contact,[11] must be distinct from what occupies and fills it; we therefore speak

6. These "parts" are elements (cf. I.4.244 and I.5.8) in an "ordering" (*taxis*; cf. *diataxis* at II.1.399) in the cosmos; see lines 116–119 below (with n. 43), and I.5.126–137.

7. This is elsewhere referred to as the "continued stability" *(diamonē)* or "preservation" *(sōtēria)* of the cosmos; see I.2.2–3, I.8.98–99, and II.1.399.

8. Sympathy in Stoic cosmobiology is a psychophysical interaction between bodies; that is, sympathy is not a metaphorical concept but defines the physical relation between living things.

9. "Provides" *(parekhesthai)* carries an inherently teleological sense, particularly with reference to the Sun's power (e.g., I.3.86, 91, 95, 97, 104; I.4.15; II.1.362 and 371); cf. I.3 n. 17.

10. On the complementary Stoic theories of place, space and void (the most important evidence for which is at *SVF* 2.503–506) see Algra (1995), especially ch. 6 (with Cleomedes discussed at 268–270). On Cleomedes on the void see Todd (1982) and Inwood (1991) 257–259. Inwood tries to draw a distinction between Cleomedes and the earlier Chrysippean theory, but Algra (1995) 269 argues convincingly that Cleomedes is rendering orthodox Stoicism in defining place as that which is occupied by body.

11. "Has no physical contact" *(anaphēs)*; cf. Epicur. *Hdt.* 40 and *Pyth.* 86 for the same term used to define incorporeality.

of such a state of subsistence[12]—namely, a capacity to receive body and be occupied—as void.[13]

25: That bodies are present in such a thing can be seen primarily in the case of liquids (i.e., all liquid substance).[14] *(a)* When, for example, we extract the solid from a vessel that contains liquid and some solid body, the liquid converges on the place of the object that has been extracted, and is no longer seen at the same level, but is reduced by an amount equal to the size of the object that was extracted. *(b)* Conversely, if a solid is placed into a vessel full of liquid, the amount of liquid that overflows is equal to the volume of the solid imported, and that would not happen unless the liquid had been present in something that had been filled by it and was capable of being occupied by body.[15] *(c)* In the case of air, too, the same thing must be understood to occur. In fact, air is forced out of the place it occupies whenever a solid occupies that place: when, for example, we pour anything into a vessel,[16] we in return

12. "State of subsistence" *(hupostasis)* is more specific than "existing"; see LS I.164. It is therefore appropriately used of the void, which is "something" (line 57 with n. 22), although conceived of as incorporeal (lines 65–66). Its subsistence comprises positive properties that constitute the way in which it "exists"; see line 68 below, where the generic verb *huparkhein* is used. Elsewhere, the verb related to *hupostasis*, *huphistasthai*, is used both of corporeal existence (II.1.338, 364 and 402), and of a geometrical abstraction (I.3.33).

13. To call void "occupied" here (as at line 18 above) is to treat it as a general concept of space, rather than as the capacity to be occupied by body, its formal definition. See Algra (1995) 69–70, and 269 n. 26.

14. Because of this reference to any type of liquid, we have translated *hudōr* in the passage that follows (as at lines 74–78 below) as "liquid" rather than "water."

15. *(a)* and *(b)* recall the standard argument for place as the "extension" *(diastēma)* between the limits of a container; see Arist. *Phys.* 211b14–17. This commonsense theory, which Aristotle rejected, was widely discussed by his commentators. Philoponus (e.g., *In phys.* 582.19–583.12) essentially adopted the Stoic concept of place as occupied void, although he rejected extracosmic void.

16. Something "solid" can be poured, inasmuch as water is "somewhat compact" (see II.4.36); it is unlikely (cf. n. 18 below) that pliable solids (e.g., salt) are also envisaged.

perceive the air[17] inside it escaping, and especially when the aperture is narrow.[18]

39: We can also conceive of the cosmos itself moving from the place that it currently happens to occupy, and together with this displacement of it we shall also at the same time conceive of the place abandoned by the cosmos as void, and the place into which it is transferred as taken over and occupied by it. The latter [place] must be filled void. If, according to the doctrine of the most accomplished natural philosophers, the whole substance [of the cosmos] is also reduced to fire, it must occupy an immensely larger place, as do solid bodies that are vaporized into fumes.[19] Therefore the place occupied in the conflagration by the substance [of the cosmos] when it expands is currently void, since no body fills it.[20] But if anyone claims that a conflagration does not occur, such a claim does not confute the existence of the void. For even if we merely conceived of the substance [of the cosmos] expanding, that is, being further extended (granted that there is no possible obstacle to such extension), then this very thing into which it would be conceived as entering in its extension would be void, just as of course what it also currently occupies is filled void.

55: So those who claim that there is nothing outside the cosmos are

17. "Air" here translates *pneuma*, which is clearly being used in a non-technical sense; on its technical sense see I.5 nn. 34–35.

18. The vessel involved here may be a clepsydra, which had a narrow aperture at the top and small perforations at the bottom. A liquid in which it was immersed entered through the perforations until the aperture was plugged, and air escaped at the top as this process occurred. Its exit could thus be "perceived in return" (the verb is *anti-lambanesthai*) by touch, and perhaps aurally, if it made, say, a whistling sound.

19. Cf. I.8.83–90 for this principle applied to the Earth.

20. Algra (1995) 321–336 makes the case for Posidonius having believed that the extracosmic void was only large enough to accommodate the expansion of matter caused by the conflagration.

talking nonsense.[21] The very thing that they term "nothing" obviously cannot stand as an impediment to the substance [of the cosmos] as it expands. As a result, when the substance expands, it will occupy something,[22] and what is on each occasion occupied in a natural [process] will be filled by the object that occupies it, and will become its place, which is void that is occupied (i.e., filled) by body. This [filled void][23] will duly become void when the substance [of the cosmos] is again compressed (that is, contracted into a smaller volume).

62: Now just as there is that which has received body, so also there is that which is capable of receiving body; the latter, which can both be filled and abandoned by body, is void. Now it is necessary that the void possess a state of subsistence. But our way of conceiving void is entirely without qualification, since void is incorporeal and without physical contact, since it neither possesses a shape nor has one imposed on it, and is neither acted on in any way nor acts,[24] but is without qualification capable of receiving body.

68: Since the void exists in this way, it is also not present at all within the cosmos. This is clear from the phenomena. For if the substance of the whole cosmos were not naturally linked throughout, then: *(a)* the cosmos could neither be held together and administered throughout by Nature, nor could its parts have any sympathy relative to one another;[25] *(b)* we would also be incapable of seeing and hearing, if the cosmos were not held

21. For Aristotle's position (cf. n. 29 below) reformulated in this way see Alex. Aphr. *Quaest.* 3.12, 105.30–35 and 106.32–107.4 (tr. Sharples [1994] 74–75), and Simplic. *In de caelo* 285.21–24 and *In phys.* 468.1–3.

22. For the void as "something" *(ti)* see *SVF* 2.331, and Brunschwig (1988) 96–99. Lines 64–67 below define its ontological status.

23. Here again (cf. n. 13 above) "void" is being used in the sense of space.

24. "Acting" and "being acted on" *(poiein* and *paskhein)* define body for the Stoics; see *SVF* 1.90 and 2.363. The sense of "acting" here is that of causing an effect; see I.3 n. 17.

25. There is a similar argument at *SVF* 2.543.

together by a single tension (that is, if the pneuma were not naturally linked throughout); for if there were intervening void spaces, our senses would be impeded by them;[26] *(c)* vessels with narrow apertures would also, when inverted in liquids, be filled when the liquid passed through the void spaces; but this does not in fact occur, because the vessels are full of air, and this air cannot be extruded because their apertures are enclosed by liquid.[27] And there are countless other [phenomena] that we need not mention now by which this [thesis] is demonstrated. It is therefore impossible that there be void present within the cosmos.

81: Aristotle and the members of his school do not admit void even outside the cosmos.[28] "The void," they argue, "must be a container of body; but no body exists outside the cosmos; so neither does void."[29] But this is simplistic, and exactly like someone saying that since water cannot be present in places that are dry (i.e., lack water), there can also be no container capable of receiving water.[30] So it should be admitted that "container of a body" is used in two senses: as that which holds body and is filled by it, and as that which is capable of receiving body.

89: "But," they say, "if there were void outside the cosmos, the cosmos would move through it, since it would have nothing that could hold it together and support it."[31] But our response will be that the cosmos

26. Alex. Aphr. *De an libr. mant.* 139.14–17 argues that if light is a body, then if there were an interstitial void, the air would be unevenly illuminated by the light present in the pockets of the void. On pneuma in vision see II.6.178–187, where its peculiar tenuity (or inherently rarefied state; cf. 1.5 n. 34) allows it to be refracted by the water in a full container without causing any overflow.

27. This would not be a problem for liquid poured into an upright vessel, or entering a clepsydra, of which the latter is more likely the situation being described at lines 33–37 above.

28. See line 55–61 above with n. 21.

29. See Arist. *De caelo* 279a12–14, and Simplic. *In de caelo* 284.21–24.

30. On the validity of the notion of an unactualized possibility underlying this counter-claim see Sorabji (1988) 133–135.

31. For this argument see Alex. Aphr. at Simplic. *In de caelo* 286.6–10 and Simplic. *In phys.* 671.4–13 (= *SVF* 2.552). The fact that here and below (see n. 34)

cannot "move" through the void, since it tends toward its own center, and has as down the [direction] toward which it tends;[32] for if the cosmos did not have its center and down as identical, it would "move" through the void[33] (as will be demonstrated in our discussion concerning motion toward the center).

96: They also claim that "if there were void outside the cosmos, the substance [of the cosmos] would, by expanding through it, be scattered and dispersed to an unlimited extent."[34] But our response will be: *(a)* The [substance of the cosmos] cannot be acted on in this way, since it has a holding power that holds it together[35] and thus preserves it. Also, *(b)* the

---

Cleomedes responds to the main thrust of arguments that we know were formulated by Alexander of Aphrodisias (fl. ca. A.D. 200) could not in itself prove that he was a contemporary of this Peripatetic (*pace* Algra [1988] 169–171 and [1995] 269; now retracted at Algra [2000] 171–172), since such arguments could have predated Alexander. Cleomedes' connection with Alexander confirms only that he belongs to an era terminating around A.D. 200, when polemics between Stoics and Peripatetics were common; see Introduction with n. 4.

32. See further lines 161–175 below. On this argument see Hahm (1977) 119, Algra (1988) 169–170 and Wolff (1988); cf. also n. 57 below.

33. This conditional sentence (lines 92–94) has sometimes been deleted because of incomplete knowledge of the manuscript tradition; see *Caelestia* Todd ed. ad loc. But it is integral to the reasoning, though when the received text at line 93 says that the cosmos would be borne "downwards" *(katō)*, that qualification can be omitted, since the void has no downward direction; see lines 150–151 below.

34. Cf. Alex. Aphr. at Simplic. *In de caelo* 286.10–23, and at Simplic. *In phys.* 671.8–13 (= *SVF* 2.552); also, derivatively, Themist. *In phys.* 130.13–17 (= *SVF* 2.553). Arist. *Phys.* 215a22–24 had cited dispersal "in all directions" *(pantēi)*, though only Themistius uses the phrase "without limit"/"to an unlimited extent" *(eis apeiron)* found in Cleomedes (at lines 97 and 113–114 below). Alexander did concede that the Stoic "holding power" *(hexis)*, introduced in Cleomedes' response, might prevent the cosmos from splitting into pieces, but thought that power no help when the cosmos was being displaced.

35. As at lines 70 and 72 above, and 98–99 below, *sunekhein* (literally "to hold together") also means "to make continuous" through the physics of the Stoic dynamic continuum, conveyed here by the *hexis* ("the holding power").

enclosing void does not act at all,[36] whereas this [substance] conserves itself through the exercise of its surpassing power,[37] as it is compressed, and again as it expands, in the void, in accordance with its natural changes—expanding into fire at one time, while setting out for the generation of the cosmos at another.[38]

104: Simplistic too is the [Aristotelians'] claim that: "if there is void outside the cosmos, it will have to be unlimited; but if the void outside the cosmos is unlimited, then there will also have to be unlimited body."[39] The reasons are: *(a)* the void's being unlimited does not imply that body is also unlimited, since the concept of the void does not cease anywhere, whereas being limited is in fact included in the notion of body;[40] *(b)* there also cannot be a "holding power" for what is unlimited: for how could something unlimited be held[41] by anything? The [Aristotelians] also make other similar claims.

36. Cf. lines 66–67 and n. 24 above.

37. This power *(dunamis)* surpasses those attributed later to intracosmic bodies such as the Earth (I.8.82–95), the Sun (II.1.357–403), and the Moon (II.3.61–65).

38. The Stoic cosmogony is the result of the evolution of a creative originative fire (e.g., *SVF* 1.107, 171; 2.774, 1027; see Pease [1955] on Cic. *De nat. deor.* 2.57); as such, it can be described as the manifestation of a biological "impetus" (for which Cleomedes uses the verb *hormān*).

39. See Alex. Aphr. at Simplic. *In de caelo* 285.32–286.2 (= *SVF* 2.535), and cf. Arist. *Phys.* 203b25–30.

40. This notion *(ennoia)*, like others in the treatise (I.8.21–26; II.1.155–170; II.3.1–4 and 34–35), is a *preliminary* idea, reached by rudimentary reasoning and requiring further refinement. Thus here we cannot naturally form a notion of body that is not finite (Sorabji [1988] 140 suggests Arist. *Phys.* 204b5–7 as a precedent), but need a demonstration (at lines 133–139 below) that an infinitely enlarged cosmos is inconceivable. In the *Caelestia* Cleomedes shows no interest in the origin of such notions; that is, he does not identify them as "natural" or "common," or see them as "preconceptions" *(prolēpseis)*. (On this feature of Stoic epistemology see Todd [1973] and Scott [1988].) They are comparable to other arguments in the *Caelestia* that are of limited value because they are based only on observations; see Introduction n. 31 and n. 54 below.

41. Perhaps *ekhesthai* (line 110) should be emended to *sunekhesthai*, given that a *hexis* is what makes [the cosmos] continuous (cf. n. 35).

112: That it is necessary that there be void outside the cosmos is evident from what has already been demonstrated. But that it is absolutely necessary that this void extend without limit in every direction from the cosmos, we may learn from the following [principle]: everything that is limited has its limit in something different in kind, different, that is, from the thing that is limited.[42] To take an obvious example: in the whole cosmos air, because it is limited, ceases [to be air] at two bodies different in kind, aether and water.[43] Similarly, the aether ceases at both the air and the void, the water at both the earth and the air, and the earth at the water. Our bodies too are similarly limited by something different in kind, their surface, and this is incorporeal.[44] It is, then, necessary that if the void enclosing the cosmos is limited rather than unlimited, it ceases [to be void] at something different in kind. But nothing different in kind from the void, at which the void ceases, can be conceived of.[45] Therefore the void is unlimited.

123: For even if we did conceive of something different in kind from the void, by which it will be limited, this [other void] will have to be filled, and what fills it will be body. And in this way there will have to

42. In the Atomist-Epicurean tradition this principle is used to demonstrate that extracosmic void is infinite, and that it is occupied by infinitely numerous bodies; see Epicur. *Hdt.* 41, Lucr. 1.958–967, and cf. Arist. *Phys.* 203b20–22.

43. By "ceasing" here Cleomedes means that the elements cease to be called by their given names. However, this does not imply that these four elements are distinct homogeneous bands or layers; in fact the Moon, which is at the "juncture" *(sunaphē)* of air and aether (I.2.37–38; II.3.83–84), has its appearance affected by not being in the "pure" part of the aether (II.3.88–90). On the gradations in the density of the Stoic aether see Todd (2001).

44. On the incorporeality of surface *(epiphaneia)* see Plut. *De comm. not.* 1080E, and the discussions at LS I.163, 165, and 301, and by Brunschwig (1988) 28–30.

45. Another Peripatetic argument (cf. lines 110–111 above) is that such a void is simply an imaginary conception; see Alex. Aphr. *Quaest.* 3.12, 105.27–35, Alex. Aphr. at Simplic. *In de caelo* 285.26–27 and 286.23–27, and Todd (1984). Thorp (1990) 159–164 discusses the Aristotelian basis for this position.

be body outside the cosmos—something that physical theory does not suggest, since all bodies are enclosed by the cosmos.[46] From this it is evident that the external void cannot be limited anywhere. Therefore it is unlimited.

130: Indeed, just as it is thought that everything that is limited is enclosed by something (otherwise it would not be limited), so too the void, if limited, is necessarily enclosed by something. What could this be? A body? Impossible, since there is no body outside the cosmos. But even if there were a [body], it again, since it is limited, will have to be enclosed by a void. And again this void would, if it is not going to be unlimited, be enclosed by another body that would itself in turn be enclosed by another void, since this body too must have boundaries. This [process would go on] to an unlimited extent, and so bodies will come into existence that are unlimited both in number and in size. None of this is possible.[47]

139: Thus if the extracosmic void is limited, and at all events enclosed by something, yet not enclosed by *body,* it will be enclosed by something *in*corporeal. So what will this be? Time? Surface? A *lekton?*[48] Something else just like them? But it is implausible that the void be enclosed by any of these. Indeed, there will have to be another void that encloses it, and this, if it is not unlimited, will have to be enclosed by another void, and this by another to an unlimited extent. So by refusing to admit that the extracosmic void is unlimited, we shall be brought round to the necessity of admitting an unlimited number of distinct voids! That is utterly

46. Cf. lines 5–7 above.

47. For the Epicureans an infinite number of bodies was, of course, possible (see Epicur. *Hdt.* 41–42), although an infinitely large body was not (Epicur. *Hdt.* 57).

48. On extracosmic time as inconceivable see already Arist. *De caelo* 279a14–18. The surface is presumably to be distinguished from a surrounding void, which has been excluded by the preceding argument. An extracosmic *lekton* (literally "the expressible," the carrier of the meaning of a proposition, and an incorporeal; see II.5 n. 18) would have nothing for which it could express a meaning.

absurd.[49] So it is necessary that we agree that the void beyond the cosmos is unlimited.

150: Since the void is unlimited, as well as being incorporeal, it will not have an upwards or a downwards [direction], nor a front, back, right, left, or center, for these directions[50] (seven in number) are observed in relation to bodies. Thus while none of them exists in relation to the void, the cosmos itself, being a body, necessarily has both an upwards and downwards [direction], as well as the remaining directions. So they say that the west is its "front," since its impetus is westward, and that the east is its "back," since it is from there that it proceeds forward.[51] Thus the north will be its "right," and the south its "left."[52]

158: There is nothing obscure about these directions in the cosmos, but the remaining ones [namely, up and down] confused the earlier natural philosophers considerably, and numerous errors occurred in this area, since they were unable to grasp that in the cosmos, which is spherical in shape, the exact center is necessarily downwards from every direction, whereas what extends from the center to the limits and right up to the surface of the sphere is upwards. The two directions coincide in the cosmos (that is, both the center and the downwards direction are identical), though in bodies that are made oblong in shape they are separated, whereas this is not the case with spherical bodies, where instead they coincide. This is because [bodies] with spherical holding powers necessar-

49. This is because the concept of the void is "without qualification" (line 67 above); this would be untrue if blocks of it were separately distinguished.

50. "Directions" *(skheseis);* the term literally means a relational state, appropriately enough since the directions are defined from an arbitrarily defined point.

51. The impetus (*hormē;* cf. n. 38 above) of the cosmos here is that of the sphere composed of aether, which contains all the heavenly bodies; *kosmos* is thus being used, as it often is, in the same sense as *ouranos* (the standard term for "heavens"). This impetus involves a daily rotation from east to west around a fixed central Earth; cf. I.2.1–4.

52. See I.6 n. 7 on these six non-central directions and the arguments for the centrality of the Earth in the cosmos.

ily tend[53] in the direction of their center from their surface, and so have as downwards the [direction] toward which they tend. So the cosmos too, since it is spherical in shape, has the same property, that is, its downwards direction and center are identical because these directions coincide in it at the same [point]. [The sphericity of the cosmos] will be the primary aim of our demonstration in our discussion concerning motion toward the center, but for now we shall demonstrate it in simpler terms, on the basis only of what is presented to us in perception.[54]

176: *(a)* All of us, at whatever latitude of the Earth we may be,[55] clearly see the heavens located above our heads, while everything around them appears to us to be sloping away. Then as we proceed to any other ter-

53. This links the geometry of the sphere and Stoic physical theory. A spherical body will have the kind of "holding power" *(hexis)* that endows it with this shape, and makes it "tend" (*neuein;* cf. lines 91–92 above) in a centripetal direction. For some indication of the physics that applies this principle to the cosmos see I.5.126–138.

54. See I.5.1–6 where the limitations of this argument are noted in more elaborate terms. Proposition *(a)* is a *conclusive* argument only for the sphericity of the *Earth*, based on the evidence of changing horizons (see I.5.107–108 with I.5.49–54), but it can only "suggest" (cf. I.5.1) the sphericity of the cosmos, or, in effect, provide a preliminary notion of it (see n. 40 above).

55. Strictly speaking, the latitude *(klima)* of an observer's locality is defined by the elevation (or inclination) of the north celestial pole above the northern horizon. But (see Figure 1) this *klima* is equal to angle ZOQ, which measures the distance along the observer's meridian from his zenith to the celestial equator. This angle is in turn equal to angle OTE, the observer's latitude, that is, the angular distance measured along a meridian of longitude from the terrestrial equator to the parallel circle passing through the observer's locality; cf. Ptol. *Alm.* 2.6. In standard usage among geographers, the distance from this parallel to the equator can be called a *klima*, and is usually given in stades (e.g., Strabo 2.5.7, 2.5.15–16, and 2.5.39–40), although for Ptolemy it is an arc given in degrees. In his *Geographia* (or *Geographical Directory*) Ptolemy advanced the study of geography by applying scientifically the terms *mēkos* and *platos* (longitude and latitude; the same terms he used in the *Almagest* for celestial coordinates) as coordinates to represent, respectively, east-west and north-south distances reckoned in degrees on (a map of) the Earth; cf. Neugebauer (1975) 934.

restrial latitude at all, what until then appeared to be sloping away is over our heads. This would not occur unless the heavens were located above the Earth in every direction (that is, unless the exact center of the cosmos were downwards, while the [direction] extending from it to the heavens were upwards). *(b)* Also, when on a sea voyage we have no land in sight, the heavens appear to us to be touching the water in a circle at our horizon. But on reaching the place where the heavens appeared to us to be touching the water, they are instead visibly located overhead, and this occurs continuously throughout the voyage. So if it were possible to sail around the whole Earth, or go around it by some other means (assuming that no part of it is uninhabitable),[56] we would learn that the heavens are located above every part of it. And so the center of the cosmos is at once downwards as well as a center. But our lesson concerning the motion of heavy bodies to the center[57] will establish this more effectively.

193:[58] Five parallel circles are drawn in the heavens: one, which we call the equinoctial circle, divides the heavens into two equal parts, while on either side of this are two that are smaller than it, but equal to one another. They are called "tropics," since we draw them through the tropical points of the Sun. Two others are also drawn on either side of these, of which the northern is called "arctic," and the one opposite to it

56. For Cleomedes (lines 210–211 and 266–267 below) the torrid zone *is* uninhabitable.

57. Only in this reference (cf. lines 94–95 and 173–174 above) to this forthcoming demonstration is the phrase "of heavy bodies" added, although it appears in other references to centripetal motion as the physical principle underlying the sphericity of the Earth and the cosmos; see Gem. *Isag.* 16.2, Strabo 2.5.2, and Theon *Expos.* 122.11–16. (Ptol. *Alm.* 1.7, 22.22–23.9 is a related, but special, case; see Wolff [1988] 499 n. 31.) But Cleomedes and these other authors must still mean that *all* the elements in the cosmos have a centripetal motion, i.e., that they are all "heavy" in that sense. See further I.5 n. 38.

58. I.1.193–273 formed a separate chapter in all earlier printed editions. But they are a geographical appendix to ch. 1, while lines 270–273 below form an obvious conclusion, and lines 262–269 complement I.1.3–16 by introducing a teleological principle.

"antarctic." These differ for different [observers] depending on differences in latitude, since they become larger and smaller, and ultimately disappear.[59] Where they do not [both] disappear, one of them must be out of sight, while the other is always visible. Five parts of the Earth are located below the intervals in the heavens that are distinguished by the circles just described:[60] First, the one enclosed by the arctic circle; second, that located below the interval between the arctic circle and the summer tropic; third, the one between the two tropics, which has the equinoctial circle located above it at its exact center; fourth, that between the winter tropic and antarctic circle; fifth, that enclosed by the antarctic circle.

209: The natural philosophers call these parts of the Earth "zones," and say that while each of the outer ones is uninhabitable because of icy cold, the one at the exact center is uninhabitable because of blazing heat, and those on either side of it are temperate since they are each tempered by the torrid and frigid zones adjacent to them. So by further dividing each of these temperate zones into two with respect to the hemisphere thought to be the upper [region of the] Earth, and the one thought to be the lower, they say that there are four inhabited zones.[61] We humans, of whom there are direct reports, inhabit one of these,[62] and the people called "circumhabitants" *(perioikoi)* another. Though the latter are in the same temperate zone as us, they inhabit the region that is thought to be

59. Here, in the next paragraph, and at I.2.79–80 it is assumed that the arctic and antarctic circles are defined; that is, that the observer is not at either pole or at the equator. See Figure 2. Note that Figures 2(a)–(f) are not true perspective drawings. We have instead in this case and others like it chosen to adopt a style of representing the sphere that is traditional in classical astronomy, and distorts perspective in order to facilitate comprehension.

60. On these zones see Arist. *Meteor.* 362a32–362b9, *SVF* 2.649, Ach. *Isag.* 62.20–63.5, *Aratea* 96.23–97.6, and Plin. *NH* 2.172. Posidonius challenged this folk geography; see I.4.90–146.

61. On these see Ach. *Isag.* 65.15–66.25, *Aratea* 97.7–23, and Gem. *Isag.* 15.1–2. See Figure 3.

62. Gem. *Isag.* 16.1 calls such inhabitants "co-habitants" *(sunoikoi).*

below the Earth.[63] The "contrahabitants" *(antoikoi)* inhabit a third zone, and those antipodal to us a fourth, and while they [both] occupy the contratemperate zone, some of them (our contrahabitants, also called "dressed by the shoulder" [*antōmoi*]),[64] occupy the region above the Earth, whereas those occupying the region below the Earth are antipodes.[65] To explain: the footprints of all who walk the Earth must face directly toward the center (that is, the exact center) of the Earth, given that the exact center of the Earth, because of its spherical shape, is downwards. Hence it is not our circumhabitants who become our antipodes, but inhabitants of the contratemperate zone in the region below the Earth—the ones who are located directly opposite us, with their footprints directly opposite ours.[66] The footprints of our circumhabitants do not, however, point toward ours, but toward those of our contrahabitants, so that again these [two groups] become antipodal to one another. Our antipodes become contrahabitants of our circumhabitants, since such relations resemble those of friends and brothers, rather than those of fathers and children, or slaves and masters; that is, they

63. This "region" (*klima* being used in the sense of a broad band of latitudes) is "thought" *(dokoun)* to be "below the Earth" from the perspective of the northern temperate zone. Based on lines 153–158 above, an observer in this zone who looks toward the west has his own region and that of the contrahabitants "above the Earth" (or to his right and left), while the other two regions are below it.

64. A contrahabitant who (following lines 157–158 above) faces west will (to use a military term) be "dressed" (or aligned) to "our" (southern European) left, and we to that person's right, assuming geographical symmetry (see n. 66 below).

65. The use of the Greek "antipodes" as a collective noun for these inhabitants seems preferable to "contrapodes."

66. But while the theory of Nature (see lines 262–269 below) may require *some* equivalently inhabited antipodal zone, it could not guarantee that its geography, and the footprints of its residents, would correspond, unless land masses were suitably distributed in each hemisphere by some teleological principle, as may have been the case for the Stoics (see I.4 n. 32). On the enduring illusion of exact antipodality see Quine (1987) 121–124.

convert,[67] in that we become circumhabitants of our circumhabitants, antipodes of our antipodes, and similarly contrahabitants of our contrahabitants.

235: Yet in relation to each of these [groups] we have something in common, as well as distinct. In relation to our circumhabitants we have in common, first, inhabiting the same temperate zone; second, having winter, summer, and the other seasons at the same time, that is, having identical lengthening and shortening in daytimes and nighttimes.[68] But there is a difference in their daytimes and nighttimes: when it is daytime in our zone, it must be nighttime in theirs, and vice versa, although this is put too loosely. For it is not by precise reckoning that the Sun begins to rise in their zone when it sets in ours, since in that case the nighttime in their zone would be long when the daytime in ours was long, and their seasons, that is, the lengthening and shortening of their daytimes and nighttimes, would be the reverse of ours. But in fact, as the Sun goes round (i.e., encircles) the Earth, which is spherical, it shines its bright light on the [parts] on which it casts its rays each time its course takes it over the Earth's curvatures. So while it is still visible above the Earth in our zone, the circumhabitants necessarily see it rising, given that it goes round an Earth that is spherical in shape, and, as it goes over the Earth's curvatures, it rises at different times for different [observers].

252: In relation to our contrahabitants we have in common: first, that we both occupy the upper hemisphere of the Earth; second, that we have daytimes and nighttimes at the same time, although this is also put too

67. The logical conversion implied here is that of strict reciprocity, whereby aRb is true if and only if bRa. Contrast Arist. *Cat.* 6b28–7a5, where the reciprocity of relatives *(ta pros ti)* can cover the relation, excluded here, between slave and master.

68. "Daytime" *(hēmera)* and "nighttime" *(nux)* will be used for the intervals from sunrise to sunset, and sunset to sunrise, respectively. Cleomedes uses the term *nukhthēmeron* ("interval of a nighttime and a daytime") for the interval from one sunrise to the next; see 1.4 n. 16.

loosely,[69] since there is shortest daytime for them when there is longest daytime for us, and vice versa,[70] given that in relation to them our seasons, that is, the lengthening and shortening of daytimes and nighttimes, are reversed. But we have nothing in common with the antipodes. Instead, everything is reversed: we occupy one another's lower regions of the Earth, and their seasons, that is, their daytimes and nighttimes, in terms of the lengthening and shortening of daytimes, are the reverse of ours.

262: The theory of Nature teaches us that circumhabitants, antipodes, and contrahabitants must exist, since none of these [groups] are described by direct reports.[71] We simply cannot travel to our circumhabitants because the Ocean separating us from them is unnavigable and infested by beasts; nor to the inhabitants of the contratemperate zone, since we cannot traverse the torrid zone. Yet the regions of the Earth that are equally temperate are necessarily inhabited to an equal extent, given that Nature loves Life, and Reason requires that all [parts] of the Earth, where possible, be filled with animal life, both rational and irrational.

270: To be demonstrated next is what causes different parts of the Earth to be frigid, torrid, and temperate;[72] and why for inhabitants of the contratemperate zone the seasons, that is, the lengthening and shortening of the daytimes, are reversed.[73]

69. Cf. lines 240–241 above.

70. It was commonly assumed in antiquity that the shortest nighttime is equal to the shortest daytime, although, as Neugebauer (1969) 158 n. 1 points out, atmospheric conditions make the shortest daytime longer than the shortest nighttime.

71. This "theory of nature" *(phusiologia)* includes the explanation for the cosmos being spherical and geocentric (lines 191–192 above; also I.5.126–138 and II.6.41–43), which in turn determines the Earth's relation to the Sun's motion in the ecliptic (I.2.73–80; cf. II.1.361–386). Cf. also Gem. *Isag.* 16.19–20 on the symmetry of habitable zones in a spherical Earth.

72. See I.2.80–82 where this project is completed.

73. This is the program for I.3; see I.3.111–114; see I.4 n. 1 on its continuation.

CHAPTER TWO

1: As the heavens revolve in a circle above the air and the Earth, and effect this motion as providential for the preservation and continuing stability of the whole cosmos, they also necessarily carry round all the heavenly bodies that they encompass.[1] Of these, then, some have as their motion the simplest kind, since they are revolved by the heavens, and always occupy the same places in the heavens.[2] But others move both with the motion that necessarily accompanies the heavens (they are carried round by them because they are encompassed), and with still another motion based on choice[3] through which they occupy different parts of the heavens at

1. The "heavens" here consist of the element *aithēr* (cf. I.4.87–88), that is, they are a band of rarefied matter with a natural tendency to sphericity (cf. I.5.134–137) and circular motion (*SVF* 2.642). (This sentence therefore supplements the description of the *diakosmēsis* at I.1.3–16.) The term *astra*, which can refer only to the fixed stars, will be translated as "heavenly bodies," when, as here, it refers generically to both the fixed stars and the planets.

2. Hence they are used to define celestial latitudes; see I.3.44–51.

3. This motion is "based on choice" *(kinēsis proairetikē)* because in the Stoic system planets are endowed with reason and thus self-motivated (cf. Cic. *De nat. deor.* 2.43 and 2.58), though such choice is exercised strictly within the limits of the "providential motion" (*kinēsis pronoētikē;* line 2) of the heavens; this, as it were, allows planets some independence relative to the motion of the fixed stars. See further Todd (2001).

different times. This second motion of theirs is slower than the motion of the cosmos, and they also seem to go in the opposite direction to the heavens, since they move from west to east.

12: The first [set of bodies] is called "fixed," but the second "planets," since these appear at different times in different parts of the heavens.[4] The fixed bodies might be likened to passengers who are borne[5] along by a ship, yet remain in their assigned places in relation to the overall space.[6] The planets, by contrast, are like passengers who move in an opposite direction to the ship (toward the stern from positions at the prow) with a relatively slower motion. They could also be likened to ants creeping on the basis of choice on a potter's wheel in a motion opposite to [that of] the wheel.[7]

20: The total number of the fixed bodies is immense,[8] but only seven planets have become known to us,[9] although it is unclear whether there

4. For this general distinction see also *SVF* 2.650. This elementary definition, given the Stoics' preoccupation with etymology (cf. II.5.82–86 and 92–101), was probably designed to draw attention to the literal meanings of *aplanē* and *planōmena/planētai*, "non-wanderers" (sc. fixed) and "wanderers" (sc. planetary).

5. The stars "are borne" (*pheresthai*, in a passive sense), since, unlike the planets (cf. n. 3 above), they do not initiate their own motion. *Pheresthai* switches back to its middle voice sense ("move") in the comparison with the planets.

6. "Space" *(khōra)* is defined by the Stoics (*SVF* 2.503–506) as the larger area within which something has its "place" *(topos)*. See Algra (1995) 263–281 on this concept.

7. For the comparison with displacement in a moving vessel see Ach. *Isag.* 39.16–20 and Hygin. *De astron.* 4.6; and with ants crawling backwards on a wheel Ach. *Isag.* 48.16–18, *Aratea* 97.33–98.1, and Vitruv. *De arch.* 9.1.15. Bodnár (1997) 200 n. 29 links Cleomedes' analogies with his acceptance of planetary motion in eccentric circles (cf. II.5.139–140).

8. For this conventional claim see Arist. *De caelo* 292a11–12, Ps.-Arist. *De mundo* 392a16–17, and Sext. Emp. *PH* 2.90 and 97.

9. For the number of planets as a basic assumption of astronomy see Dercyllides at Theon *Expos.* 200.2–3. (This passage could reflect Posidonian ideas, as I. G. Kidd [1978a] 11 suggests, particularly if Dercyllides can be dated to the first decades of the first century A.D., as Tarrant [1993] 11–12 and 72–76 argues.) On the "Chaldean" planetary order followed here (as opposed to the "Egyptian," in

are still more. The one held to be farthest away, the star of Saturn, named *Phainōn* (The Shining One), completes its own circuit in a period of 30 years in accordance with its motion that is based on choice. Below it is the star of Jupiter, named *Phaethōn* (The Radiant One), which completes its own circuit in a period of 12 years. Below this is *Puroeis* (The Fiery One), the star of Mars, which has a relatively disorderly motion,[10] although it too is held to complete its own circuit in 2 years 5 months.[11] The Sun is thought to be below this, and thus at the center of the other planets.[12] By going round its own circuit in 1 year it demarcates the seasons by this motion, while it provides the days by the motion that accompanies the heavens. Below this is the star of Venus, and it too has a period of 1 year; when it sets later than the Sun, it is called *Hesperos* (Evening Star), but when it rises before it, *Heōsphoros* (Dawn-Bringer), which some also like to call *Phōsphoros* (Light-Bearer). Below Venus is the star of Mercury, named *Stilbōn* (The Gleaming One); they say it goes round its particular circuit in 1 year. Below this is the Moon,[13] closest to the Earth of all the heavenly bodies, in that accepted theory places it at the junction of the air and the aether, which is why its own body is also visibly murky.[14] The illuminated part of it has its luminance from the Sun,

---

which the Sun immediately follows the Moon rather than being centrally located) see Préaux (1973) 213–217 and Aujac (1975) 124 n. See also II.7 n. 2. Cleomedes gives the *sidereal* period for each of the seven planets, that is, the time it takes for each planet as seen from the Earth to return to a given star. For the *synodic* planetary periods, that is, the periods from one conjunction (or opposition) with the Sun to the next, see II.7.8–10. On the epithets applied to the planets see Pease (1955) on Cic. *De nat. deor.* 2.52–53.

10. Cf. Plin. *NH* 2.77. This motion is only "relatively" disorderly, given lines 43–46 below, though Cleomedes does not explain why he regards it as such.

11. Gem. *Isag.* 1.26 gives 2 years 6 months for Mars' sidereal period. Others (Cic. *De nat. deor.* 2.53 and Theon *Expos.* 136.8) give just under 2 years. On Copernicus' use of Cleomedes' report see Rosen (1981) 452–453.

12. On the teleological implications of its location see II.1.396–403.

13. On the Moon's proximity to the Earth see II.3.81–101.

14. See II.3.83–84 and II.5.1–7.

since the hemisphere of the Moon that is turned toward the Sun always gets illuminated. The Moon completes its own circuit in 27½ days, and is in conjunction with the Sun in 30 days.

43:[15] All these planets have a motion opposite to the heavens, and so are seen in different [positions] at different times, yet they do not effect a disorderly course, that is, they do not go through random parts of the sky but through what is called "the zodiac," though without going beyond it.[16] The band of the zodiac is at an oblique angle because it is positioned between the tropical circles and equinoctial circle, touching each of the tropical circles at one point, while dividing the equinoctial circle into two equal [parts]. This zodiac has a determinable width, with [parts] in the north, the south, and in between.[17] That is why it is also described by three circles: the central one is called "heliacal,"[18] and the two on either side of it "northern" and "southern."[19] Whereas the other planets approach the northern and southern circles at different times in accordance with the motion based on choice through this zodiac, only the Sun

15. In earlier divisions of Book I, the first sentence of this paragraph (lines 43–46) marked the end of chapter 2, and chapter 3 began with a remark about the zodiacal circle in line 46. But the whole of lines 43–82 offers an effective complement to the preceding account of planetary motion, while returning the discussion to the topic of differences in temperature between terrestrial zones (cf. 73–82 with I.1.270–271). As such, this passage belongs with chapter 2.

16. On the zodiac see Gem. *Isag.* 5.51–53 and Ach. *Isag.* 52.25–55.6. Ptol. *Alm.* 1.8, 27.20–29.16 is less elementary.

17. In I.2 (lines 45, 46, 49, 53, 61, 70 and 73) the "zodiac" is a circular "band" (as we have translated *kuklos* at line 46), whereas *zōidiakos* (with or without *kuklos*) usually refers to the heliacal circle within that band (see next note).

18. Cleomedes does not use the term "ecliptic" *(ekleiptikos);* at II.5.146 it appears in a gloss; see II.5 n. 32. He calls this circle the "zodiacal circle" (I.4.30–31), the "circle that passes through the middle of the zodiacal constellations" (II.5.144–145; II.6.32; II.7.1–2), and "the circle at the middle/center of the zodiacal [band]" (I.4.53–54; II.6.5, 12). At I.4.49–71 he explains that the "heliacal circle" *(hēliakos kuklos)* is properly the circle on which the Sun travels, and that the circle termed "heliacal" (here at line 51) is its trace.

19. See Figure 4(a).

moves exclusively through the central circle, and does not approach either the northern or southern circles.[20] It does approach both the north and south of the heavens as it goes from one solstice to another, but approaches neither [extreme] of the zodiac. Instead, in its course it follows the circle at the exact center of the zodiac; that is why this circle has the name "heliacal."

60: The remaining planets approach north and south both of the heavens, and of the actual zodiac, because they move in it as in a spiral. That is, when they go down from the northern to the southern circle, and from there go back up toward the northern circle, they effect a motion through the zodiac that is neither straight, nor even simple like the Sun, but is like a spiral.[21] And when they move from the northern to the central circle they are said to lower themselves relative to their high position, and when they go through the central circle and approach the southern circle are said to lower themselves relative to their low position.[22] But on going back up from there to the central circle they are said to elevate themselves relative to their low position, and on crossing the central circle and approaching the northern circle, are said to elevate themselves relative to their high position.[23] Because the heavens slope from north to south in the zone that we inhabit,[24] the northern [parts] of the zodiac

20. For a more elaborate analysis of the Sun's motion see I.4.30–43.

21. This spiral-like course, which is confined to the zodiacal band, is the result of the planet's eastward motion along the heliacal circle and of its north-south motion with respect to this same circle.

22. The terms used here for the low and high positions (i.e., the extremal latitudes) of planets in the zodiac are *tapeinōma* and *hupsos.* They sometimes also identify the varying distances to the Earth of planets in their eccentric orbits (II.5.133–141). In astrological contexts, they designate planetary exaltations and depressions, that is, the positions in the heavens from which the planets have their greatest and least effect on an individual (cf., e.g., Ptol. *Tetrabibl.* 1.19.41–42 and Sext. Emp. *AM* 5.35).

23. See Figure 4(b).

24. On this northern latitude see I.3 n. 9. On the northern elevation of the zodiac see I.3.35–43.

are as a result elevated far above the horizon, whereas the southern [parts] are much closer to the horizon.

73: This is how the planets move in the zodiac, but the Sun, by moving in the heavens through the band between the tropics, necessarily makes torrid the terrestrial band below the interval (described above)[25] between the tropics. But when the Sun comes back from the south to the north, it does not go beyond the summer tropic, nor, on going from the summer tropic to the south, does it go beyond the winter tropic. The result is that the zones in the extreme regions[26] are frigid, since they are at the greatest distance from the Sun, whereas the zones below the band between the tropical and arctic [and antarctic] circles are temperate. This [motion of the Sun] is what causes some [regions] of the Earth to be frigid, others torrid, and still others temperate.[27]

25. I.1.194–196.

26. The phrase actually used by Cleomedes at line 78 is "the zones below the Bears" (*αἱ ὑπὸ ταῖς ἄρκτοις ζῶναι*). But since the Bears are a northern constellation, the inclusion of the antarctic zone can perhaps only be justified (as it is by Goulet 192 n. 99) in light of Arist. *Meteor.* 362a32, where the phrase "the other [of the two] Bears" (*ἡ ἑτέρα ἄρκτος*) means "the antarctic region." On the other hand, if *ὑπὸ ταῖς ἄρκτοις* were emended to *ὑπὸ τοῖς ἀρκτικοῖς* ("below the arctic [circles]"), then the reference to both arctic and antarctic zones would be consistent with the unusual use of *ἀρκτικοί* to describe both the arctic and antarctic circles later in the sentence at line 80. We have translated it as "arctic [and antarctic]" there only to prevent confusion, and not as an essential supplement to the text. But, however the text is emended, the meaning intended by Cleomedes in this sentence, and reflected in our translation, is clearer than the language in which it is expressed.

27. Further on the Sun's power see II.1.361–375.

CHAPTER THREE

1: The following is essentially what causes the seasons (i.e., the lengthening of daytimes), to be reversed in the temperate zones. The Earth is spherical in shape, and thus [located] downward from every part of the heavens;[1] as a result its latitudes do not have an identical position relative to the zodiac, but different ones are located below different parts of the heavens. (That is why, as has been demonstrated,[2] they differ also in their temperatures.)

6: So in the mid-torrid zone,[3] which occupies the [latitude] of the Earth at the exact center, the heavens slope neither to the north nor to the south, but maintain a position of complete equilibrium such that each of the poles is observed on the horizon, with no arctic circles existing at this latitude; instead, all the stars set and rise again, and not a single one can be always visible there.[4] But when someone goes from this latitude to the

1. See I.1.159–175.
2. See 1.1.209–269.
3. The phrase "in the torrid zone" *(en tēi diakekaumenēi)* is best translated as "in the mid-torrid zone," in view of the qualifying phrase that follows it, which shows that it refers to the latitude elsewhere defined as "below the equinoctial circle" (lines 52–53 below; cf. I.1.205–206 and I.4.158), i.e., the terrestrial equator.
4. For a formal demonstration of the observational situation at the equator see Theod. *De hab.* 2.16.8–33, and also Figure 2(d).

temperate zones the position of the heavens appears increasingly different: one of the poles becomes concealed, while the other becomes elevated (that is, raised above the horizon). So for anyone coming to our temperate zone from the mid-torrid zone the south pole would go out of sight, since in the course of the journey it would be obstructed by the curvature of the Earth. The north pole, by contrast, would be elevated high above the horizon. But if we hypothesize someone traveling from the mid-torrid zone to the contratemperate zone, the opposite would occur: the south pole would be elevated above the horizon, and the north pole would go out of sight.[5]

22: So let us hypothesize someone coming from the mid-torrid zone[6] to our temperate zone. Now when that person is still [directly] below the equinoctial circle, each of the poles will be seen on the horizon, and no star will be either out of sight or always visible, and so there will also be no arctic circles. (An arctic circle must exist at our latitude to enclose the stars that are always visible, and an antarctic one to enclose those that are [always] concealed.) But someone who initiates the process of reaching here[7] from the south will necessarily have the south pole concealed by the curvature of the Earth, and the north pole proportionately elevated. And in this way the heavens assume for that person a slope from the north to the south: that is, of the stars near the poles some will go out of sight, others will be always visible, and the arctic circles enclosing these stars will exist with their slope necessarily changing in accordance with the forward direction of the journey. Since en route the heavens continually assume a position that is increasingly sloped, the northern parts of

5. Cf. the earlier account of the observations of changing horizons at I.1.176–182, where they served as prima facie evidence of the sphericity of the cosmos. At I.5.107–111 they help to demonstrate conclusively the sphericity of the Earth; see Figure 5.

6. Because this zone is uninhabitable (cf. I.1.187–189 and 265–267), this has to be a hypothesis.

7. "Here" is the latitude of Greece (cf. line 39 with n. 9 below), used to identify the north.

the zodiac will be seen as high (that is, elevated well above the horizon), and the southern parts as low (that is, much closer to the horizon).[8] Also, for someone who goes to the north from the south in this way and arrives at the terrestrial latitude of Greece, at which Aratus also composed his poem *The Phaenomena*, "the head of Draco" and "the feet of Helice" will be touching the horizon.[9] Also, the circle enclosing the stars that are [always] concealed will necessarily become equal to the size of the arctic circle.

44: Given that the heavens slope in this way, we must next imagine that each of the fixed stars, as it is revolved along with the heavens around its own center, describes a circle.[10] Now these circles are all parallel, and while the equinoctial circle is the largest of them, the smallest are those around the poles of the cosmos. Thus the circles proceeding from the [poles] to the equinoctial circle will become larger in proportion to their distance from the poles, whereas those proceeding from the equinoctial circle to the poles will become smaller in proportion to their distance from the equinoctial circle. At the latitude below the equinoctial circle all these circles (the greatest, smallest, and intermediate ones) have half-sections above as well as below the Earth.[11]

8. These definitions may be needed because at I.2.62–69 (cf. I.2 n. 22) "highness" and "lowness" described planetary motion in the zodiac rather than the changing appearance of the zodiac at different latitudes. I.2.69–72 only briefly foreshadowed this additional sense.

9. At Arat. *Phaen.* 58–62 (cf. *Schol. in Arat. vet.* 98.7–8) Draco is said to look "as if it is inclined towards the tip of Helice's tail: the mouth and the right temple are in a very straight line with the tip of the tail. The head of Draco passes through the point where the end of settings and the start of risings blend with each other" (tr. D. Kidd [1997] 77). Draco is at 54°N and so the latitude in question here is 36°N, that of Rhodes (cf. Ptol. *Alm.* 2.6, 109.5–10), the fourth of the seven canonical locations that mark latitudes in the northern hemisphere; see further Neugebauer (1975) 44, and cf. D. Kidd (1997) 199–200.

10. Each of these circles is called a "day-circle."

11. See Figure 2(d).

54: When someone reaches our zone from that latitude, then just as the north pole is elevated and the heavens slope, so these circles too no longer maintain the same position relative [to the Earth], but the equinoctial circle (a great circle that divides the heavens into two equal [parts]), has precisely half remaining above the Earth, and half below it. (That is because every circle that divides the heavens into two equal parts is either the horizon, or is divided into two equal [parts] by the horizon, so that it has half always visible above the Earth, and half concealed.) Thus since the equinoctial circle is a great circle, it also maintains the same position relative [to the Earth] even in the temperate zones, whereas the [successive circles] that proceed from it toward the poles do not. Instead, all the larger sections of the [circles] that proceed toward the north pole are necessarily above the Earth, since they are more elevated in our temperate zone, whereas the smaller sections are below the Earth. All the circles that proceed toward the south pole, by contrast, have larger sections below the Earth and smaller ones above it, at least [in our zone] where the whole antarctic circle <must> also <be> concealed, whereas the arctic circle is always visible.[12]

69: That is the situation in our temperate zone, but in the contratemperate zone the situation is reversed: that is, what is low [on the horizon][13] for us is high there, and vice versa, since they have the heavens sloping from the south to the north. Someone traveling there from the [latitude] below the equinoctial circle has the north pole going out of sight, and the south pole elevated, and so what is high for them is low for us, and vice versa. Thus they also have the arctic circle concealed, while the opposite circle is elevated in an amount equal [to the lowering of its counterpart].[14]

12. Our translation follows a Byzantine paraphrase; see the apparatus criticus at *Caelestia* Todd ed. I.3.67–68. See also Figure 5 (b).

13. See the definitions of "high" and "low" at lines 36–38 above, and cf. n. 8 above.

14. See Figure 5(c).

76:[15] Given all this, the Sun will obviously touch[16] all the [day-]circles between the tropics as it effects its course through the zodiac from one solstice to another. So when it touches the winter tropic as it goes from north to south, it causes our shortest daytime.[17] That is because of all the circles which the Sun touches this one has [for us] the largest section below the Earth and the smallest above it, and so [on touching it] the Sun necessarily causes the shortest daytime and longest nighttime in our temperate zone. But when, after touching the winter tropic, it again turns back toward us, then as it goes up toward the more elevated parts of the heavens, it continually encounters circles that have sections above the Earth larger than the section of the winter tropic [that is above the Earth]. In this way it provides a daytime that proportionately increases, while still remaining shorter than the nighttime, as long as the Sun's course is toward the equinoctial circle. But when the Sun touches the equinoctial circle, where the [sections] above and below the Earth are equal, it causes the [vernal] equinox. Finally, as the Sun goes up from the equinoctial circle to the summer tropic, it also necessarily provides daytimes that are longer than nighttimes when it encounters circles that have larger sections above the Earth. Such lengthening proceeds until the Sun approaches the summer circle, which, of all the circles that the Sun touches at our latitude, has the largest section above

15. This paragraph enlarges on I.2.73–78.

16. Since the heavenly bodies appear to be equidistant from the Earth, they also appear to move in the same plane, and on, as it were, the inner surface of a relatively small hemisphere (cf. II.1.85–86 with II.1 n. 20). It is within this conceptual framework that the Sun can be said to "touch" the day-circles.

17. Here and elsewhere we translate *poiein* as "cause." This is justified by the association of this verb with *aitia* (the standard term for "cause") in an identical context dealing with solar "power" at II.1.365–367; cf. also Posid. F18.26 EK for the related phrase *poiētikē dunamis* similarly associated with *aitia* in a more general context (see Appendix n. 23). The causality in question is also teleological (cf. F18.22 EK), as the use of "provide" *(parekhetai)* (cf. I.1 n. 9), and "bring to completion" *(epitelein)* in this context (e.g., I.2.31 and I.4.2) indicates. On Posidonius' interest in aetiology see, for example, T85EK with Kidd *Comm.* 72–74.

the Earth. In this way it provides our longest daytime at the summer solstice.

95: But <when> it goes down to the south from here, and encounters circles that have sections above the Earth that are proportionately smaller than that of the summer tropic, it provides a shorter daytime, although the daytime remains longer than the nighttime until the Sun approaches the equinoctial circle. But when it causes the autumnal equinox by touching the equinoctial circle, it immediately goes through it[18] and touches circles that have smaller sections above the Earth, and so after the autumnal equinox nighttimes become longer than daytimes. The daytime gets continually shorter until the Sun approaches the winter tropic, while the nighttime remains longer than the daytime until, after the solstice at the winter tropic, the Sun provides a daytime that gets longer by turning back from this tropic to approach the equinoctial circle and causes the vernal equinox.

107: This is the situation with the parallel circles just described, and since those circles that are low [on the horizon] for us are high (i.e., elevated) for those in the contratemperate zone, and vice versa, our summer tropic is thus also their winter tropic through having its smallest section above the Earth, whereas their summer tropic is our winter tropic.[19] This [contrast] is the cause of the seasons' (that is, of the lengthenings and shortenings of daytimes) being reversed in the contratemperate zones, and it is the general cause of the universal lengthening and shortening of daytimes and nighttimes.[20] Nothing like this, however, occurs in the mid-torrid zone where instead there is a permanent equinox, since equal parts of all the parallel circles are above and below the Earth.[21]

18. At I.4.37–43 this motion is linked with the angle of the ecliptic at the equinoctial circle.

19. See Figure 5(b)–(c).

20. This completes the plan outlined at I.1.270–273. It will be refined in I.4 (cf. especially lines 232–239) with reference to the relation between nighttimes and daytimes considered over the period of the whole year.

21. Cf. lines 51–54 above, and I.4.234–235.

CHAPTER FOUR[1]

1: By effecting its motion based on choice through the zodiac the Sun occupies different parts of it at different times; in that way it completes [the cycle of] the seasons. It causes the summer solstice when, in very close proximity to our habitation,[2] it describes its northernmost circle and causes the longest daytime and the shortest nighttime. It causes the winter solstice when, on getting farthest from our habitation (that is, in its lowest position in relation to our horizon), it describes its southernmost circle, and causes the longest nighttime of the year and shortest daytime. It causes the vernal equinox when, in its course from the winter solstice to the north and the summer tropic, and located precisely halfway between both of them in its course, it describes a circle that divides the heavens into two equal parts and causes daytime to be equal to nighttime. It causes the autumnal equinox when, on turning back from the

1. This chapter recapitulates I.3.76–106 before offering a more elaborate analysis (lines 18–89) of what causes variations in daytimes and nighttimes, a topic it later (lines 147–196) pursues with special reference to the northern hemisphere. Despite a digression (lines 90–146) on Posidonius' account of terrestrial zones, this discussion is as coherent as, for example, II.1, and is kept as a single unit here instead of being divided, as it has traditionally been, into four separate chapters.

2. On the latitude of this habitation see I.3 n. 9.

summer tropic to the south and the winter tropic, and likewise on getting precisely halfway between the two, it describes the same equinoctial circle. [To sum up], the Sun provides daytimes that increase in length when it turns back from the winter tropic to the northern [parts] of the heavens, and [daytimes] that decrease in length when it goes down in the opposite direction from the summer tropic to the south and the winter tropic.

18:[3] The lengthening of daytimes and nighttimes does not add and subtract an equal amount during each [complete] day,[4] but when the daytime starts to be lengthened, then in the first month it grows longer by 1/12 of the whole amount by which the longest daytime exceeds the shortest; in the second by 1/6 [of that amount]; in the third by 1/4; in the fourth by 1/4 again; in the fifth by 1/6; and in the sixth by 1/12. Thus if the longest daytime exceeds the shortest by 6 hours,[5] then in the first month 1/2 hour will be added to the daytime; in the second 1 hour; in the third 1 1/2 hours, so that the addition amounts to 3 hours in the three-month period; and in the fourth month 1 1/2 hours will again be added; in the fifth 1 hour; in the final month 1/2 hour. In this way the 6 hours by which the longest daytime exceeds the shortest will reach their total.

30:[6] The cause of these additions' being unequal is the following. The zodiacal circle[7] through which the Sun effects its course is slanted and so

3. With this paragraph cf. Gem. *Isag.* 6.29–50 and Aujac (1975) 38 n. 1.

4. This "day" *(hēmera)* is the period defined by two successive sunsets, what the Greeks usually called a *nukhthēmeron* (the term used at lines 72–89 below). Our terms "daytime" and "nighttime" for *hēmera* and *nux* (cf. I.1 n. 68) distinguish the periods of solar illumination and darkness within any such "day."

5. This is the situation at the Hellespont (54°N); see Ach. *Isag.* 57.2–6, Hipparch. *In Arat. et Eudox. Phaen.* 1.3.7 (26.16–23), *Schol. in Arat. vet.* 304.3–5 and 305.9–306.5, and Ptol. *Alm.* 2.6, 109.17–18. Cf. II.1.442, where 9 hours is given for the shortest nighttime at the Hellespont.

6. With this paragraph cf. Gem. *Isag.* 6.34–39.

7. That is, the central circle of the zodiacal band, as is explained at lines 52–53 below. The Sun appears to go "through" it, in the sense that it follows it in its course. That is, this zodiacal circle is the trace of the heliacal circle on the zodiacal band, and is sometimes itself called the "heliacal circle"; cf. I.2.46–52, and I.2 n. 18.

intersects with the equinoctial circle at two points, while touching each of the tropics at one point. It intersects with the equinoctial circle and adjacent parallel circles more directly,[8] but abuts on the tropical circles more obliquely, that is, at a more inclined [angle]. Because it produces acute angles in this latter way, it becomes the cause of [the Sun's] approaching and distancing itself from the tropics more slowly. The Sun, that is, as it effects its lengthy course through the zodiacal circle, distances itself from the tropics more slowly, whereas at the equinoctial circle, where the zodiacal circle is more upright, it effects its approach and withdrawal from it more abruptly.[9] Providence, in other words, has marvellously fashioned the relation of the zodiacal circle to the tropics in such

8. Here we have revised Todd's text by deleting at line 34 a phrase seemingly designed to explain "more directly" (*ὀρθότερος*) in the same way as "more obliquely" (*πλαγιώτερος*), which is explained in the next clause by the phrase "at a more inclined [angle]" (*ἐπὶ πλέον ἐγκλινόμενος*). The deleted phrase is *καὶ ὀλίγου δεῖν πρὸς ὀρθὰς γωνίας*, "i.e., almost at right angles." But this is obviously untrue of the angle in question, which is approximately $23\frac{1}{2}°$. Neugebauer (1975) 961 n. 4 recognized the problem, and proposed deleting *γωνίας* ("angles") and translating the phrase as "almost as a straight line." He explained this as meaning "under a constant angle, in contrast to a changing direction farther away until tangential contact with the tropics," and noted an identical Latin expression ("paene directim") at Mart. Cap. 8.878. However, the common phrase *πρὸς ὀρθάς* would then have to be used unprecedentedly to mean *πρὸς ὀρθὰς γραμμάς*, whereas *γωνίας* is normally understood in this phrase, as at *Caelestia* II.5.70. Also, the Greek for "in relation to a straight line" would presumably be *πρὸς εὐθεῖαν γραμμήν*, by analogy with the standard phrase for "in a straight line," *ἐπ' εὐθείας* (*γραμμῆς*) (e.g., II.6.182). The adjective *ὀρθός* also normally refers to lines that are perpendicular to one another. We suspect that *καὶ ὀλίγου δεῖν πρὸς ὀρθὰς γωνίας* was derived from a gloss, designed to balance the genuine gloss on *πλαγιώτερος*, its intrusive status suggested by its awkward location after the main verb *τέμνει*. (For another deleted gloss see II.5.145–146, with II.5 n. 32.) The glossator was perhaps either impressed (consult Figure 6) by the near equality of the angles of intersection at C and D, or else was exaggerating the angle of intersection at these points in order to sharpen the contrast with what happens at A and B. In either case, the thought is so poorly expressed that we decline to attribute it to Cleomedes.

9. See Figure 6.

a way as to ensure that changes in the seasons occur imperceptibly rather than abruptly.[10]

44: The time intervals between the tropics and the equinoctial circle are not equal either: from the vernal equinox to the summer solstice there are 94½ days; from the summer solstice to the autumn equinox 92½ days; from this equinox to the winter solstice 88; and from the winter solstice to the vernal equinox 90¼.[11]

49: So the problem arises: given that the four quarters of the zodiacal circle are equal, why does the Sun not complete its passage through them in an equal time? Now the answer to be given is that if the Sun effected its course through the zodiacal circle itself, it would go through all its parts in an equal time. But the heliacal circle is in fact located below the central circle of the zodiacal band, and at a position much closer to the Earth. Yet if, despite being located below the zodiacal circle, the heliacal circle had the same center as the zodiacal circle, the Sun would also go through the four parts of its own circle in an equal time, since then the diameters drawn out from the tropical and equinoctial [points of the zodiacal circle] would also divide the heliacal circle into four equal parts. But in fact [these two circles] do not happen to have the center. Instead, the heliacal circle is eccentric [relative to the zodiacal circle], and for this reason is not divided into four equal [parts] by the diameters just mentioned. Rather, its arcs are unequal, since only circles with identical centers have arcs equally divided by diameters, whereas circles that do not have the same centers do not.[12]

62: So since the heliacal circle is eccentric, then if it is divided into

10. Cf. II.1.396–399, and Cic. *De nat. deor.* 2.49.

11. See Figure 7(a). Other sources report a year of 365¼ days, with 88⅛ days in the period from the autumn equinox to the summer solstice, and 90⅛ days in that from the winter solstice to the spring equinox. See Gem. *Isag.* 1.13–16, Ptol. *Alm.* 3.4, 233.21–24 and 237.20–238.4 (with Hipparchus' prior authority invoked at 238.3–4), and Theon *Expos.* 153.6–12.

12. For similar demonstrations see Gem. *Isag.* 1.31–34, Theon *Expos.* 153.16–158.11, and Mart. Cap. 8.848–849.

twelve [parts], just like the zodiacal circle, there will be unequal sections of the heliacal circle located below equal sections of the zodiacal circle. Its largest section will be that located below Gemini, its smallest that below Sagittarius:[13] that is also why [the Sun] goes through Sagittarius in the shortest period, but through Gemini in the lengthiest, since it is at its greatest height in Gemini, but closest to the Earth in Sagittarius, while proportionately [distant] in the other signs. So consequently the Sun's circuit is also eccentric since it does not always move at the same height,[14] but in accordance with its course it moves both on high and back toward [points] closer to the Earth.[15]

72:[16] Nor are all the intervals of a nighttime and a daytime equal to

13. A *zōidion* may be either a zodiacal constellation or a zodiacal sign (a *dōdekatēmorion* or one-twelfth part of the zodiacal circle; see n. 17 below). Here Gemini and Sagittarius are zodiacal signs.

14. "Height," here and elsewhere, translates *hupsos,* which in cases such as this refers to the distance of a celestial body from the center of the cosmos, the Earth.

15. See Figure 7(b).

16. As noted in the Introduction (nn. 11 and 12) in connection with the dating of the *Caelestia,* the argument at lines 72–80 here parallels Gem. *Isag.* 6.1–4, and both passages offer a less sophisticated analysis of day lengths than Ptolemy's at *Almagest* 3.9. Also, Geminus uses the periphrasis "a nighttime and a daytime added together" *(to sunamphoteron nux kai hēmera)* instead of Cleomedes' term "interval of a nighttime and a daytime" *(nukhthēmeron).* His passage is as follows: "[1] 'Day' is used in two senses: in one sense it is the interval of time from a rising of the Sun to a setting, but in the other sense 'day' is used for the interval of time from one rising of the Sun to the next. [2] In the second sense the day is [identical with] the revolution of the heavens *and* the rising of the arc that the Sun traverses as it moves in the direction opposite to the heavens during their revolution. [3] That explains why a nighttime and a daytime added together is also not precisely equal to every nighttime and daytime. Instead, while their lengths are equal relative to perception, relative to precise reckoning there is some slight and imperceptible variation. [4] The reason is that while the revolutions of the heavens are in equal intervals of time, the risings of the arcs that the Sun traverses during the revolution of the heavens are not. That explains why a nighttime and a daytime added together is not <equal> to every nighttime and

one another by precise reckoning, as is supposed, but only in relation to perception. That is because the revolution of the heavens themselves is necessarily less than every interval of a nighttime and a daytime, given that in a whole course the heavens complete their own circuit more quickly than in the interval of a nighttime and daytime that the Sun proceeds through as it goes in the opposite direction to the heavens. For when the heavens have come right round to the same point, the Sun is not yet observed in the east; instead it is only when the arc of the circle that the Sun in accordance with its motion based on choice completes during the interval of a nighttime and a daytime is elevated that it too is seen in the east. So if all the *dōdekatēmoria*[17] of the zodiacal circle, which are equal, also rose in an equal time, every interval of a nighttime and a daytime would consequently be equal as well. But in fact the summer signs rise upright and set obliquely, and as they rise upright the period of their rising is longer, and so the parts of them through which the Sun goes in the interval of a nighttime and a daytime rise proportionately more slowly.[18] But the opposite occurs with the winter signs.[19] Thus the revolutions of the aether are equal, but the intervals of a nighttime and a daytime are not, at least on the most precise reckoning.

90: The Sun, as we have said,[20] approaches the tropics and withdraws from them rather slowly, and for that reason spends a longer time near

---

daytime added together." In the final sentence of sect. [4] we delete *pasa* ("every") before the first use of "a nighttime and a daytime added together" to create the required parallel between this sentence and the first sentence of sect. [3]. Also, the supplement "<equal>" should have the form *isē* in Greek, not *ison* (Aujac), again as in the first sentence of [3].

17. As the technical term for the zodiacal signs, that is, for those twelfth parts of the zodiacal circle that are named after certain constellations, we leave *dōdekatēmorion* transliterated, as also at I.7.21, I.8.38, and II.1.319 and 328. For the definition and identification of the *dōdekatēmoria* as zodiacal signs see Gem. *Isag.* 1.1–4.

18. See Figure 8.

19. See Ptol. *Alm.* 2.7 on rising times.

20. Lines 30–40 above.

them. Also, the parts below the tropics are not uninhabitable, nor are those still further south. Syene[21] is in fact located [directly] below the summer tropic,[22] and Ethiopia farther south than this.[23] Taking his key from this [evidence], Posidonius believed that the whole latitude below the equinoctial circle was also temperate.[24] And where the reputable natural philosophers had claimed that there were five terrestrial zones, he alone claimed that the one they called "torrid" was inhabited and temperate. That is, he argued that[25] *(a)* if the [latitudes] below the tropics are not uninhabitable, nor those still farther within them,[26] despite the Sun's spending longer there, how could the [latitudes directly] below the equinoctial circle not be much *more* temperate, since the Sun approaches this circle rapidly and again distances itself at an equal speed, and does not spend a prolonged time at that latitude, when moreover, as he says, *(b)* the nighttime there is always equal to the daytime, and for this reason has a length appropriate for cooling <the air>? *(c)* Since this air is also in the exact center (i.e., most voluminous [part]) of the [nocturnal] shadow, there will be rains and winds that can cool the air, because even in Ethiopia rains reportedly fall continuously in the summer, and espe-

21. Its contemporary name is Aswan. The same location is assigned it at Plut. *De def. or.* 411A and Strabo 2.5.7. See also I.7.71–72.

22. See I.5.59–60; I.7.71–72; II.1.211–212 and 270.

23. The Ethiopians (located in modern Sudan) were often identified (e.g., by Homer *Odyssey* 1.22–24; cf. *Iliad* 1.423) as the most equatorial race in the known world, just as they had been the most peripheral on a flat Earth; see Gem. *Isag.* 16.28 with Aujac (1975) 152 n. 3.

24. Posidonius, in addition to lines 90–131 here (= F210EK), addressed issues involving zones at F49.1–145 EK; cf. also F209 and F211EK. His treatise on this subject, *On the Ocean*, was known to Strabo.

25. The three explanations that follow may be "alternative possible hypotheses" (Kidd *Comm.* 136), but they are not mutually incompatible as are the explanations of paradoxical eclipses anonymously proposed at II.6.168–177 (see II.6 n. 21). I. G. Kidd (1978a) 14 mistakenly tries to use such multiple explanations to support his interpretation of Posidonius' philosophy of science in F18EK as an activity based on hypotheses. See Appendix, especially n. 7.

26. I.e., closer to the equinoctial circle.

cially at its height, and these are also thought to be the source from which the Nile floods during the summer.[27]

109: That, then, is Posidonius' position. And if this is the situation with the regions below the equinoctial circle, then the seasons will have to occur there twice a year, since the Sun is certainly at their zenith twice, inasmuch as it causes two equinoxes.[28]

113: Those who oppose this opinion of Posidonius argue [as follows]. *(a)* As far as the Sun's spending longer at the tropics is concerned, his doctrine would have to be sound. Yet, in addition, the Sun distances itself a considerable amount from the tropics, and so the air below them also cools off a considerable amount. As a result, those latitudes can be inhabited. But it distances itself just slightly from the equinoctial circle (which is halfway between the tropics), and effects a rapid reversal of direction toward it.[29] *(b)* The [latitudes] below the tropics receive annual winds from the frigid zones, and these moderate the extreme heat from the Sun by cooling the air.[30] But they cannot penetrate as far as the

27. On *(c)* see Strabo 2.3.3 (F49.51–61 EK) and 2.3.3 (F49.134–135 EK).

28. This parenthesis could be an implication that Posidonius himself drew, or Cleomedes' interjection, based on I.3.88–89 and 98–106, and lines 8–13 above, where the vernal and autumnal equinoxes are distinguished. The reasoning is that although the "seasons" at the terrestrial equator are not distinguishable by the usual criterion of the length of the daytimes and nighttimes (defined by the Sun's return overhead and by its departure; cf. lines 234–235 below), they are definably different in temperature, and in this respect form two pairs of seasons which are defined by the Sun's return to an equinox and departure for a solstice. This claim reinforces the arguments for equatorial habitability in *(a)*–*(c)* that precede, but in such general terms that it is probably best taken as a Cleomedean aside.

29. Since the distances involved here depend on the size of the Earth, this observation may reflect some debate about the distance between the terrestrial tropics and equator; see Strabo 2.2.2 (Posid. F49.10–36 EK) with Kidd *Comm.* 223–228.

30. See Strabo 3.2.5 (Posid. T22EK) on Posidonius' account of annual ("etesian") winds. At *Aratea* 97.1–6 this argument is attributed to an earlier Stoic, Panaetius (ca. 185–109 B.C.).

equinoctial circle. If they do, then the Sun, given the length of its course, will make them hot, indeed flaming hot. *(c)* The equality [below the equinoctial circle] of nighttime to daytime would not by itself have the power to cool the air there, given that the Sun has an indefinable power, and at all times sends out its ray toward that latitude in a perpendicular and intense form, since it certainly does not significantly slope away from that latitude. (The natural philosophers suppose that most of the Great Sea, which is centrally placed in order to nourish the heavenly bodies,[31] is inserted at this latitude.)[32] Thus in this case Posidonius does not seem to adopt the correct view.

132: But on [Posidonius'] hypothesis[33] that the whole Earth is inhabited, some habitations will be encircled with shadows *(periskioi)*, others shadowed unidirectionally *(heteroskioi)*, and others shadowed bidirectionally *(amphiskioi)*. Encircled with shadows are those habitations below the poles where the year will be [equally] divided into daytime and nighttime, the equator will be the horizon, and the same six signs of the zodiac will always be above and below the Earth.[34] Shadows there accordingly describe a circle, and make people "encircled with shadows," since in latitudes below the poles the heavens revolve just like millstones.[35] Shadowed unidirectionally are the temperate habitations, since whenever the Sun is in the south, people in the northern zone have shadows that slope toward the north, while those in our contratemperate zone have them sloping toward the south.[36] Shadowed bidirectionally will be those

31. On the equatorial ocean as excluded from Posidonius' geography see F118EK with Kidd *Comm.* 459–461.

32. "Inserted" *(hupoballesthai)* may have teleological implications linked to the geographical symmetry between terrestrial zones (on which see I.1 n. 66).

33. F210EK ends at line 131, but Theiler (F284) correctly includes lines 132–146 as a logical extension of lines 90–131. For its classification of zones see Strabo 2.5.37 and Ach. *Isag.* 66.28–67.6; also, cf. Ptol. *Alm.* 2.6, at 102.9, 107.14, and 114.23, and see I.6 n. 10.

34. See further lines 225–231 below.

35. See Figure 9(a).

36. See Figure 9(b).

below the equinoctial circle: for when the Sun leaves the equinoctial circle for the south (that is, the winter tropic), their shadows slope toward the north but when it travels to the summer tropic from the equinoctial circle, they would point toward the south.[37] That, then, is [Posidonius'] distinction between the zones of the Earth.

147: But what must also be realized here[38] is that while daytime lengthens and shortens over the same time period for everyone occupying our temperate zone, the addition and subtraction [of daytime] is still not equal at all [latitudes]. Instead, there is a major contrast between them, with some having a minimal addition and subtraction, others a very large one, and still others an intermediary one. This is caused by the heavens' not sloping equally at all [latitudes], and the north pole not being elevated [everywhere] an equal number of degrees from the horizon, but just minimally for anyone living in the south, more for anyone in the north, and an intermediary amount for anyone in between.

156:[39] To explain. Those traveling to the south from the north necessarily have the north pole at a lower elevation and the heavens with less of a slope, while the opposite holds for those going away from there to the north. That is because on the Earth [directly] below the equinoctial circle each of the poles, as we have said,[40] is observed on the horizon, and all the parallel circles have equal sections above and below the Earth. There the axis [of the cosmos] is also the diameter of the horizon, and there are no stars that are either always visible or always out of sight. But because of the spherical shape of the Earth, those who come toward us from the mid-torrid zone have our pole elevated, have altered horizons,

37. See Figure 9(c).

38. "Here" must refer to lines 1–89 above, the opening discussion of the lengths of daytimes. Lines 147–231 form in effect a supplement to lines 18–29 above, based on the scheme of terrestrial latitudes recapitulated from I.3.1–68.

39. Cf. I.3.6–12 and 15–18, where the situation at the equator is also taken as the point of reference. Only here (lines 161 and 164) is the axis of the cosmos (on which see Gem. *Isag.* 4.1) mentioned.

40. At I.3.9, 23–24 and 52–54.

and have the axis [of the cosmos] no longer as the diameter of any [horizon] because the slope that develops as the heavens are elevated above the plane depends on the elevation of the pole.

166:[41] There are also different arctic circles at different [northern] latitudes depending on alterations in horizons. This is because the arctic circles enclosing the stars that are always visible at each latitude must be described by the pole and the distance [from the pole] <at> each latitude's horizon.[42] So for people living near the mid-torrid zone (that is, south of us), the arctic circles are very small because the heavens slope minimally, and the [north] pole appears minimally elevated above the horizon. But for people in the north (that is, near the frigid zone) it is necessary that the arctic circles be very large, the pole have a significant elevation above the horizon, and the heavens consequently slope to an extreme degree. But for people right in between the north and the south (and this includes the Greeks and everyone at the same latitude),[43] all the [phenomena] just described occur to an intermediary degree.

179: Those parallel circles that are divided by the horizons at each latitude are, therefore, also divided equally in the mid-torrid zone, but unequally at the other latitudes. The larger and smaller sections above

41. Cf. with this paragraph I.3.32–35 and 44–68.

42. The idea behind "be described" *(graphesthai)* seems to be that, as the celestial sphere makes its daily rotation, the arctic circle is traced on the celestial sphere by the northernmost point of the horizon. The text that we have translated restores a deletion in Todd's edition and adds a supplement, so that at line 169 we now read *γράφεσθαι πόλῳ καὶ διαστήματι <πρὸς> τῷ παρ' ἑκάστοις ὁρί-ζοντι*. That is, we take the distance between the pole and the horizon to be defined "at" the horizon, and for "at" we have supplied a preposition that defines locality much as it does in the standard phrases "in the north/south" (*πρὸς ἄρκτῳ/νότῳ*). In so emending we reject the possibility that the Greek as it stands can mean that the arctic circles are "drawn by the pole and the distance to the horizon" ("dessinés par le pôle et la distance à l'horizon," Goulet). That is the kind of meaning clearly needed by this text, and we have tried to elicit it with minimal intervention, though we concede that the solution may not be entirely satisfactory.

43. Given I.3.39–42 (cf. I.3 n. 9), this latitude would be 36°N.

and below the Earth will balance out at those latitudes, and in this way the lengthenings and shortenings of daytimes will also balance out proportionately.[44]

184: Those living adjacent to the mid-torrid zone do not have daytimes that significantly lengthen and shorten, since there the heavens slope minimally, and cause a slight variation in the division of parallel circles into unequal [sections] by the horizon. But people at the latitude adjacent to the frigid zone have an extreme variation in the lengthenings of daytimes and nighttimes, since there the heavens have an extreme slope, and also because the pole has a significant elevation above the horizon and for this reason makes the arctic circle there extremely large, resulting in its not being far distant from the summer tropic. The horizons at those latitudes consequently divide the parallel circles unequally in so extreme a disproportion that the lengthenings and shortenings of daytimes also involve an extreme variation.

197: For example, in Britain, when the Sun is in Cancer and causes the longest daytime, the daytime reportedly consists of 18 equinoctial hours and the nighttime of 6,[45] so that over this time interval there is light at this latitude during the nighttime, since the Sun runs alongside the horizon and sends out its rays over the Earth. (This of course also happens at our latitude, when the Sun approaches the horizon, since its light is well in advance of its rising.) So in Britain too there is light during the night, so that even reading becomes possible. In fact this is said to be absolutely necessary because, due to the section of the summer tropic below the Earth being minimal at this latitude, the Sun at this time effects its course alongside the horizon without going too far below the Earth.

44. On this "balancing out" *(summetria)* see lines 232–239 below. Over the course of a year each latitude will receive in principle (cf., however, I.1. n. 70 and n. 51 below) an equal amount of light and darkness.

45. See II.1.444. At Ptol. *Alm.* 2.6, 113.6–9 the parallel at 58°N ("through the southern part of Ireland") is assigned a longest daytime of 18 hours. For less precise data see Gem. *Isag.* 6.7–8, and Plin. *NH* 2.186.

208:[46] In the island called Thule (said to have been visited by Pytheas, the philosopher from Marseilles) the whole summer tropic is reportedly above the Earth, and in fact it becomes the arctic circle at that latitude.[47] When the Sun is in Cancer there, the daytime will last 1 month, if all the parts of Cancer are also always visible there; but if not, then to the extent that the Sun is present in the parts of that sign that are always visible.

213:[48] For those proceeding to the north from this island other parts of the zodiacal circle in addition to Cancer will become always visible in due proportion, and this means that as long as the Sun is going through those parts of the zodiacal circle that are [always] visible above the Earth at each latitude, it will be daytime.

218: And necessarily there exist latitudes of the Earth where the daytime is 2 and 3 months long, and 4 and 5 months, too.[49] Also,[50] since directly under the pole there are 6 signs of the zodiac above the Earth, then as long as the Sun is going through these signs, which are always visible, it will be daytime, since here the same circle is the horizon, the arctic circle, and the equinoctial circle. That is, people at Thule have the summer tropic coinciding with the arctic circle, but people still further north have the arctic circle exceeding the summer tropic relative to the parts [of the heavens] leading to the equinoctial circle, and this occurs proportionately. But for those directly under the pole the equinoctial cir-

46. At Ptol. *Alm.* 2.6, 114.9–11 Thule, an island north of Britain, perhaps identifiable with Iceland or part of Scandinavia, is assigned a latitude of 63°N, with a longest daytime of 20 hours. The latitude at which the longest daytime is a month is 67°N (Ptol. *Alm.* 2.6, 115.8–16). For the claim by Pytheas (probably late fourth century B.C.) that the arctic circle and summer tropic coincide at Thule see Strabo 2.5.8, Kidd *Comm.* 745–746, and Roseman (1994) 104–109, who translates and discusses Cleomedes' report (lines 208–210 = Pytheas T27 Roseman) and its context.

47. See Figure 10(a).

48. Cf. with lines 213–219 Ptol. *Alm.* 2.6, 115.8–116.20.

49. See Figure 10(b).

50. Cf. with lines 219–231 Ptol. *Alm.* 2.6, 116.21–117.9.

cle assumes all three relations: it is the arctic circle because it encloses the stars that are always visible, since at this [latitude] absolutely no star sets or rises; it is the horizon because it separates the hemisphere of the heavens that is above the Earth from the one below the Earth; and it is the equinoctial circle because it alone divides daytime and nighttime equally at that latitude, and while everywhere else too it is the equinoctial circle to an equivalent degree, it is no longer either a horizon or an arctic circle.

232: These, then, are the variations in the lengthenings and shortenings of nighttimes and daytimes, although the darkenings and illuminations of the air are made equal at all latitudes. In the mid-torrid zone, that is, nighttimes are always equal to daytimes, but at the other latitudes this kind of equalization is achieved differently: that is, the longest daytimes at each latitude are made equal to the longest nighttimes,[51] and neither the darkenings nor illuminations of the air exceed one another, but the year as a whole divides them equally.

239: The cause of the whole variation in the cases just described is the Earth's shape, which is spherical, as a fortiori is the whole cosmos itself. In other words, none of the [phenomena] just described could occur with other kinds of shapes. We shall demonstrate next that both the whole cosmos, and its most significant parts, do have this shape.[52]

51. Since atmospheric conditions make the longest daytime longer than the longest nighttime (see I.1 n. 70), this assumption of exact symmetry does not in fact hold true.

52. The Earth's sphericity had previously been assumed; see I.1.224, 245–246, I.3.2–3, and I.4.163.

CHAPTER FIVE

1: Now sight alone seems to suggest that the cosmos is a sphere,[1] but *that* must not be made a criterion for its shape, since everything does not in fact usually appear to us as it [really] is.[2] It follows that it is on the basis of what is "clearer" (that is, what is presented cognitively to us)[3] that, in accord with what is patently implied,[4] we should aim to arrive at things

1. Cf. I.1.176–191, where cosmic sphericity was demonstrated on the basis of "what is presented in perception," or, as here, "sight" *(opsis)*.

2. On the Stoic concept of the criterion of truth, and the methodology summarized in this opening paragraph, see also the Introduction.

3. The terminology in the phrase *apo tōn enargesterōn kai kataleptikōs hēmin phainomenōn* is used elsewhere by Posidonius; see T83.2, F159.2 and F169.45 EK, and cf. Kidd *Comm.* 74 who rightly calls it a "criterion." The words *katalēptikōs phainomena* that gloss "clearer" also embody the technical Stoic term for a cognitively reliable sense presentation, the *kataleptikē phantasia* (see Introduction with n. 29).

4. This translates *kata tēn phainomenēn akolouthian*, a phrase difficult to interpret without a context for its use. Presumably the implication *(akolouthia)* is "apparent" (the literal meaning of *phainomenē*) because human reason recognizes it as clear or evident, and our translation "patently" is designed to reflect this through alternative language. Thus at lines 126–138 below (which are anticipated at lines 6–9 here) there are two inferences (or "transitions"; cf. *metiontes*, line 8)

that are not in and of themselves fully displayed. Accordingly, if we demonstrate that the most solid (that is, the most compact) part of the cosmos, the Earth, has a spherical shape, we could easily learn by a transition from this to its remaining parts that they are all spherical, and in this way that the whole cosmos too has this sort of shape.[5]

10: There have been numerous differences among earlier natural philosophers about the shape of the Earth: some, by following only the sense presentation based on sight, claimed that its shape was flat (i.e., a plane);[6] others, who supposed that water would not be stationary on an Earth that did not have a "deep" (i.e., concave) shape, said that it had just that shape;[7] others claimed that the Earth was shaped like a cube (i.e., square); others that it was shaped like a pyramid.[8] Our school, as well as

---

from the sphericity of the Earth to that of the air and aether (and thus the whole cosmos), and these are grasped via physical theory and its geometrical manifestations. In the main body of this chapter too (lines 20–113) implications are grasped as "evident" within more restricted contexts in the arguments for the Earth's sphericity, and so the phrase *kata tēn phainomenēn akolouthian,* despite its use here in connection with the larger cosmological picture, has completely general application.

5. The structure of the cosmos is itself unobservable, that is, it is one of the things "not in and of themselves fully displayed" *(mē autothen ekphanē).* It is displayed to the extent that we can see the air and aether, but their sphericity can only be inferred with the help of physical theory (see lines 128–134 below); it cannot be inferred from the phenomena alone, as can the sphericity of the Earth (see lines 104–113 below).

6. While the flat-Earth cosmology is associated with Homer and early Greek thought (e.g., by Gem. *Isag.* 16.28), the present text may apply to the Epicureans (on whom see Furley [1996]), who at II.1.2–5 are similarly described as having inferred the Sun's size from its "visual sense presentation"; see also II.1. n. 101.

7. Democritus (DK 68A94) thought that there was a central hollow on a flat and oblong Earth.

8. For other shapes see Ptol. *Alm.* 1.4, 15.23–16.7 for the cylinder, and Theon *Expos.* 120.23–121.1 and Eucl. *Phaen.* 4.26–6.14 for the cone and cylinder. See also the Epicurean polemic at Cic. *De nat. deor.* 1.24.

all the scientists and most of the Socratic school, affirmed that the shape of the Earth was spherical.[9]

20: Now since no other shape beyond those mentioned could be appropriately attached to the Earth, the following disjunction would be necessarily true: *the Earth is either flat (i.e., a plane), or concave ("deep"), or square, or pyramidal, or spherical in shape.* So having posited this disjunction as true, then by going forward on the basis of what the logicians call "the fifth undemonstrated [argument constructed] through multiple [disjuncts],"[10] we shall demonstrate that the Earth has a spherical shape. That is, we shall state, as indeed we shall demonstrate, that *the Earth is neither flat, nor concave, nor square, nor pyramidal.* We shall then conclude that *it is absolutely necessary that the Earth be spherical.*

30: That the Earth is not a plane we may learn from the [following arguments]. *(a)* If it were flat (i.e., a plane) in shape, then there would be a single horizon for all peoples: for it is inconceivable how, if the Earth had such a shape, horizons would alter. And given a single horizon, the Sun's risings and settings, and thus also the beginnings of daytimes and nighttimes, would occur at the same time everywhere. But this does not in fact happen; instead, in terrestrial regions the variation in the [phenomena] cited is clearly very considerable, in that the Sun sets and rises at different times at different places. For example, the Persians who live in the east are said to encounter the onset of the Sun 4 hours earlier than the Iberians who live in the west.[11] This [variation] is also proven

9. For the Stoics ("our school"; cf. Posid. T85EK) see *SVF* 2.648 and Posid. F18.20 and F49.6–7 EK; for Plato see *Phaedo* 108e4–109a6 (with Furley [1989b] 23–26). The "scientists" *(hoi apo tōn mathēmatōn)* include, but are not restricted to, mathematical astronomers (see II.6.122).

10. On this argument see *SVF* 2.241 (80.17–20) and 2.245 (82.34–83.3), and Sext. Emp. *PH* 2.158. It has the form "either p or q or r," "not p and not q," "therefore r." The way that "not p" and "not r" are demonstrated in this chapter is via the Stoics' "second undemonstrated argument" (e.g., *SVF* 2.242, p. 81.1–12), their equivalent of *modus tollens;* see n. 28 below.

11. For the Iberians as the westernmost inhabitants of the known world see lines 42–43 and 78 below, and II.1.459–461; also Strabo 1.4.5 and Aujac (1966)

from other [phenomena] but particularly from the eclipses of the heavenly bodies, which, though they are eclipsed everywhere over the same time period, are still not detected at the same hour. Instead, the [heavenly body] that is eclipsed among the Iberians in the 1st hour is detected as undergoing an eclipse among the Persians in the 5th hour, and proportionately elsewhere.[12] *(b)* If the Earth's shape were flat, the pole would be seen by everyone at an equal distance from the horizon, as would the same arctic circle. Nothing like this is present in the phenomena; instead, the elevation of the [north] pole is seen as very small for the inhabitants of Syene and for the Ethiopians, yet as very great among the Britons, and proportionately at intervening latitudes.[13] *(c)* For someone leaving for the north from the south some of the stars that are seen in the south are concealed, and others previously out of sight come to be seen in the north, and the opposite happens to anyone who

---

184. Strabo 1.4.5 and others usually cite India rather than Persia as the easternmost location.

12. The sphericity of the Earth is best demonstrated from the observation of lunar eclipses at different times and locations; see Ptol. *Alm.* 1.4, 15.12–13 and Theon *Expos.* 121.5–12. Plin. *NH* 2.180 uses both lunar and solar eclipses in his argument for the Earth's sphericity, but because of lunar parallax, a solar eclipse will differ for different observers "at the same hour," and in some cases will not be seen at all. It is unusual in antiquity to find comparisons of the Sun's rising times at different longitudes. It is worth noting that Cleomedes' estimate of a 4-hour difference between Persia and Iberia (Spain) may be based on reports of a lunar eclipse that was observed simultaneously at Arbela on Sept. 20, 331 B.C., during Alexander's campaign and at Carthage. Ptolemy (*Geog.* 1.4) says that the eclipse occurred in the 5th hour (of night) at Arbela and in the 2nd hour at Carthage, which would make the difference between times of sunrise there roughly 3 hours (though it should only be about 2¼ hours). Pliny's report (*NH* 2.180), which is better so far as it goes, maintains that the eclipse took place in the 2nd hour at Arbela, but says only that it was seen at the same time that the Moon rose (that is, at sunset) in Sicily. See Neugebauer (1975) 667–668 and 938.

13. Syene is at the latitude of the northern tropic (I.4.93, and lines 59–60 below), while the Ethiopians are probably at the latitude of Meroe, the southernmost latitude identified in lists of latitudes; see II.1.440–441.

might go south from the north.[14] None of this would happen if the Earth had a flat shape that caused a single horizon. Therefore the Earth does not have this shape. *(d)* Daytimes would also turn out to be of equal length for everyone,[15] although entirely the opposite is present in the phenomena.

57: *(e)* Indeed, if the Earth had a shape that was flat (i.e., a plane), the whole diameter of the *cosmos* would be 100,000 stades![16] Look.[17] *(1)* People at Lysimachia have the head of Draco overhead, while Cancer is located above the area of Syene. *(2)* The "arc [of the meridian]" between Draco and Cancer is 1/15 of the "meridian" passing through Lysimachia and Syene (as is demonstrated by sundials). [*(3)* Syene is 20,000 stades from Lysimachia.][18] *(4)* 1/15 of the whole "circle" is 1/5 of the "diameter."[19]

14. At I.3.12–43 the sphericity of the Earth was identified as the cause of these variations.

15. This is a consequence of there being a single horizon; see lines 30–39 above.

16. This argument concerns the size of the *cosmos*, not the Earth (see Goulet 200 n. 162 against Neugebauer [1975] 961–962 plus fig. 97 at 1406), and must originally have been a *reductio ad absurdum* that invoked a bizarre cosmology (familiar to Ptolemy; see *Alm.* 1.3, 11.14–24 and cf. n. 21 below) in which a flat Earth has the heavens parallel to it, rather than encircling it. This is the structure implied by assumptions *(3)* and *(4)* below; i.e., if 1/15 of the flat Earth's extent is 20,000 stades, its total extent is 300,000 stades, or the same as that of the heavens, which must therefore be parallel to it. Hence "arc," "meridian," "diameter," and "circle" need to be used in quotation marks in this passage to reflect their etiolated meaning when applied to such a nonspherical cosmology. However, by having the flat Earth surrounded by a sphere in this argument (see n. 21 below), Cleomedes, or his source, has misrepresented this reasoning. Hence the figures in this passage should be used with caution in historical or geographical reconstructions; for attempts at these see, for example, Neugebauer (1941), Collinder (1964), and Goulet 200 n. 162.

17. What follows is a "procedure" (*ephodos;* see Introduction n. 38), with four overt assumptions and an implicit axiomatic principle (see n. 22 below).

18. This is to anticipate the information given at line 68 below.

19. This is because the diameter of a circle is taken to be a third of the circumference, as at I.7.118–120, and II.1.299–300 and 321–322.

63: So if, by hypothesizing that the Earth is a plane,[20] we produce perpendiculars to it from the extremities of the "arc" extending from Draco to Cancer, they will [by *(1)*] touch the "diameter" [of the Earth] that measures the "meridian" through Syene and Lysimachia.[21] Thus the distance between the two perpendiculars will be 20,000 stades, since [by *(3)*] there are 20,000 stades between Syene and Lysimachia. So since [by *(2)* and *(4)*] this distance is 1/5 of the whole "diameter," the whole "diameter" of the "meridian" will be 100,000 stades. So if the cosmos has a "diameter" of 100,000 stades, it will have a "great circle" of 300,000 stades[22]—in relation to which the Earth, although a point [in relation to the cosmos], is 250,000 stades [in circumference], while the Sun is much larger than this, although it constitutes a minimal part of the heavens.[23]

20. In the calculations of the size of the Earth or the Sun in I.7 and II.1, the sphericity of the Earth and the cosmos are, of course, implicitly assumed.

21. See Figure 11. In order to reflect the misleading reasoning in this passage (see n. 16 above), the shapes of the flat Earth and flat heavens in this figure are finite and circular. They should be indefinite in shape and extent. As Ptolemy, *Alm.* 1.3, 11.14–24, argues, if *both* the Earth *and* the heavens are flat, the heavenly bodies will move in a straight line "to infinity," with no plausible explanation available for their rising and setting, or for their rectilinear motion ever being reversed. Thus the problem that Cleomedes should be posing a flat-Earther is that the cosmos is infinite, rather than, as is concluded here, minuscule.

22. The calculation is implicitly based on the axiom that two quantities have the same ratio to one another as the equimultiples of these quanta taken in corresponding order (cf. Eucl. *El.* 5 prop. 15). Thus if *d* is the diameter of a circle, and *c* is its circumference, then $d:c :: 1:3 :: 5 \times 20{,}000:15 \times 20{,}000$. The subsequent calculations of the circumference of the Earth, and the size of the Sun, also rely on this axiom (see I.7 nn. 9 and 21; II.1 nn. 58 and 73), which is known to have interested Posidonius (see I.7 n. 9). Perhaps the present argument, more appropriately articulated (see previous note), was originally Posidonius' *reductio ad absurdum* of the theory of a flat Earth.

23. The comment following the dash may be Cleomedes' insertion, since if he, rather than any source, is responsible for misunderstanding the cosmological assumptions of this argument (see n. 21 above), he could now be drawing on Eratosthenes' calculation of the size of the spherical Earth (I.7.109–110), and alluding to the later calculation of the size of the Sun (cf. II.1.324–325).

Surely it is obvious from this [argument] too that the Earth cannot be a plane?

76: That the Earth also does not have a "deep" (i.e., concave) shape may be seen from the following [arguments]. *(a)*[24] If its shape were like this, daytime would begin for the Iberians before the Persians, since the protrusion [of one side] of the Earth would obstruct [the Persians] who are close to it, without impeding the line of sight for [the Iberians] who are at a greater distance away. After all, when any hollow object is out in the Sun, the part of it on the Sun's side is in a shadow when the Sun rises, whereas the part directly opposite is illuminated.[25] So if the Earth's shape were concave, the outcome would be same throughout the whole world too: that is, people in the west would encounter sunrises earlier. But in fact the opposite is present in the phenomena. *(b)* Also, if the Earth had this sort of shape, then for people in the south, the north pole would be visible at a significant distance above the horizon, while it would be obstructed for people in the north by the protrusion [of the side of the Earth closest to them]. And by the same token, given this shape, more stars would always appear visible to people in the south, and consequently their arctic circle would be larger. Entirely the opposite to this is present in the phenomena. *(c)* Those living in the deepest part of the concavity would be unable to see the six zodiacal signs above the Earth, and thus not even half of the equinoctial circle. For example, when we descend to a deeper place and look back toward the heavens, we see a small part of it, not the whole hemisphere.[26] *(d)* Nighttimes would also always exceed daytimes, since the arc of the heavens located below the convex part [of the Earth] would far exceed the

24. Ptol. *Alm.* 1.4, 15.16–17 briefly notes this argument.

25. This use of any hollow object to illustrate the consequences of the Earth being hollow is in line with Stoic accounts of concept formation "by analogy"; see, for example, *SVF* 2.87 (with a parallel at 29.15–16).

26. As at I.8.130–132, valleys rather than subterranean locations (cf. II.1.38–44) must be intended here.

arc located above the concave part, assuming that the Earth is at the exact center of the cosmos.

98: If the Earth were cuboid (i.e., square), the result would be a daytime consisting of 6 hours and a nighttime of 18, given that each side of the cube would be illuminated for 6 hours. But if the Earth were like a pyramid too, each side of it would be illuminated for 8 hours.[27]

102: So if the phenomena demonstrate that the Earth has none of these shapes, it is necessarily spherical in accordance with the fifth [undemonstrated argument] from multiple [disjuncts]. But it can also be demonstrated directly that the Earth is spherical by starting out in the same way from the phenomena, given that precisely the same [phenomena] that demonstrated that the Earth had none of the [non-spherical] shapes listed above demonstrate that it is spherical.[28] That is, horizons on the Earth alter, and the following are not observed as being identical everywhere: the stars in the north and the south, the elevation of the pole, the size of the arctic circle, and the lengths of nighttimes and daytimes. All these phenomena clearly demonstrate that the shape of the Earth is spherical. For it is impossible for any of them to occur with a different shape,[29] but it is possible for properties of this kind to be displayed only in association with a sphere.

114: Also,[30] when at sea we are about to approach land, our line of

27. Ptol. *Alm.* 1.4, 15.19–23 also mentions a triangular and square shape, and notes that they would be subject to the arguments against the Earth's having any polygonal shape.

28. The reasoning at lines 30–101 has been "indirect." That is, if "p entails q" is the direct argument (i.e., given phenomena entail that the Earth is spherical), then previously the argument form was "not-q entails not-p; but p; therefore, not not-q," i.e., *modus tollens* (see n. 10 above). In other words, nonspherical shapes are eliminated because they entail impossible phenomena.

29. For these arguments used to eliminate nonspherical shapes see lines 30–39, 44–49, and 54–56 above, and cf. I.4.239–242.

30. This is an elementary application of the argument from changing horizons at lines 30–39 above. Cf. I.1.183–190 where it is used in a slightly different form as a preliminary proof of the sphericity of the cosmos.

sight first encounters mountain peaks, while everything else is obstructed by the Earth's curvature.[31] Next, as we go over the curvatures, we encounter in the course of the journey the sides and spurs of mountains. And within the boats themselves, when we ascend the mast and get above the obstructing curvatures, we invariably see those parts of the land that are invisible from the decks and the hold. Also, when a ship leaves land, the hulls disappear first, although the masts are still visible; but when it approaches land from the sea, then, by the same token, the masts are seen first, while the hulls are still obstructed by the curvature of the water. All these [phenomena] indicate through virtually geometrical demonstrations that the shape of the Earth is spherical.[32]

126: It is necessary, then, that the air enclosing the Earth also be a sphere, since exhalations from the whole of the Earth rise and flow together, and thus fashion the same shape for the air too.[33] (Solid bodies can also, of course, have numerous configurations, but in the case of a substance that is composed of pneuma or of fire,[34] wherever [such types of matter] exist independently, nothing like this can happen. This is because they reach the shape proper to their nature by being "tensionized," that is, by being stretched equally in all directions from the exact center [of the Earth], since their substance is malleable, meaning that nothing

31. If the Earth has a "curvature" (*kurtōma;* here and at lines 118, 120, and 124 above), then, since curvatures define sphericity (cf. I.1.247 and 250), this argument contains a *petitio principii.* Ptol. *Alm.* 1.4, 16.13–18 argues more carefully that the surface of the sea causes the land to appear to "rise up" as we approach it. For other versions see Plin. *NH* 2.164, Strabo 1.1.20 (= Posid. 3b Theiler), and Theon *Expos.* 122.17–123.4.

32. Cf. also I.8.17–18 on geometrization.

33. On terrestrial exhalations cf. *SVF* 2.527 (168.25–26), and see I.8.79–88. The argument from here to line 138 was outlined at lines 6–9 above; see notes 4 and 5 above.

34. Pneuma is a compound of air and fire (e.g., *SVF* 2.442), and so the two elements identified here, the air and the aether, can also form a realm of matter describable as "composed of pneuma or fire," since they represent pneuma in its purest form, unaffected by solid matter.

solid exists which would configure them differently.[35]) Since the air is spherical, so too is the aether, since [the aether], being in turn able to enclose the air,[36] and neither being bent into angles by anything solid, nor having anything forcing it into some shape with uneven lengths,[37] is itself necessarily a sphere. So it is absolutely necessary that the whole cosmos too have such a shape.[38]

139: It is also entirely plausible that the most complete of bodies has the most complete of shapes. And the cosmos is the most complete of all bodies, while the sphere is the most complete of all shapes. For the sphere can enclose every shape that has the same diameter as it, but no other shape can enclose a sphere that has a diameter equal to it.[39] So it is absolutely necessary that the cosmos be a sphere.

35. Pneuma holds the Stoic cosmos together by "tensional motion" (see I.1.72–73). Hence non-solid matter is next said to have a sphericity that results from being "tensionized" *(tonousthai)*, i.e., from the oscillation of pneuma from the center to the periphery of the cosmos and back to produce a "stretching effect" (the literal meaning of *tonos*).

36. Ptol. *Alm.* 1.3, 13.21–14.16 argues that the aether (for him immutable) is spherical just because of its homogeneity, and because it is occupied by observably spherical bodies.

37. See I.1.166–168 for the contrast between sphericity and shapes with uneven lengths.

38. See also *SVF* 2.547 and 681, and Posid. F8EK, for this conclusion. Although the present argument emphasizes the relative density of the elements, and the capacity of the lighter ones for pneumatic, or tensional, motion, it need not conflict with the earlier principle (I.1.94–95, 173–174, and 191–192), that all the elements are centripetal. Furley (1993) defends such a system of dynamics. Wolff (1988) 497–533 and 539–542, however, suggests that the lighter bodies push the heavier ones to the center and move around them in a vortex-like motion.

39. For this claim see Pl. *Tim.* 33b1–7, Arist. *De caelo* 286b25–33, Cic. *De nat. deor.* 2.47, *SVF* 2.1009 (299.15–19), and cf. Ptol. *Alm.* 1.3, 13.14–20. Cleomedes' version holds for the five regular solids; see Eucl. *El.* 13 props. 13–18.

CHAPTER SIX[1]

1: We shall establish that the Earth occupies the exact center of the cosmos, by which it is enclosed, by again[2] starting out from the procedure based on the fifth undemonstrated [argument constructed] through multiple [disjuncts]. The following disjunction, that is, is necessarily true: *The Earth, which is encompassed by the cosmos, is either in its east, or its west, or its north, or its south, or higher than its center or lower—or it occupies its exact center.* But, as we shall demonstrate, *none of the [propositions] prior to the last is true.* Therefore, *it is necessary that the Earth occupy the center of the cosmos.*

9:[3] That it is not in the east is clear from the following. *(a)* If it were in the east, then shadows from objects illuminated at sunrise would be

1. This whole chapter can be compared with Ptol. *Alm.* 1.5, which is diagrammatically analyzed at Pedersen (1974) 39–42. The arguments here also assume the thesis of I.8, that the radius of the Earth is negligible in relation to the distance from the Earth to the Sun, or to the edge of the cosmos, i.e., that the observer is effectively at the center of the Earth. The theses of I.6 and I.8 are thus conventionally linked; see Eucl. *Phaen.* prop. 1 (10.11–12), Gem. *Isag.* 16.29, and Theon *Expos.* 120.11–12.

2. See I.5.23–26.

3. At Ptol. *Alm.* 1.5, 17.4–5 such displacement is described as a case of the Earth's not being on the axis of the cosmos, but still equidistant from both poles.

shorter, but at sunset these would be sent out farther; that is because when objects that give out light are nearby, shadows [cast by them] are shorter, but when they are more distant, shadows are invariably enlarged in proportion to the distance.[4] *(b)* If we were closer to the east, all the [heavenly bodies] would appear larger to us as they rise, while on setting (that is, as they continually went farther away), they would appear smaller.[5] *(c)* The first six hours of the daytime would also be very short, since the Sun would reach the zenith rapidly, whereas those hours after the sixth would be lengthened, given the greater distance from the zenith to the west.[6] None of these [observations] is present among the phenomena. Therefore the Earth is not farther east. Yet by the same token neither is it farther to the west; otherwise all the [observations] would as a result be the opposite to those just described.

22:[7] But if the Earth were farther north, then whenever the Sun rises, shadows from objects illuminated by it would as a result extend in the direction of that region. And if it were in the south, shadows would also slope southwards, both when the Sun rises and when it sets. In fact none of this occurs, but at the equinoxes, when [the Sun] rises, shadows slope toward where it sets at the equinoctial setting [point], and when it sets, they slope toward where it rises at the equinoctial rising [point]. But at

4. The longer shadows will, given *(c)* below, also be cast for a longer period of the daytime.

5. Cf. Ptol. *Alm.* 1.5, 18.5–8, and in the *Caelestia* cf. I.8.32–37.

6. Ptol. *Alm.* 1.5, 17.9–18.4 notes only the elimination of the equinox, here left implied.

7. Ptol. *Alm.* 1.5, 18.12–19.8 describes this as a case of the Earth's being on the axis of the cosmos but displaced toward one of the poles. He rejects it by using the same evidence of the altered visibility of zodiacal signs used here at lines 33–35 against the Earth's being "higher" or "lower" than its central position. The latter cases are identical with the present case of displacement toward one of the poles, and this seems to be acknowledged in the summary at line 40 below when only "four regions" of displacement are noted. The six catalogued in this chapter were perhaps intended to correspond to the six noncentral directions identified at I.1.150–158.

winter solstices when [the Sun] rises, [shadows slope] toward where it sets at the summer solstitial setting point, and when it sets, [they slope] toward where it rises at the summer solstitial rising [point]. But when, <on going> back <up>[8] from the south, it rises [at summer solstices],[9] shadows slope toward where it sets at the winter solstitial setting [point], but when it sets, they slope toward where it rises at the winter solstitial rising [point]. Thus [over the course of a year] the shadows form a chiasmus.[10] Therefore the Earth is not in any of these regions.

33: If the Earth were higher than the center, then the half of the heavens above the Earth would not be visible, nor would the six zodiacal signs (i.e., 180 degrees), nor half of the equinoctial circle, but less than all these [would be visible].[11] Hence nighttimes too would as a result always be longer than daytimes. But if the Earth were lower than the center, the result would be the complete opposite to what has just been described, since the hemisphere above the Earth would be larger.[12] [This does not happen.] Therefore the Earth is at neither a higher, nor a lower, position.

8. This supplement translates ἀνατρέχων, proposed for a lacuna in *Caelestia* ed. Todd, apparatus criticus at 1.6.29. A verb is needed here to complement ἐνθένδε ("thence," i.e., back again from the south).

9. This gloss is needed to balance "at winter solstices" (line 28), and is implied by "back up."

10. This chiasmus, or "decussation," involves two diagonals for the solstitial shadows; see Figure 12. The illustration could be Posidonian, given that at I.4.132–146 (cf. Figure 9) zones are defined in terms of shadows in a passage that can be attributed to him; see I.4 n. 33.

11. Throughout this paragraph the observer is assumed to be at ground level. On the special conditions under which less than 180 degrees are seen in this position from a centrally located Earth see I.8.127–132 and I.8 n. 33.

12. More than 180° would be visible from a centrally located Earth if the Earth were large enough, or the cosmos sufficiently small; see I.8.134–139 and I.8 n. 34, and II.6.122–138 and II.6 n.16. Also, under normal conditions, the refraction of light by the air at the horizon enables any terrestrial observer to see slightly more than half of the celestial sphere at any time.

40: It has been demonstrated that the Earth is not in any of the four regions [mentioned above].[13] Therefore it is necessary that it occupy the exact center of the cosmos. (In addition, the Earth is the heaviest of the bodies in the cosmos, and so must occupy the [place] farthest downwards, which [in a sphere] is also identical with the exact center.)[14]

13. See n. 7 above.

14. This refers to the centripetal motion of the elements (the demonstration of which was postponed at I.1.94–95, 173–174, and 191–192; see Introduction n. 6), which will establish that the Earth is stable at the center of a spherical and stable cosmos, and so rule out saving the phenomena introduced in this chapter (like many of those in I.5 and I.8) by a different cosmological arrangement (see Introd. n. 23). Ptol. *Alm.* 1.5–6 (his equivalents to *Caelestia* I.6 and I.8) are immediately followed by a demonstration (1.7) that the Earth is immobile by a physical theory broadly similar to the Stoic theory of centripetal motion; see Wolff (1988) 498–502.

CHAPTER SEVEN

1: Natural philosophers have held numerous doctrines about the size of the Earth, but two of these are superior to the rest. Eratosthenes' doctrine demonstrates its size by a geodesic procedure,[1] while Posidonius' is less complicated. Each [philosopher] takes certain assumptions [as being the case], and then arrives at demonstrations via the implications of the assumptions. The first [doctrine] that we shall discuss is Posidonius'.[2]

7: He states that *(P1)* Rhodes and Alexandria are located below the same meridian.[3] *(Def. 1)* Meridians are the circles drawn through the poles of the cosmos, and through a point that lies at the zenith of each of those

1. Our translation reflects Gratwick (1995) 178 n. 1. While the procedure here is called *geōmetrikos,* that term cannot mean "geometrical," since Posidonius as well as Eratosthenes (ca. 276–194 B.C.) employs geometry.

2. For literature on this calculation see Kidd *Comm.* 728–729; add Taisbak (1973–74) and Gratwick (1995). See also Figure 13. In this calculation and that of Eratosthenes below the unit of measurement used is the stade. Since it was never standardized, its value is difficult to determine from the ancient sources, and it cannot be readily translated into modern values for comparative purposes. For further discussion see Lloyd (1987) 233–234, and the literature cited at González (2000) 216–217.

3. This was the standard view (Ptol. *Alm.* 5.3, 364.9–10 and Strabo 2.5.7; cf. Toomer [1984] 255 n. 16), although Rhodes is in fact 1°50' west of Alexandria.

[observers] who stands on the Earth. (Thus, while the poles are the same for everybody, the point at the zenith is different for different [observers], which is why infinitely numerous meridians can be drawn.) Rhodes and Alexandria, then, are located below the same meridian, and *(P2)* the distance between the cities is held to be 5,000 stades.[4] Let it be assumed that this is so. *(Def. 2)* All the meridians are also included among the great circles in the cosmos, since they divide it into 2 equal parts by being drawn through its poles.[5]

18: Now with this assumed to be the case, Posidonius next *(P3)* divides the zodiacal circle (which, since it too divides the cosmos into 2 equal parts, is [by *def.* 2] equal to the meridians) into 48 parts by dividing each of its *dōdekatēmoria*[6] into quarters. Now if the meridian through Rhodes and Alexandria is also divided into the same 48 parts as the zodiacal circle, then its sections will be equal to the sections of the zodiacal circle just identified. The reason is that *(def. 3)* when [2] equal magnitudes are divided into equal [parts], their parts must also be equal to the parts of what has been divided.[7]

4. "Is held" (δοκεῖ) (14–15; also line 43 below) here implies "generally believed on good authority"; cf. I.2.26–27 and II.3.18 for this sense, and I.4.197–199, II.1.441, II.3.10, and II.7.1 for figures cited on external authority, without reservations. Kidd *Comm.* 726, however, calls the number given in *(P2)* "hypothetical," and claims that the whole Posidonian calculation reported here is designed to illustrate "the hypothetical method rather than the accuracy or certainty of the figures." But even though Posidonius uses a method of calculation that can be identified independently of the numbers used (see n. 9 below), that does not make the numbers cited "hypothetical." There is a significant difference between using numbers that have some reasonable credibility, and arbitrarily (or hypothetically) stipulating numbers to facilitate a calculation (as at II.1.274–275 in Posidonius' measurement of the size of the Sun). The number in *(P2)* is clearly of the former kind. See also n. 11 below.

5. On the definition of great circles see I.3.56–62.

6. On this term see I.4 n. 17.

7. Compare Eucl. *El.* 1, *Not. comm.* 3 (5.11–12): if equals are subtracted from equals, the remainders are equals.

27: Now with [*(P1)–(P3)*] assumed to be the case, Posidonius next says that the star called Canobus, which is located in the south at the rudder of Argo, is very bright.[8] (This star is not seen at all in Greece; that is why Aratus does not mention it in his *Phaenomena*.) But for people going to the south from the north, the star starts to be seen at Rhodes, and once seen on the horizon immediately sets along with the revolution of the heavens. But when we reach Alexandria by sailing the 5,000 stades from Rhodes, this star, when precisely at the meridian, is determined as being elevated above the horizon ¼ of a zodiacal sign, that is, [by *(P1)* and *(P3)*] 1/48 of the meridian through Rhodes and Alexandria.

38: Now it is necessary [by *(P1)* and *(P3)*] that the section of the same meridian located above the distance separating Rhodes and Alexandria also be 1/48 of that meridian, because the Rhodians' horizon is also distant by 1/48 of the zodiacal circle from that of the Alexandrians.[9] So since [by *(P2)*] the portion of the Earth located below this section is held to be 5,000 stades, the portions located below the other sections also consist of 5,000 stades. And in this way the circumference of the Earth is determined as 240,000 stades[10]—*if* [by *(P2)*] there are 5,000 stades between

8. On Canobus (sometimes spelt Canopus) (α Carinae), the brightest star after Sirius, see Aujac (1975) 132 n. 6 (on Gem. *Isag.* 3.15), and Kidd *Comm.* 725–726.

9. In Figure 13, since RV is parallel to AW, angle APV = α (Eucl. *El.* 1 prop. 29). Since triangle APV is similar to triangle RTV, angle RTA = angle APV = α. The three ratios that are identical in this calculation (cf. I.5 n. 22) are 1:48 :: ¼ *dōdekatēmorion*:full circle :: 5,000:240,000 stades (on the numbers in the latter see n. 4 above). Since the same basic principle or axiom underlies Eratosthenes' calculation (n. 21 below), Posidonius could be the source for both reports in this chapter. He was certainly interested in relations between ratios; see Galen *Inst. log.* 18 (cf. F191.13–15 EK), and on the axiomatization of such relations, see Hankinson (1994) 73–74.

10. Strabo 2.2.2 reports "about *(peri)* 180,000 stades," i.e., 500 stades per degree of the circumference. This implies a distance of 3,750 stades between Rhodes and Alexandria, the figure that Strabo 2.5.24 says that Eratosthenes calculated by the use of sundials.

Rhodes and Alexandria. Otherwise, [it will be determined] in proportion to the [true] distance.[11] That, then, is Posidonius' procedure for dealing with the size of the Earth.

49: Eratosthenes' [calculation], by contrast, involves a geodesic procedure, and is considered to possess a greater degree of obscurity.[12] But the following [assumptions], when stated by us as presuppositions, will clarify his account.

51: Let us first assume here too [cf. *(P1)*] that *(E1)* Syene and Alexandria are located below the same meridian; [second] that *(E2)* the distance between the two cities is 5,000 stades.[13] Third, [assume] that *(E3)* the rays sent down from different parts of the Sun to different parts of the Earth are parallel, as geometers assume to be the case.[14] Fourth, let the following assumption demonstrated by geometers be made: that *(E4)* straight lines intersecting with parallel lines make the alternate angles equal.[15] Fifth, [assume] that *(E5)* the arcs [of a circle] standing on equal angles are similar, that is, have the same proportion (namely, the same

11. The gloss "true" is justified here, since the figure of 5,000 stades is only "held" (i.e., widely believed) to be true (see lines 14–15 and 43). Kidd *Comm.* 726–727 again (cf. n. 4 above) sees the *caveat* in line 46 as evidence of Posidonius' use of "hypothetical method." But aside from an uncertainty as to whether this caveat was entered by Posidonius or Cleomedes, for Posidonius to admit quantitative corrigibility (cf. also I.4.208–213 and II.3.40) would scarcely show that he "was not primarily interested in figures or in accuracy" (Kidd *Comm.* 727). In fact, since the same basic principle or axiom underlies the two calculations reported in this chapter (see n. 21 below), it is even conceivable that Posidonius was the source for both of them.

12. On Eratosthenes' calculation see Figure 14. For discussions see the literature cited at Lloyd (1987) 231 n. 55; note especially Taisbak (1973–74), Newton (1980), Rawlins (1982a) and (1982b), and Gratwick (1995). For further literature see González (2000) 214–217.

13. Syene is in fact about 3° east of Alexandria. For *(E1)* and *(E2)* see also Strabo 2.5.7, where *(E1)* is qualified as "approximate."

14. Cf. also II.1.197–209 for this assumption.

15. Eucl. *El.* 1 prop. 29.

ratio) to their own circles, as is also demonstrated by geometers.[16] (For example, when arcs stand on equal angles, then if one of them is one-tenth part of its own circle, all the remaining arcs will also be one-tenth parts of their own circles.)

64: Someone who has mastered these [assumptions] would have no difficulty in learning Eratosthenes' procedure, which is as follows.[17] He says that *(E1)* Syene and Alexandria are located below the same meridian. So since *(def. 2)* the meridians are included among the great circles in the cosmos, the circles of the Earth located below them are necessarily also great circles. Thus the size that this procedure demonstrates for the [arc of the] circle of the Earth through Syene and Alexandria will be in a ratio with the great circle of the Earth.[18]

71: Eratosthenes says, and it is the case, that *(E6)* Syene is located below the summer tropical circle.[19] So when the Sun, as it enters Cancer and produces the summer solstice, is precisely at this meridian, the pointers on the sundials are necessarily shadowless, since the Sun is located vertically above them. (This [shadowless area] is reportedly 300 stades in diameter.)[20] But in Alexandria at the same hour pointers on sundials do cast a shadow, since this city is located further north than Syene. Now

16. Compare Eucl. *El.* 3 def. 11, which concerns segments.

17. As befits a pedagogical treatise, the basic method of calculation is summarized in advance, as at lines 21–26 above.

18. Lines 104–106 below offer the justification for translating the correlatives here (ἡλίκος . . . τηλικοῦτος) as referring to a ratio.

19. For this location see I.4 n. 21. For the absence of shadow at Syene at the summer solstice see Ach. *Isag.* 67.5–6, Plin. *NH* 2.183, Ptol. *Alm.* 2.6, 107.16–20, and Strabo 2.5.7.

20. Only Cleomedes, here and at II.1.211–213 and 270–273, provides this value. Three hundred stades are $\frac{1}{800}$ of 240,000 stades, or $\frac{1}{600}$ of 180,000 stades, the two figures for the Earth's circumference attributed to Posidonius (see n. 10 above). Neugebauer (1975) 655–656, however, suggests that the figure of 300 was derived from the demonstration by water clocks that the Sun is $\frac{1}{750}$ of its own orbit (see II.1.184–191) with 240,000 stades / 750 = 320 stades, rounded off to 300 stades.

since [by *(E1)*] the two cities are located below a meridian (a great circle [by *(def. 2)*]), if we draw an arc from the tip of the pointer's shadow on the sundial at Alexandria round to the base of the pointer, this arc will be a section of the great circle in the sundial's bowl, since the sundial's bowl is located below a great circle.

84: If we next conceive of straight lines produced through the Earth from each of the pointers, they will coincide at the center of the Earth. So since [by *(E6)*] the sundial at Syene is located directly below the Sun, then if we also conceive of a straight line going from the Sun to the tip of that sundial's pointer, the line going from the Sun to the center of the Earth will be a single straight line. If we conceive of a second straight line drawn from the bowl at Alexandria, that is, from the tip of the pointer's shadow up to the Sun, this line and the first one will be parallel [by *(E3)*], since they extend from different parts of the Sun to different parts of the Earth.

94: Now a [third] straight line extending from the center of the Earth to the pointer at Alexandria meets these parallel lines, and as a consequence [of *(E4)*] makes the alternate angles equal. One of these [angles] is at the center of the Earth where the lines drawn from the sundials to the center of the Earth coincide. The other is where the tip of the pointer at Alexandria and the line drawn from the tip of the pointer's shadow up to the Sun through the point where the line touches [the tip] coincide. The arc drawn from the tip of that pointer's shadow round to its base stands on this second angle, while the arc extending from Syene to Alexandria stands on the angle at the center of the Earth.

103: Now the arcs are similar to one another, since [by *(E5)*] they stand on angles that are equal. Thus the ratio that the arc in the bowl [at Alexandria] has to its own circle is the same as the ratio of the [arc] from Syene to Alexandria [to its own circle].[21] The arc in the bowl is certainly determined as one-fiftieth part of its own circle. So the distance from Syene

21. There are in fact three ratios here that are the same: 1:50 :: $1/50$ circle:full circle :: 5000:250,000 stades. Cf. n. 9 above.

to Alexandria is necessarily one-fiftieth part of the great circle of the Earth: namely [by *(E2)*] 5,000 stades. Therefore, the [great] circle [of the Earth] totals 250,000 stades.[22] And that is Eratosthenes' procedure.

111: Also, at winter solstices sundials are positioned in each of these cities, and when each sundial casts shadows, the shadow at Alexandria is necessarily determined as the longer because this city is at a greater distance from the winter tropic. So by taking the amount by which the shadow at Syene is exceeded by that at Alexandria, they also determine this amount as one-fiftieth part of the great circle in the sundial. So it is evident from this [calculation] too that the great circle of the Earth is 250,000 stades.[23] Thus the diameter of the Earth will exceed 80,000 stades, given that it must certainly be ⅓ of the great circles of the Earth.[24]

121: Those who say that the Earth cannot be spherical because of the hollows occupied by the sea and the mountainous protrusions, express a quite irrational doctrine.[25] For neither is there a mountain determined higher, nor a depth of sea [greater], than 15 stades. But 30 stades has no

22. Only Arrian at Philoponus *In Arist. meteor.* 15.13–15 also gives this figure as 250,000 stades. For 252,000 stades see Galen *Inst. log.* 12.2 (26.21–27.1), Plin. *NH* 2.247, Strabo 2.5.7 and 2.5.34, Theon *Expos.* 124.10–12, and Vitruv. *De arch.* 1.6.9.

23. See Figure 15. Bowen (2002) notes that Cleomedes, in relying on Eratosthenes' figure for the size of the Earth in other arguments (I.5.72; II.1.294–295), may be indicating that he considered it superior. If so, the point of his mentioning this new computation (the work of an unidentified "they") may be that it replicates and confirms the one attributed to Eratosthenes at lines 64–110, thus making the procedure common to both more reliable than Posidonius', which (see n. 11 above) used reported and questionable information for a crucial premise.

24. For the diameter as ⅓ of the circumference of the circle see I.5.62–63, and II.1.295–301 and 320–323; also Gem. *Isag.* 16.6. The figure of 22/7 for π, introduced by Archimedes (second century B.C.) is respected by Theon *Expos.* 124.12–17. Cf. also II.1 nn. 69 and 77.

25. On the Earth's sphericity being unaffected by uneven surface conditions see Strabo 2.3.3 (= Posid. F49.131–132 EK), Plin. *NH* 2.160 with 162, Plut. *De fac.* 924A, Sen. *NQ* 4B.11.3, and Theon *Expos.* 124.7–127.19.

ratio to over 80,000 stades,[26] but is just like a speck of dust would be on a ball. The protrusions on the rondures[27] from plane trees also do not stop them from being rondures. Yet these protrusions have a ratio to the total sizes of the rondures greater than that of the hollows of the sea and the mountainous protrusions to the total size of the Earth.

26. That is, there is "no ratio" that affects observations, the sense of this phrase at I.8.1–18.

27. "Rondures" for *sphairia* (literally "little spheres," or here berries) preserves the analogy with a spherical Earth.

CHAPTER EIGHT

1: While the Earth has the size demonstrated through the procedures just described, there are several ways of proving that it has the ratio of a point not only to the total size of the cosmos, but also to the height of the Sun, which the sphere that encloses the fixed stars far exceeds.[1] So even if 100,000,000 pitchers of water may amount to a significant[2] number when conceived of in isolation, they still have no ratio to the sea, not even to the Nile, or any other river of significant size: by the same token, the Earth, with its diameter of over 80,000 stades, appears to have a significant size when assessed in isolation, yet clearly does not have any [significant] ratio to the height of the Sun, still less to the total size of the cosmos.[3] This is because one magnitude must have a ratio to another magnitude when the larger can be measured by the smaller[4]—when, say,

1. This enclosing sphere is the "enclosure" *(periokhē)* or "circumference" of the cosmos; see line 85 below and II.3.49.

2. "Significant" *(axiologos)*, i.e., "expressible in a ratio."

3. Thus in this context the size of the Earth is, as Ptol. *Alm.* 1.6, 20.5–6 says, "relative to perception" *(pros aisthēsin)*; see n. 8 below.

4. This is true, though it is also true that two magnitudes can be in a ratio when the smaller does *not* measure the greater. For the general definition of a

it is 10 times larger or, if you like, even 10,000 times so. But the [single] pitcher of water would not measure the sea, not even the Nile. So just as the pitcher has no [significant] ratio to the [quantities] mentioned, so too the size of the Earth has no [significant] ratio to the size of the cosmos. This is proven by innumerable [arguments] that involve essentially geometrical demonstrations.[5]

19: *(a)*[6] Although the Sun is much larger than the earth and sea combined, it sends out to us (as will be demonstrated in what follows) an appearance of being about 1 foot wide, despite being very bright. We can thus form the notion that the Earth, if we should look toward it from the height of the Sun, would either not be seen at all, or be seen with the size of a minuscule star;[7] but if by hypothesis we were elevated to a distance far beyond the Sun, and right up to the sphere of the fixed stars, the Earth would not be seen by us at all, not even if imagined as having a brightness equal to [that of] the Sun.[8] Hence the [fixed] stars too must be larger than the Earth, in that they are visible from it, whereas the Earth could not be seen from the height of the sphere of the fixed stars. The Earth is certainly far smaller in size than the Sun, since the Sun itself too, if imagined at the height of the fixed stars, will perhaps appear as large as a star.

32: *(b)* That the Earth is a point in relation to the size of the cosmos

---

ratio see Eucl. *El.* 5 defs. 3 and 4. On the idea of measurement see Eucl. *El.* 5 defs. 1 and 2 (for magnitudes), and *El.* 7 defs. 3 and 5 (for whole numbers).

5. Cf. I.5.124.

6. Cf. II.3.51–59 on the size of heavenly bodies, and lines 158–160 below.

7. Under such conditions the Earth would appear to be moving in the same orbit as the Sun moves when observed from the Earth.

8. This argument is irrelevant to the main thesis of the chapter, which concerns the smallness of the Earth's radius in relation to the distances of bodies such as the fixed stars, in other words, the absence of parallax in observations made from the Earth of such distant bodies (see Figure 16). It is the relative distance of those bodies, not the Earth's inherent size, that determines the observational situation, namely, that the observations made from the Earth's surface are the same as if taken from the center of the Earth.

is also evident from observing the heavenly bodies: that is, they are not only visibly equal in size from every part of the Earth, but also similar in shape.[9] Neither [observation] would result unless the straight lines falling from every part of the Earth to all parts of the heavens were equal to one another. The Earth must, therefore, have the ratio of a center[10] to the whole cosmos.

37: *(c)*[11] The *dōdekatēmoria* of the zodiac also prove this: that is, precisely six of them are seen above the Earth, and the bulk of the Earth[12] does not conceal a single degree, indeed not even a small fraction of a degree, since precisely 180 degrees are always detected above the Earth. That is, half of the equinoctial circle is always above the Earth, as is evident from the equinoxes, where nighttime exceeds daytime not even by a hair's breadth;[13] and that would not happen if the Earth's bulk cut off any part of the equinoctial circle—that is, if the 80,000 stades of that bulk were in some [significant] ratio to it.

46: *(d)* Something like the following is seen among the phenomena.[14]

9. At I.6.13–15 the Earth's centrality in the cosmos is demonstrated from the observation that heavenly bodies do not appear unequal from different terrestrial locations. Cf. also II.5.112–114 where the equal distances of the signs of the zodiac are noted.

10. "Center," *qua* geometrical center, is frequently used in the same sense as "point" *(sēmeion)* in this chapter.

11. The arguments of *(c)* can also demonstrate the centrality of the Earth (see I.6.33–34), and be used in estimating the size of the Sun (see II.2.7–10).

12. The Earth's "bulk" *(bathos)* is its third dimension (cf. the specification of its diameter at line 45 below), and equivalent to *onkos*, "volume" (see lines 82 and 92 below, and cf. I.1.32 and 61).

13. On the day of the equinox, nighttime and daytime are observed to be equal; in other words, the Sun's day circle is then the equinoctial circle, and is divided in half by the observer's horizon. Cleomedes is assuming that the Sun is at the equinoctial point when it rises above the horizon rather than reaching this point during the course of the daytime or nighttime.

14. This indefinite expression *(toiouton ti)* alone suggests that the observations that follow are not precise; see further n. 16 below.

There are two stars,[15] identical in both color and size, and directly opposite one another. One occupies the 15th degree of Scorpio; the other, which belongs to the Hyades, occupies the 15th degree of Taurus.[16] Their color resembles that of Mars, and they are always observed on the horizon at the same time, with the one rising as the other is setting. This would not occur if the bulk of the Earth could obstruct any part of the zodiacal circle: for although one star rises and the other sets at the same time, the setting of the one that has risen would anticipate the rise of the

15. These are Antares (α Sco) and Aldebaran (α Tau). These stars are exactly opposite in longitude, but not in latitude, although they are also both close to the zodiacal circle. See Ptol. *Alm.* 7.5 and 8.1.

16. Neugebauer (1975) 960 notes that since Cleomedes' values for the longitude of Aldebaran and Antares differ by 2⅓° from those given by Ptolemy (who likewise puts the stars opposite to one another), and then, assuming *(a)* a constant of precession of 1 degree per century, and *(b)* the accuracy of Cleomedes' values, he argues that the *Caelestia* would have to be dated to the fourth or fifth century A.D., some 233 ± 50 years after A.D. 138, the epoch of Ptolemy's star catalogue. But this argument takes for granted that the values of Cleomedes' longitudes are tropical, or determined in relation to a fixed vernal point, in the same way as are Ptolemy's, and ignores the possibility that they are sidereal longitudes determined in relation to the fixed stars in the Babylonian style. Yet it is now clear that "Babylonian" methods were not supplanted in Greco-Latin astronomy by Ptolemaic methods immediately on the publication of the *Almagest* and *Handy Tables*, but continued to flourish for at least several centuries more; see, for example, Jones (1997) and (1999). Moreover, *(b)* is questionable on grounds of language (see n. 14 above), and from the general quality of Cleomedes' astronomical evidence. Cleomedes' error may in fact be much larger than the ½° that Neugebauer envisages, especially if Cleomedes is making the vague claim that the stars are in the middles of their signs more precise than he ought. There is certainly precedent for this sort of misleading precision in the history of astronomy. Hipparchus himself (*In Arat. et Eudox. Phaen.* 2.1.20–22 [132.10–134.2]; cf. ibid. 1.2.18–20 [20.4–22.9]) claimed that Eudoxus put the tropic and equinoctial points at the midpoints of their respective zodiacal signs, though to judge from Hipparchus' own quotations of Eudoxus, Eudoxus made only the much less precise remark that these cardinal points are in the middles of their respective constellations. See further Bowen and Goldstein (1991) 241–245.

one that is setting by the total interval of time in which it was necessary for the one rising over the part of the heavens obstructed by the bulk of the Earth to become visible on the horizon.

57: *(e)* Sundials also offer a major proof that the Earth has the ratio of a center to the heliacal sphere,[17] since the shadow of the Earth revolves along with the Sun, as Homer clearly indicates when he says: *The shining light of the Sun fell in the Ocean, dragging black night over the fertile land.*[18] Since the Earth's shadow is always in opposition to the Sun and is conical in shape, the very tip [of the shadow] is necessarily opposite the center of the Sun.[19] Sundials that have the shadows of their pointers completing a circular course along with the shadow of the Earth are, therefore, marked out on the Earth by experts.[20] So since no sundial can be marked out at the exact center of the Earth, although one can be marked out in every part of it, obviously the whole Earth has the ratio of a point to the height of the Sun, and thus to the sphere conceived of from it.[21] <Also, every tip on a pointer must have the ratio of a center to the heliacal sphere>,[22] since there obviously cannot be several centers belonging to a single sphere. This means that the pointers on all the sundials that can be marked out on the Earth have precisely the ratio [to the heliacal circle] they would have even if they were con-

17. See also Ptol. *Alm.* 1.6, 20.13–19, Theon *Expos.* 128.5–11 (after emendation), and Calcid. *In Tim.* 111.8–17.

18. *Iliad* 8.485–486; cf. also II.6.22–23.

19. See II.6.4–8.

20. They represent the diurnal course followed by the shadow cast by the unilluminated portion of the Earth, and so, in the literal sense of the verb used *(sumperinostein)*, they "come back home [to their original position] by going round with" that larger shadow.

21. The sphere is "conceived" or "imagined" *(nooumenē)* (cf. line 77 below, and II.1.72), since it is like the limit of a body, which in Stoic ontology "subsists in mere thought" *(kat' epinoian psilēn)*; cf. *SVF* 2.488.

22. The sentence in angle brackets has been transposed from lines 66–67 above, where it interrupted the argument.

tracted into a single point.[23] So since there is no part of the Earth on which a sundial could not be set up, the whole Earth has the ratio of a point to the Sun's height and to the sphere conceived of from it.

79: No problem need be raised here about how the Earth, with the status of a point in relation to the size of the cosmos, sends nutriment up to the heavens, as well as to the bodies encompassed by them, despite the heavenly bodies being so large in number and size.[24] That is because the Earth, while minuscule in volume, is vast in power in that virtually alone it comprises most of the substance [of the cosmos]. So if we imagined it totally reduced to smoke or air, it would become much larger than the circumference of the cosmos, and not only if it became smoke, or air, or fire would it become much larger than the cosmos, but also if it were reduced to dust. (We can, for example, see that even wooden objects that disintegrate into smoke expand almost without limit, as does vaporized incense, and every other solid body that is reduced to vapor.) And if we imagined the heavens, along with the air and the heavenly bodies, contracted to the compactness of the Earth, they would be compressed into a volume smaller than it. Thus while in volume the Earth may be a point in relation to the cosmos, since it has an indefinable power (that is, has a natural [capacity] to expand almost without limit), it does not lack the power to send nutriment up to the heavens and to the bodies in them. And this [process] would not cause the Earth to be totally expended, since the Earth itself also acquires something in turn from the air and the heav-

23. Goulet (204 n. 206) claims that here Cleomedes has forgotten the principle underlying Eratosthenes' calculation of the size of the Earth in I.7: that shadows vary depending on the latitudes at which sundials are located. But the present argument turns on the observation that in defining the *paths* of shadows in sundials, no account need be taken of the size of the Earth; see Ptol. *Alm.* 1.6, 20.16–19.

24. Terrestrial exhalations are celestial nutrition in Stoic cosmobiology; see, for example, *SVF* 2.650 (196.8–11), 663, and 690, Posid. F10EK, and cf. Cic. *De nat. deor.* 2.83.

ens. For "the way up is the way down" (to quote Heraclitus),[25] given that <the> substance [of the cosmos] is naturally disposed to be completely transformed and changed by yielding in every way to the artificer for the administration and continuing stability of the whole cosmos.[26]

100: So while the Earth has the ratio of a point to the height of the Sun, some use arguments of the following kind to establish that it does not have the ratio of a point to the sphere of the Moon.[27] *(a)* The distances of the [lunar sphere] from the fixed stars are said not to appear equal at every latitude, but larger and smaller at the same hour for different [observers]. Yet this would not be the result if the straight lines drawn from the Earth to the height of the Moon were equal, for then the distances of the [lunar sphere from the fixed stars] would also appear equal. *(b)* The eclipse of the Sun is also adduced as a sign[28] of this [conclusion], in that it is not eclipsed to an equal extent for all [observers], but it is frequently eclipsed totally, partially, or not at all for different [observers].[29] This would not result if the Earth were a point in relation to the height of the Moon rather than being at an [observationally] significant distance.

25. See Heraclitus at DK 22B60.

26. This "craftsman" *(dēmiourgos)* is fire, or an "artificer" *(tekhnikos)* (e.g., *SVF* 2.1032). The active principle in Stoic physical theory also "crafts"; see *SVF* 2.300.

27. See Plut. *De fac.* 921D (cf. Cherniss [1951] 138) on "certain mathematicians" who were said to argue that there is no lunar parallax. The question of lunar parallax (that is, whether observations taken on the Earth's surface can for practical purposes be treated as though they were taken at the Earth's center) is different from the question about whether the Earth is at the center of the lunar sphere. Note that Cleomedes speaks of the Earth as having the ratio of a center to the solar sphere, but does not think that the Earth lies at the center of this sphere (cf. I.4.53–62). Cf. also Ptol. *Alm.* 4.1, esp. 266.1–4. Aristarch. *De magn.* prop. 2 (352.5–6) does not allow for lunar parallax.

28. A sign *(sēmeion)* is an observable indication of something that is not directly observable (cf. *sēmeioutai* at II.1.216). On the elaborate theory of signs in Stoic epistemology see Burnyeat (1982).

29. That is, it is often the case that a *single* solar eclipse is observed as total or partial by different observers, while being invisible to everyone else.

And this is why the Moon conceals [the Sun] either totally, partially, or not at all for different [observers].[30]

113: Some use the following arguments to claim that the Earth does not have the ratio of a point [to the size of the cosmos].[31]

114: *(a)* They say that when elevated our sight observes objects that are not observed at ground level but are concealed beneath the horizon, and it does this to an increasing extent the higher it is elevated; therefore the heavens are not divided equally from every part of the Earth, and this is considered evidence that the Earth does not have the ratio of a point [to the size of the cosmos]. Our response must be that this [kind of observation] is caused by the sphericity of the Earth's shape. Thus even if the Earth were one stade in size, the result would be identical, just as long as the Earth is centrally located [in the cosmos] and has a spherical shape. (And of course it could not be claimed that an Earth of such limited size did not have a ratio of a point to the cosmos!) So in this case the Earth's shape must be held to be the cause.

124: Also, if someone extended in thought a plane from every point on the Earth,[32] there would not be more, or less, of the heavens seen above the Earth, but an equal amount at an elevation as well as at ground level. Of course the size of the heavenly bodies appears equal at an elevation as well as at sea level.

127: But if someone at this stage said that the half of the heavens above the Earth was not observed at ground and sea level, but only at extreme heights, their claim might have some rationale, since [in some cases] the heavens are certainly divided into two equal parts at extreme heights, but not at ground level, where instead less [than half] is visible above the Earth.[33] But, in fact, it is irrelevant to our argument whether *more* [of

30. See II.3.71–75.

31. The arguments that follow were obviously created for dialectical purposes, not as the record of positions actually adopted.

32. Cf. Figure 1 or Figure 5(a).

33. Both the language and logic of this sentence are awkward. The point seems to be that the contrast in the extent of horizons between a plane and an elevated

the heavens] is observed above the Earth when our sight is elevated, since this necessarily results from the Earth's shape being spherical. What must therefore be offered is evidence that the Earth is not a point in relation to the whole cosmos—[that is], not as to whether it is possible to see more than half the heavens when our sight is elevated, but as to whether an equal amount of them is *not* seen above the Earth [when viewed] from plane surfaces,[34] given that the horizons at ground level are planar, while those seen from an elevated position are, and are called, conical.[35]

140: There is the additional claim *(b)* that different parts of the Earth would not be frigid, torrid, and temperate, unless the Earth maintained distances from the sphere of the Sun that were [observationally] significant; that instead, if the Earth were a point [in relation to that sphere], the Sun would not even be described as approaching us and as withdrawing again. Now here, as with *(a)*, the response must be that the Earth's shape causes all these [phenomena] too: that is, some locations are torrid, frigid, or temperate depending on how the Sun's rays are sent down to the latitudes of the Earth.[36] (This is observed even in relatively small subdivisions that are also a short distance from one another. Certainly, some [places] in Elis are parched, while the adjacent [part] of Achaea has no extreme heat at all.) So even if the Earth were of minuscule size, the result would be the same, in that the Sun's rays are not sent

---

position may on occasion be such that a horizon of 180° can only be observed from a relatively elevated position. "Ground level" *(ek tōn khthamalōn)* here may therefore be a valley (cf. the example at I.5.92–94) rather than the planar horizons envisaged later in the paragraph. However, the special conditions involved are not clearly identified.

34. For this to occur, the Earth would either have to be displaced from the center of the cosmos (cf. I.6.33–35), or the cosmos be much smaller, or the Earth much larger. This is also a consequence of the theory refuted at II.6.123–145 (see II.6 n. 16), that more than 180° is seen from ground level over the curvatures of the Earth.

35. Strabo 1.1.20 infers the Earth's sphericity from the enlarged horizon seen from an elevated location.

36. See I.2.73–80.

down to all its latitudes in the same way, but in a perpendicular and intense form in the case of some latitudes, while obliquely and in a diminished form in the case of others. Also, the Sun's approach to us and its subsequent withdrawal are identified with respect to its relation to our zenith, since the straight lines produced from the Earth to Cancer and Capricorn are equal to one another.[37]

157: That the Earth has in fact the ratio of a center [to the cosmos] is demonstrated by these and many other [arguments]. But having in our opening argument[38] stated that the Sun sends out to us an appearance of being about 1 foot wide, despite its being much larger than the Earth, it is this very [claim about its size] that we must demonstrate next by offering in sufficient number for the present introduction [arguments] derived from a group of authors, including Posidonius,[39] who have written treatises exclusively on this subject.

37. That is, the Sun at the summer and winter solstices is not at determinably different distances from the Earth, but is in a different relation to the Earth considered as a sphere.

38. Lines 19–21 above.

39. Posidonius reportedly demonstrated that the Sun was larger than the Earth in a book of one of his treatises (F9EK; cf. Romeo [1979] 14–15), but Cleomedes' reference to other authors implies that not the whole of *Caelestia* II.1 is derived from Posidonius. Theiler nonetheless makes this chapter his F290a.

Book Two of Cleomedes'
The Heavens

OUTLINE

Book Two

Chapter 1. The Sun is not the size it appears to be, as the Epicureans believe, but has a size calculable as far larger than that.

Chapter 2. The Sun is larger than the Earth.

Chapter 3. The Moon, other planets, and stars are not the size they appear to be.

Chapter 4. The Moon is illuminated neither by inherent light, nor by reflection, but by the mingling of the Sun's light with the Moon's body.

Chapter 5. The phases of the Moon are caused by its motion in relation to the Sun and the Earth.

Chapter 6. Lunar eclipses are caused by the Moon falling into the Earth's shadow.

Chapter 7. *Appendix:* Data on the extremal latitudes of the planets, the maximum elongation of the inner planets, and the planetary periods; Conclusion.

CHAPTER ONE

2: Epicurus and most of his school[1] claimed that the Sun was the size it appeared to be[2] because they followed only the sense presentation caused by sight: that is, they made this presentation a criterion of its size.[3] We can therefore see what follows from their claim: namely, that if the Sun is the size it appears to be, it is quite clear that it will have in total more than one size, in that it appears larger as it rises and sets, but smaller as it culminates, while from the highest mountains it appears extremely large

1. Since the Epicureans are mentioned collectively at lines 414–415 and 418–419 below without any qualification, this phrase (like "most of the Socratic school" at I.5.17) must refer to the whole sect.

2. See Epicur. *Pyth.* 91, Lucr. 5.564–573, Cic. *Acad.* 2.82, Philodem. *De signis* col. 9 (sect. 14 De Lacy [1978]), and Demetr. Lacon (ed. Romeo [1979]) *passim.* The latter two Epicureans both flourished close to the time of Posidonius. On this theory see Sedley (1976) 48–53, Romeo (1979), and Barnes (1989).

3. On the principle underlying this claim, *viz.* that all sense impressions are true, and on the criterion in Epicurean epistemology, see LS sects. 16–17; Cleomedes may have addressed this issue elsewhere (cf. line 408 with n. 95 below). (At I.5.11–13 the belief in a flat Earth is similarly traced to exclusive reliance on visual appearance; cf. I.5 n. 6.) For a defense of the Epicureans against what is obviously Cleomedes' polemical oversimplification of their position see Algra (2000) 181–182.

when it rises.[4] Now either they will have to say that in total it has more than one size, or, if this is obviously absurd, it is absolutely necessary that they concede that it is not the size it appears to be.[5]

13: Some [Epicureans] say that the Sun appears larger to us as it rises (and sets)[6] because its fire is widened by the air through the force of its rapid ascent.[7] But this involves utter ignorance.[8] The Earth, after all, is located at the exact center of the cosmos, and has the ratio of a center to it; it is therefore equidistant from the sphere of the Sun in all directions,[9] and so the Sun does not come near the air either at its rising and setting, or in any other part of its course. In fact the Sun does not even rise everywhere at the same time, but, given the Earth's spherical shape, it rises, sets, and culminates at different times in different places. So since it can be both rising and culminating at different places, it will be in total both larger and smaller: larger for those for whom it is rising, smaller for those

4. At lines 47–48 below the same location is associated with an illusion of the Sun's greater distance. As Ross (2000) 868 notes, Cleomedes is mistaken here, since the Sun's size is reduced when viewed from a height; she suggests that his extromission theory of the visual ray (cf. n. 15 below) may have led him to assume that in such cases there was greater refraction through the atmosphere at the horizon. On the general issue of celestial illusions and their interpretation see Ross and Plug (2002).

5. If an object has conflicting appearances, then it is not seen as it really is; see Burnyeat (1979) on this principle in ancient epistemology and cf. n. 16 below.

6. There is no subsequent reference to the Sun's being flattened by the pressure of setting, but that can perhaps be taken as implied.

7. On the possibility of the Sun's being reconstituted by the confluence of fiery particles see Lucr. 5.660–665, who does not refer to the air having the causal role assigned it here. Cleomedes probably has in mind the theory of a diurnally rekindled Sun that is attacked later; see lines 426–466 and n. 105 below.

8. Epicurus was often charged with "ignorance" *(apaideusia);* in addition to line 452 below, see LS sect. 25E-H and Pease (1955) on Cic. *De nat. deor.* 1.72.

9. This only appears to be the case; in reality the Sun's orbit is eccentric in relation to the Earth. See I.4.49–71 and II.5.103–132.

for whom it is culminating—at the same hour of day![10] Nothing is more absurd than this.

26: These kinds of suggestions, then, are utterly meaningless and futile. The Sun appears larger to us as it rises and sets, and smaller at its culmination, because we see it at the horizon through air that is denser and damper (that is what the air closer to the Earth is like), while we see it culminating through less adulterated air. So in the latter case the ray sent out toward it from the eyes is not refracted, whereas the ray sent out to the horizon whenever the Sun rises or sets is necessarily refracted on encountering air that is denser and damp.[11] And in this way the Sun appears larger to us (as, of course, objects in water also appear to us other than they are because they are not seen along a straight line).[12] Therefore all such states must be held to be conditions affecting our line of sight, not as properties associated with the objects that are being seen. (When the Sun is observed from [within] deep wells, at least where this is possible,[13] it also reportedly appears much larger because it is seen through air within the well that is damper. And in this case the Sun cannot, of course, be said to be enlarged for those looking in its direction

10. See II.1.430–437 and 448–452 for analogous reasoning used against the Epicurean explanation of the Sun's orbit as a daily extinction and rekindling at sunrise and sunset.

11. On such atmospheric conditions affecting the appearance of the Sun and its size see Arist. *Meteor.* 373b12–13, *Schol. in Arat. vet.* 419.6–420.2, Posid. F119.12–20 EK (in which Posidonius questions this explanation), Sen. *NQ* 1.6.5, and Ptol. *Alm.* 1.3, 13.3–9. For a discussion, with a translation, of II.1.27–75 in relation to the history of the analysis of the "solar illusion," i.e., the apparent enlargement of the Sun near the horizon, see Ross (2000). For a general explanation of what is known as the celestial illusion, and includes the lunar and solar illusions, see Plug and Ross (1994).

12. Such a case of refraction is analyzed at II.6.178–187, though with reference to the visibility, not the enlargement, of the submerged object.

13. That is, where these (vertical) wells are located at latitudes on or between the tropics, so that the Sun can be observed at zenith from the base of the wells at some point during the year.

from [within] the well, but diminished in size for those doing so from above, but quite clearly the humid darkness of the air within the well causes it to appear larger for the [former] observers.)

45: The Sun's distance [from the Earth] also appears larger and smaller to us: as it culminates it appears very close, but as it rises and sets it appears farther away, while from the highest mountains it appears at a still greater distance. And when it appears close by, it also appears very small; but the more distant it appears, the larger it also seems. The quality of the air causes all such [appearances]: that is, when seen through damp and denser air, the Sun appears larger to us and more distant, but when seen through clear air it appears smaller in size and closer in distance. (Thus Posidonius claims that if it were possible for us to see through walls and other solid bodies, as Lynceus could in the legend, then the Sun, when seen through these, would appear much larger to us and as removed to a much greater distance.)[14]

57: While the Sun appears larger and smaller to us, as similarly do the distances involving it, the [visual] cone that impinges in reality on it from the rays that flow out <from> the eye[15] is necessarily very large. But since the size and distance of the Sun are both contracted to what appears to be an extremely small quantity, we can conceive of two cones: one that impinges in reality on the Sun, the other that does so in appearance.[16]

14. Lines 51–56 = Posid. F114EK; see Kidd *Comm.* 442–443. Lynceus was a legendary figure with preternatural visual powers.

15. Hahm (1978) 65–69 (who ignores Cleomedes) doubts that the early Stoic theory of vision involved rays in the form of visual pneuma flowing out from the eye, as indicated here and at II.6.181–185 (cf. I.1.72–74) where our "line of sight" is like a conical "bead" of pneuma drawn on an object. Note the verb *epiballein* ("impinge") at lines 59, 62, and 234 below, and also II.4.130. For more on the visual cone see lines 253–258 below, and II.5.110–112 (cf. II.5.49–54).

16. A "real" visual cone represents the conditions under which the Sun is seen without any conflicting appearances. See Burnyeat (1979) 73–75 on the implication that an object that is seen by different observers as having conflicting appearances must in principle be visible to all observers with the same appearance.

These will have a single vertex at the pupil of the eye, but two bases: one in reality, the other in appearance. Therefore the real distance is to the apparent distance as the real size is to the apparent size. But the bases of the [two] cones are equal to the real and apparent diameters respectively.[17] Therefore the real distance is to the apparent distance as the real size is to the apparent size. But the real distance is almost immeasurably[18] larger than the apparent one, since the Earth has the ratio of a point to the height of the Sun, and to the sphere conceived of from it. Therefore it is absolutely necessary that its real size be immeasurably larger than its apparent size. Therefore the Sun is not the size it appears to be.

76: Also, if the Sun is the size it appears to be, then if we imagine it being double its size, each of its parts, when divided into two, will appear to be 1 foot wide.[19] So if we also imagined it so enlarged as to extend over a distance of a 1,000,000 stades, each of its parts that is 1 foot wide would appear the size it is. If so, it follows that the Sun would in fact appear the size that it is, although that is clearly impossible: human sight simply cannot attain such a degree of power that objects extended over 1,000,000 stades appear the size that they are in reality, since even

17. Here the Sun is treated like a flat disk; see Figure 17. But at II.5.51–54, with the help of Eucl. *Opt.* prop. 27, it will be shown that less than half of the Moon's sphere is seen by a terrestrial observer.

18. "Immeasurably larger" *(apeirōi meizōn* or *apeiromegethēs)*, literally "larger by an unlimited [amount]," often qualified by "almost" *(skhedon)* (cf. II.1.85, 135, 241, 266, 360), presumably because attempts can be made to calculate the Sun's real size.

19. "1 foot wide" *(podiaios)*, this width being the approximate diameter (cf. the qualifications at I.8.20 and 159) of the flat disk that the Sun appears to be (see n. 17 above, and cf. lines 267–269 below); it is an informal measurement in contrast to more systematic procedures (see II.3 n. 7). The term itself was probably inherited from Heraclitus by Epicurus (see Sedley [1976] 52–53), and is frequent in descriptions of the Epicurean doctrine by secondary sources. As Algra (2000) 187 notes, these distort a claim that may have been intended as a polemical reaction to astronomers' claims about its distance and size. On Aristotle and the philosophical basis for this illusion see Schofield (1978) 113–114.

the cosmos itself, although of almost unlimited size, appears to us to be very small.[20]

87: Now since what follows from the Sun's being 1 foot wide is impossible, it is impossible for it to be 1 foot wide. For it cannot also be claimed that, when the Sun is extended over such a great distance, some of its parts that are 1 foot wide will appear the size that they are, while others will not.[21] That is because the distances from the Earth to all the parts of the Sun will be equal, since the Earth has the ratio of a center to the heliacal sphere. Thus *all* of its parts that are 1 foot wide, not some specific parts more than others, will have to appear the size that they are. So if all its parts that are 1 foot wide appear the size they are, the Sun itself, when extended that much, will appear in total the size that it is. Since this is obviously impossible, its parts that are 1 foot wide will not even appear to be the size they are—instead, they will not even appear at all! So if the Sun itself is 1 foot wide, it will not even appear. But it does appear. Therefore, it is not 1 foot wide. So it is clear from this, I think, that if the Sun were the size it appears to be, it would not appear. But since it does appear, it is not the size that it appears to be.

102: If the Sun is the size it appears to be, and if the sense presentation derived from sight is itself the criterion for the size that belongs to it, it could be said to follow that this appearance would also be a criterion for the [other qualities] that appear to belong to it.[22] Hence if it is

20. "Cosmos" here must refer to the visible celestial hemisphere, which we imagine to be smaller by assimilating the sizes of heavenly bodies to familiar bodies at shorter distances. For an analysis of this illusion see the passage translated from the Arabic as sect. 7 of Ptolemy's *Planetary Hypotheses* Book I, Part 2 at Goldstein (1967) 9.

21. Here we have an imaginary opponent who must have accepted the argument in the preceding paragraph, and is then trying to wriggle off the hook by claiming that the Sun is large in some places, but 1 foot wide in others.

22. This sentence marks a return (until line 139) to arguments based (like those at lines 5–56 above) on conflicting appearances.

the size that it appears to be, it also has the qualities that it appears to have. But it appears hollow, <spinning>,[23] and flashing,[24] although this configuration does not belong to it; certainly at other times it is seen as smooth, Moon-like, and not spinning.[25] Yet is impossible for all these [qualities] to belong to it. Therefore the Sun's being 1 foot wide, of which they are a consequence, is also a falsehood.

110: Again, if the Sun is as large as it appears, and has the qualities it appears to have, then since it also *appears* stationary, it would *be* unchanged in position.[26] Yet it is not unmoved, and so is not unchanged in position. Hence it is also not the size that it appears to be.

114: The absurdity of their claim could also be proved with the utmost

23. This supplement, proposed by R. Renehan at *Caelestia* ed. Todd II.1.106, balances "not spinning" in the next sentence (see n. 25 below). For such spinning see Arist. *De caelo* 290a12–18; it is an optical effect that can occur at both sunrise and sunset.

24. For hollowness and brightness see Arat. *Phaen.* 828–830, and for apparent hollowness caused by interposed clouds see *Schol. in Arat. vet.* 410.8–13 and 411.6–10. Such a shape could be dark, and so *marmairōn* ("flashing") (line 107) might be emended to *melainomenos* ("blackened"; cf. *Schol. in Arat. vet.* 411.8).

25. μὴ δινούμενος ("not spinning") (line 108): i.e., the Sun's appearance at times other than sunrise and sunset. Given the evidence at n. 23 above, there is no basis for emending it, either to μελαινόμενος ("blackened") (Theiler) (an epithet more appropriate in the preceding line; see n. 24 above), or μηδὲν πυρούμενος ("in no way ignited") (Marcovich [1986] 117). Marcovich finds support for his emendation in πυρωπός (line 132 below), but μηδὲν πυρούμενος must refer to an igneous constitution, not a fiery appearance, and its true parallels are elsewhere: πυρώδης at I.5.129, or πύρινος at II.3.93, II.5.4, and II.6.104.

26. For this argument, supplied with its major premise ("everything that appears also is the case"), see the Epicurean Demetr. Lacon col. 20, with Romeo (1979) 16 and n. 41, and Algra (2000) 185–186, who uses this text to defend the Epicureans, who were well able to handle this widely recognized type of illusion; see Lucr. 4.391–396, and also Sen. *Ep.* 88.26 and Alex. Aphr. *De an.* 71.17–18 with Todd (1995) 127 n. 26. For the generic case of distant objects appearing stationary although really moving see Sext. Emp. *PH* 1.118.

cognitive reliability on the basis of the following [argument]. If the Sun is indeed the size that it appears to be, it is, I think, evident that the Moon too is the size it appears to be.[27] And if it is so itself, then so too are its phases: so when it is crescent-shaped, the distance from horn to horn is also as large as it appears to be. This further implies that the distances [from the Earth] to the heavenly bodies near the Moon are also as large as they appear to be, from which it further follows that all the distances of the heavenly bodies without exception are also as large as they appear to be. Hence the whole hemisphere of the heavens above the Earth is also as large as it appears to be.[28] But this is not so. Therefore neither is the Sun the size that it appears to be. (Also, if the Moon along with its phases is the size it appears to be, then the black spots that appear in it are also the size that they appear to be. If so, the [Moon's] "mountains" must also be the size they appear to be.[29] But this is not the case. Thus neither is the Sun the size it appears to be.)

129: Now when the air is "pure" (that is, in a natural state), we cannot look back at the Sun. But when the condition of the air enables us to look at it, it appears differently to us at different times: sometimes white, sometimes palish, occasionally fire-like, and is often to be seen as red ocher in color, or blood-red, or yellow, and occasionally even multicolored, or pale green.[30] Also, we think that the pale cloud-like flecks that

27. See Epicur. *Pyth.* 91 and Lucr. 5.575–584 for this as an Epicurean claim. For more on the size of the Moon see II.3.1–33.

28. For Epicurean recognition that the Sun can appear closer than it really is see Lucr. 4.405–413, and Diog. Oen. Fr. 13 cols. I.13–II.10 Smith (1993). Lucr. 4.397–399 (on distinct mountains appearing to coalesce in the distance) could presumably be applied to the distance between heavenly bodies, even if, for an Epicurean, they will still be the size that they appear to be.

29. Goulet 210 n. 237 argues that these mountains are terrestrial. Yet although the Moon, as he notes, is a rarefied body (I.2.37–39, II.3.88–91 and II.4 *passim*), it could still appear mountainous as a result of its unevenly "turbid" appearance (see II.3.89 and II.5.2).

30. See Arat. *Phaen.* 832–879 on the Sun's varying appearances, recorded there as weather signs; cf. "palish" (*ōkhriōn*, line 132) with *Phaen.* 851.

often appear around the Sun belong to it,[31] although they are at an almost immeasurably vast distance from it. Again, often when setting or rising on a mountain peak, the Sun sends out to us the appearance of its touching the peak, although its distance from every part of the Earth is as vast as is to be expected when the Earth has the ratio of a point in relation to its height.[32]

140: So surely it is utterly stupid—is it not?—to follow this kind of sense presentation[33] instead of making something else a criterion, at least for things of such a great size,[34] bearing in mind that being misled in these cases usually causes significant damage.

144: The utter inconceivability of the [Epicureans'] claim is also very clearly proven on the basis of arguments constructed in the following way. Imagine a horse released to run along a plane in the time interval between the Sun's outer rim's emerging over the horizon and its complete emergence.[35] A fairly obvious guess would be that it would advance at

31. On "flecks" *(knēkides)*, pale spots produced by a light cloud that lacks moisture, see *Aratea* 126.24–25. They probably created a halo; see Goulet 210 n. 238, who notes Sen. *NQ* 1.2.3.

32. For the same illusion see Lucr. 4.404–413. This passage is evidence that the Epicureans could handle the issue of the Sun's vast distance, as is Diog. Oen. F13, II.1–10, adduced by Algra (2000) 185. For the constructive use of the illusion see lines 227–232 below.

33. One based just on sight (or visual observations), that is; see lines 3–4 and 102–105 above.

34. The best definition of such an alternative criterion is given at I.5.2–6. Using its language, the Sun is not "fully displayed," and therefore implications about its size must be drawn from phenomena by more elaborate forms of reasoning, initially involving the comparison of different phenomena (lines 145–224 below), and later the use of calculations (lines 225–356 below).

35. At Philodem. *De signis* cols. 10–11 (sect. 15 De Lacy [1978]) a Stoic, Dionysius of Cyrene, is reported as arguing that the Sun must be very large because it reappears slowly from behind an obstruction; the present argument is a rough attempt to measure the speed of this apparent motion. (On Dionysius and Posidonius, whose lives probably overlapped, see Romeo [1979] 14–16.) Cf. also II.2.13–18 with II.2 n. 6.

least 10 stades, whereas a very swift bird would go many times farther than the horse, and again a missile with a very swift momentum would go much farther than the bird, so that in such a period of time it would cover at least 70 stades.[36] Now on the hypothesis that the heavens travel as fast as the horse, the diameter of the Sun would be determined as 10 stades; if as fast as a very swift bird, then much larger; but if as fast as the missile, it would be at least 70 stades. Given all this, the Sun will not be 1 foot wide, that is, not the size that it appears to be.

155: Now we could form the notion that the heavens move immeasurably many times swifter than the missile from the following procedures.

156: When the Persian King was on his expedition to Greece, he reportedly stationed people at intervals from Sousa as far as Athens, so that what he accomplished in Greece could be indicated orally to the Persians through people stationed at intervals successively receiving oral messages from one another. The oral message that progressed through the stages of this relay reportedly reached Persia from Greece in two intervals of a nighttime and a daytime. Now if such a movement (or "impact") of air,[37] although extremely swift, covered a minuscule portion of the Earth in two intervals of a nighttime and a daytime, one can, I think, form a notion of what kind of speed the heavens have, and that it is immeasurably swifter than this, since in one interval of a nighttime and a daytime the heavens go through a distance immeasurably many times greater than that from Greece to Persia.[38]

36. On missile throws as units of measurement see Lucr. 4.408–409. This missile, however, cannot depend on human propulsion, since 70 stades is seven times as far as a horse gallops in roughly 2 minutes. Note that at lines 150 and 153 we have emended the manuscript reading 200 to 70. The minuscule letter used for 200 (sigma) could have been confused with that for 70 (omicron), and 70 also makes more sense of the numbers at lines 166–168 (see n. 39 below).

37. Stoics (see *SVF* 1.74 and 2.138 and 139) defined speech as an "impact" *(plēgē)* on air.

38. For the Persian relays see Hdt. 8.98 and Xen. *Cyr.* 7.6.17–18. This argument is a "procedure" *(ephodos)*, as we have defined it in the Introduction (cf.

166: Also, if we imagined the missile going through the great circle of the Earth, it would not even go through the 250,000 stades [of the Earth's circumference] in three intervals of a nighttime and a daytime![39] Yet the heavens go through the full extent of the cosmos, despite its being immeasurably larger than the Earth, in one interval of a nighttime and a daytime. Thus no notion of the speed—the rapid movement, that is—of the heavens can even be formed, and nothing like it can be interpreted in terms of a ratio. The Poet displays how great the speed of the heavens' course is through the following [verses]: *The dim distance that a man sees with his eyes when sitting on a promontory, looking upon the wine-blue water, is equivalent to one stride of the gods' loud-neighing horses.*[40] But this is expressed in an exaggerated way, and with striking expansiveness. Not only is [Homer] pleased to use the farthest extent of sight to indi-

---

n. 38), since it relies on an axiom that could be made explicit as follows: "If two bodies, *A* and *B*, move in circles around the same center, and *A* moves along its circumference farther than *B* does along its circumference in the same time period, then *A* moves faster than *B*." Not surprisingly, Cleomedes does not spell this out. Similarly, the argument that follows (lines 166–170) is an application of the axiom: "If two bodies, *A* and *B*, move in circles around the same center, and *A* moves the same distance along its circumference as *B* does along its circumference, but in a shorter time period, then *A* moves faster than *B*."

39. Since this must be the missile introduced at lines 149–153 above, it will have to cover 200 stades during sunrise if we retain the manuscript reading at 150 and 153 above. Thus if the Sun's diameter is $^1/_{750}$ of its orbit (lines 184–191 below), it will rise in $^{1440}/_{750}$ minutes (= 1.92 mins.), and the missile will cover 150,000 stades in a full day. It will thus orbit an Earth of 240,000 (or 250,000) stades in less than 2 full days. But 200 can be plausibly emended to 70 (see n. 36 above), and in that case the missile will cover 52,500 stades in a full day, and, in conformity with the text at lines 167–168 ("not even in 3 days" being taken to mean "in [significantly] more than 3 days"), take over 4 days (specifically 4 days, 18.3 hours) to orbit the Earth.

40. Homer *Iliad* 5.770–772. Milton *Paradise Lost* 8.38 is closer to Cleomedes in saying that the heavens have "Speed, to describe whose swiftness number fails." Although the Homeric verses identify only the sight line from the promontory to a distant horizon, Cleomedes adds additional upward and downward sight lines because of the comparison with celestial horses and the reference to the sea.

cate the speed involved in the rapid movement of the heavens, but also adds to it an upper distance along with the [depth of] the sea below. Yet even this description falls short of properly indicating the swiftness of the heavens. The speed that the heavens employ in their rapid movement has no limit, and no notion of it can be formed. So surely it is stupid to believe that a part of them that is 1 foot wide could rise in such an interval of time?[41]

184: The naïveté of this claim is also proved by water clocks, since these are a means of showing that, if the Sun is *1 foot* wide, then the greatest circle of the heavens will have to be *750 feet*! For when the Sun's size is measured out by means of water clocks, it is determined as 1/750th part of its own circle: that is, if, say, 1 *kuathos*[42] of water flows out in the time it takes the Sun to rise completely above the horizon, then the water expelled in the whole daytime and nighttime is determined as 750 *kuathoi.* Such a procedure was reportedly first conceived of by Egyptians.[43]

192: The [Epicurean] doctrine is also refuted by colonnades that face south, since the shadows of the columns are sent out in parallel lines.[44] That would not happen unless the Sun's rays were sent out to each column in straight (i.e., perpendicular) lines.[45] And the rays would not be sent out perpendicularly in the direction of each column unless the diameter of the Sun were coextensive with the whole colonnade.[46]

41. In, that is, the interval determined in the three "guesstimates" (cf. *stokhazesthai* at line 147) at lines 147–153 above.

42. A *kuathos* was 1/6 of a *kotulē*, which was about 1/4 of a liter, or about 1/2 of a modern pint.

43. On ancient accounts of this method see Goulet 210 n. 243 and Kidd *Comm.* 449–450 (add Sext. Emp. *AM* 5.75 to the critics). Cf. lines 297–299 below where this method is applied to the Moon. The "water clock" *(hudrologeion)* must have been a type of clepsydra (see I.1 n. 18).

44. At the latitude of Greece, that is, the shadows would always point in a northern direction.

45. See I.7.53–56 for this principle.

46. Cf. II.3.23–33 for the argument that the Sun and the Moon must both have diameters at least as large as the terrestrial shadow cast by the Moon in a

197: Streets that throughout the inhabited world are arranged facing the equinoctial rising point also reportedly become shadowless when the Sun rises at the equinoxes, a result that is conditional on the Sun's size being coextensive with the whole inhabited world, specifically with its width. Again, at midday at the equinox all the streets in the inhabited world are illuminated on both sides, so that the Sun's size is coextensive not only with the width, but also with the length, of the whole inhabited world. (The length of the inhabited world extends east-west, whereas its width extends north-south.) Thus when the Sun rises on the day of equinox and renders the streets that face it shadowless, it has its diameter coextensive with the width of the inhabited world, whereas when it culminates and illuminates all the streets on both sides, it has its diameter coextensive with its length. (But the Sun is said not to culminate for everyone at the same time, but only for those who live below the same meridian. So it must be stated that the preceding [argument] is expressed rather loosely.)[47]

211: Also, when the Sun is in Cancer, bodies illuminated at Syene become shadowless at exactly midday over [a circular area] 300 stades in diameter.[48] This clearly reveals that the Sun is not 1 foot wide. If it were 1 foot wide, none of this would happen.

216:[49] That the Sun is not 1 foot wide is also indicated by shadows: for when the Sun displays its rim above the horizon, the shadows sent out are very long, but when it is above the horizon they are contracted

---

solar eclipse. Here the colonnade is analogous to the Moon, since it is the object that creates a terrestrial shadow when illuminated by the Sun.

47. The caveat is needed, since the Sun rises and reaches culmination at different times at successive meridians (cf. I.5.37–39). We are asked to extrapolate from the situation at a given location to the whole of the inhabited world.

48. See I.7.75–76, and lines 270–273 below.

49. Cf. I.6.9–32, where the evidence of shadows is used in an elementary argument for the cosmocentrality of the Earth. Here shadows serve to indicate the Sun's distance from the Earth, and *eo ipso* the fact that its real size is greater than its apparent one.

to a much smaller size. This would not happen unless the Sun's rays were much higher than all terrestrial bodies, and that would not happen if the Sun were 1 foot wide. Therefore it has a diameter greater even than the highest mountains, since when it is completely visible above the horizon, it sends out rays higher than the peaks of mountains (that is, from a higher position).

225: The following procedure, which goes forward on the basis of the phenomena alone, demonstrates not only that the Sun is not 1 foot wide, but also that it has a prodigious size.[50] For when the Sun rises or sets in alignment with the peak of a mountain, anyone at a significant distance away from the peak sees its rim, which is observable on each side of the peak. This would not occur unless the diameter of the Sun were larger than the peak causing the obstruction. So if this peak is 1 stade in diameter,[51] the diameter of the Sun will have to be larger than 1 stade.[52] (The preceding is said to be observed among the phenomena not just in the case of a mountain peak, but also in that of the largest islands. For when our line of sight is at a significantly elevated position, and impinges on one of the largest islands from a considerable distance away, the island appears so small that here too the Sun's rim visibly protrudes on each side when it rises or sets in line with them. From this it is clear that the diameter of the Sun is also greater than the length of the largest islands.)

240: With this taken [as true] on the basis of the phenomena alone, the next stage is to demonstrate that it is necessary that the diameter of

50. This procedure could be based on Eucl. *Opt.* prop. 5 (cf. II.4.121–122 with II.4 n. 32): i.e., when two objects are aligned relative to an observer, and the closer one blocks (or, in this case, almost blocks) the more distant one, then the more distant object must be the larger of the two.

51. This is its one-dimensional appearance at such a distance; cf. *mēkos*, "length," in line 238.

52. This assumes that in the next illustration there is no distance between the Sun and the peak, or the Sun and the island; i.e., that the Sun is as close to these intervening objects as it appears. Distance will be incorporated in the next stage of reasoning.

the Sun be almost immeasurably greater than the diameters of the largest islands. This is established via the following procedure. *(1)* If an isosceles triangle has a base, say 1 stade in length, and if the sides are produced by an amount equal to those that enclose the base that is 1 stade long, the base of this [second] triangle will then be twice the base that is of 1 stade in length. Then *(2)* if we once more produce sides equal to the whole of the sides [of the second triangle], its base will be four times the base of the triangle [posited in *(1)*], and thereafter the same proportion will progress without limit.[53] Now assume *(3)* that we see one of the largest islands from a considerable distance when the Sun is rising or setting in line with it, with its rim visibly protruding on each side, and that the island is located between us and the Sun. Now *(4)* if our line of sight encompasses the island, the cone formed from the line of sight will have the diameter of the island as its base. So if its diameter is 1,000 stades, then the base of the cone will also be the same size. Now let us hypothesize *(5)* that the Sun is as distant from the island as the island is from us. So since [by *(3)*] the Sun's rim visibly protrudes on each side of the island, the rays that flow from our eyes to the Sun are [by *(1)*, *(2)*, and *(5)*] double [the length of] those that reach the island. Thus [by *(1)*] the base of this [second] triangle will also be double the diameter of the island.[54] If [by *(4)*] the latter is 1,000 stades, the diameter of the Sun will be 2,000 stades, since it is the base of the larger triangle. So since [by *(5)*] the Sun is as distant from the island as we too are on the opposite side of it, the diameter of the Sun will be 2,000 stades. But the distance [in each case] is not equal; instead, we are a short distance from the island, while the Sun is immeasurably many times farther away from us, and so the diameter of the Sun will [by *(2)*] also be almost immeasurably many times greater than the diameter of the island. How, then, could the Sun's size be 1 foot wide when it extends over such a great distance?

53. See Figure 18.
54. See Figure 19.

269: The following kind of procedure reveals better than any other the estimated value for the size of the Sun.[55] [It assumes] *(1)* Syene is located below Cancer; thus when the Sun is located in this sign and stands precisely at culmination, objects illuminated by it are shadowless in this area up to a diameter of 300 stades.[56] With this as true among the phenomena, Posidonius hypothesized *(2)* that the heliacal circle is 10,000 times greater than the Earth's circumference.[57] Starting out from this [premise], he demonstrated that the diameter of the Sun must be 3,000,000 stades. That is, if [by *(2)*] one circle is 10,000 times greater than the other, then the section of the heliacal circle that the Sun's size occupies must be 10,000 times greater than the section of the Earth that the Sun renders shadowless when located overhead. So since [by *(1)*] this section extends to a diameter of 300 stades, the section of its own circle that the Sun occupies at any time must be 3,000,000 stades.[58] But this is taken [as true] on the basis of the aforementioned hypothesis. And while it is plausible that the heliacal circle be no less than 10,000 times greater than the Earth's circumference (given that the Earth has the ratio of a

55. On this calculation see Kidd *Comm.* 443–447 and Neugebauer (1975) 655–656. On problems involved in reconciling it with other reports of Posidonius' determination of the size and distance of heavenly bodies see Kidd *Comm.* 454–456 and 464–466 on F116 and F120 EK.

56. See I.7.71–76, where this premise forms part of Eratosthenes' calculation of the circumference of the Earth.

57. Contrast this arbitrarily stipulated value (a genuine hypothesis) with the value for a terrestrial distance employed in the calculation of the circumference of the Earth in I.7; see I.7 nn. 4 and 11. It is unlikely (*pace* Heath [1913] 348 and Kidd *Comm.* 447) that Posidonius derived the ratio 10,000:1 for the heliacal circle to the Earth's circumference from Archimedes.

58. This calculation, like earlier ones (cf. I.7 nn. 9 and 21, and see n. 73 below), involves two identical ratios (cf. I.5 n. 22). Here 300:240,000 (the ratio between the diameter of the area at Syene and the Posidonian measurement of the circumference of the Earth given in I.7) (or 1:800) is multiplied by 10,000, the hypothesized figure in *(2)*, and the ratio is converted to 3,000,000: 2,400,000,000.

point to it), it is possible that we do not know that it is in fact greater, or again less, than this.[59]

286: The following procedure[60] is therefore considered to carry a greater degree of cognitive reliability.[61] *(1)* In total eclipses the Moon is said to measure out the Earth's shadow twice.[62] (This is because the time interval in which it enters the Earth's shadow is equal to the time interval in which it is concealed by the shadow, so that there are three equal intervals: the one in which it enters; the one in which it is concealed; and the one in which it exits from the shadow on exposing its outermost rim immediately after the second interval.) So since [by *(1)*] *(2)* the Earth's shadow is measured out twice by the size of the Moon, it seems plausible *(3)* that the Earth is twice the size of the Moon.[63] So since, by Eratosthenes' procedure, *(4)* the great circle of the Earth is 250,000 stades,[64] then *(5)* the Earth's diameter must exceed 80,000 stades. Thus [by *(3)*] *(6)* the diameter of the Moon must be 40,000 stades.[65] So since *(7)* the Moon (just like the Sun) is also $^1/_{750}$th part of its own circle (as is estab-

59. Even without this interjection, which is presumably Cleomedes', the hypothetical nature of *(2)* must have been obvious to Posidonius; see n. 57 above.

60. Kidd *Comm.* 444 suggests that the next calculation (lines 286–338) should not be attributed to Posidonius, but since it has an identical axiomatic basis (see n. 73 below), Posidonius could have been its source, if not its original author; cf. also I.7 n. 9.

61. Literally, it involves "some greater degree of clarity," where the clarity is attached to cognitively reliable reasoning; see Introduction 14–15 and II.6 n. 29.

62. See Aristarch. *De magn.* prop. 5 (352.13 Heath [1913]).

63. On the relative sizes of the Earth and the Moon in Stoic cosmology see II.3 n. 19.

64. See I.7.109–110.

65. This rounding off means that the Earth's circumference might as well be 240,000 stades (Posidonius' value) as the Eratosthenean value given in *(4)*. The claim at *(6)* must assume that the Earth's shadow is cylindrical, not conical (cf. II.2.27; II.6.90–108), in shape. See Figure 20. This in turn assumes that the Sun is the size of the Earth.

lished by water clocks),[66] and since *(8)* 1/6th part of its circle is the distance extending to its height from the Earth,[67] it follows [by *(7)* and *(8)*] that *(9)* this distance is 125 lunar magnitudes.[68] But *(10)* each of the 125 lunar magnitudes also has [by *(6)*] a diameter of 40,000 stades. Therefore, *(11)* there are 5,000,000 stades from the Earth to the height of the Moon (at least by this procedure).[69]

304: In addition, hypothesize that *(12)*, based on a simplified reckoning, the motion of the planets that is based on choice occurs at the same speed.[70] Since *(13)* the Moon goes through its own circle in 27½ days, while the Sun has a period of 1 year,[71] *(14)* the heliacal circle must be 13 times the lunar circle.[72] Thus *(15)* the Sun will also be 13 times the size of the Moon (since [by *(7)*] each of them is 1/750th part of its own circle.) Therefore, in accordance with the preceding assumptions, *(16)* the Sun's diameter is determined as 520,000 stades.[73]

66. See lines 184–191 above.

67. This follows, if the diameter of a circle is 1/3 of the circumference; see *(20)* below, and cf. I.7.119–120 (with I.7 n. 24). If the Earth is not "a point" in relation to the orbit of the Moon (cf. I.8.100–112 and II.3.71–80), then *(8)* is a hypothesis, albeit less radical than *(12)*.

68. That is, 750/6 = 125. This also assumes that the lunar circle is concentric with the circumference of the Earth.

69. For this value see Apollonius of Perge at Hippolytus *Refutatio* 4.8.6 (101.31). It can also be reached from *(8)* above (i.e., 750 x 40,000/6); it is, in other words, 1/6 of the lunar circle.

70. This hypothesis (cf. lines 334–338 below) is used to explain anomalous planetary motion at Gem. *Isag.* 1.19. The speed in question is not angular but linear; it is the rate at which a planet runs through the circumference of its circular course.

71. These are their sidereal periods given earlier: I.2.28–29 (Sun), and I.2.41–42 (cf. II.3.97–98) (Moon).

72. The computation here is approximate. Cleomedes is considering only the whole number of times the year is divided by 27½.

73. This calculation is again (cf. n. 58 above) implicitly based on an identity between ratios: specifically, 1:13 :: Moon's diameter:Sun's diameter :: 40,000: 520,000.

312: Since *(17)* the heliacal circle (just like the zodiacal circle) is divided into 12 parts, then *(18)* each of its twelfth-parts will [by *(16)*] comprise 32,500,000 stades,[74] and *(19)* the distance from the Earth to the Sun is [by *(8)*] 2 twelfth-parts. (Aratus[75] also states this with reference to the zodiacal circle as follows: *Six times the length of the ray from an observer's eye-glance would subtend this circle, and each sixth, measured equal, encompasses*[76] *two constellations.* Here he has called 2 *dōdekatēmoria* of the zodiacal circle "constellations" *(astra)*, and in the verses quoted reveals that the distance from the Earth to the Sun is 1/6 of the whole circle.)[77] In other words, while [by *(8)*] *(20)* the whole diameter is 1/3 of that circle, then *(21)* the distance from every part of the Earth to the Sun is 1/6 [of that circle; i.e., 65,000,000 stades] (since that circle has as its center the Earth, which is located at its exact center).[78] So since the heliacal circle is determined by this procedure [cf. *(7)* and *(16)*] as 390,000,000 stades, each of its twelfth-parts, as we said [at *(18)*], is 32,500,000 stades. So if the latter too are divided into 30 degrees, just like the *dōdekatēmoria* of the zodiacal circle, each degree will be 1,083,333 1/3 stades. There will be 720 half-degrees in the whole circle, but the Sun will [by *(7)*] be 1/750 of it, and hence less than 1/2 degree. Thus since 1/2 degree is 541,666 2/3 stades, the Sun itself is with probability determined [by *(16)*] as having a

74. That is, 750 × 520,000 / 12 = 32,500,000 (see lines 324–325 below). Here, in *(19)* and at line 326 below, *dōdekatēmoria* can be translated as "twelfth-parts," since it is not being used in its technical sense to designate the zodiacal signs (cf. I.4 n. 17 and lines 319 and 328 below) which are each 1/12 of the zodiacal circle.

75. *Phaen.* 541–543, translated by D. Kidd (1997) 113, with minor changes.

76. At *Phaen.* 543 the best manuscripts of Aratus have περιτέμνεται ("intercepts") for the verb in Cleomedes' text, περιτέλλεται ("encompasses"). See D. Kidd (1997) 113.

77. The radius of the zodiacal circle is thus taken as 1/6 of its circumference; see Hipparch. *In Arat. et Eudox. Phaen.* 1.9.12 (94.16–17) and *Schol. in Arat. vet.* 320.13–15. A scholiast in the latter collection (321.4–6), however, notes that the radius is equal to one side of a regular hexagon inscribed in the circle (cf. Eucl. *El.* 4 prop. 15 porism).

78. This is how it appears; for the reality see I.4.57–71.

diameter of 520,000 stades, in accordance with the assumptions as we have made them.

334: Yet it is certainly held to be implausible *(12)* that the planets move at an equal speed in accordance with the course based on choice. Rather, the course of the more distant planets, composed as they are of a more tenuous fire, is much swifter. For how is it possible for the Moon, whose own body is mixed with air, to have its course based on choice equal in speed to those [planets] that subsist from fire, which is tenuous (i.e., extremely light)?[79]

339: So while different claims have been made regarding the size of the Sun, none of the natural philosophers and astronomers has claimed that it has a diameter less than that given above. (Hipparchus, they say, demonstrated that it was 1,050 times the size of the Earth!)[80] So how could it be 1 foot wide when it is determined as being of immeasurable size by every method of reasoning that adopts an essentially systematic procedure?

345: Now since [by *(7)* and *(21)*] there have to be 125 solar magnitudes between the Sun and the Earth, then if the Sun is 1 foot wide (i.e., the size that it appears to be), the Earth's distance from it will have to be 125 feet![81] (That means that the Sun will be located well below the highest mountains, since some of them have a vertical height exceeding even

79. If the varying density of the planets' aethereal mediums causes them to move at different speeds, then their apparent speeds will also necessarily differ, as they would anyway because of their varying distances from the observer (cf. II.5.102–107), even if their real speeds were equal (cf. Eucl. *Opt.* prop. 54, which proves that "when objects move at an equal speed, those more remote seem *[dokei]* to move more slowly," tr. Burton [1945]).

80. On Hipparchus see Toomer (1974–75) 140.

81. Diog. Oen. Fr. 13 col. II.5–8 Smith (1993) argued that the Sun's distance was greater than it appeared; otherwise, it would ignite the Earth. Cleomedes could respond that it would then also have to be larger than it appeared. Furley (1996) 125 argues that the Epicureans are committed to a cosmos in which heavenly bodies are not disproportionately distant from an Earth to which they are proportionate in size.

10 stades.)[82] Thus, on Epicurus' doctrine, the Sun's height (in relation to which the Earth is a point, despite [by *(4)*] being 250,000 stades [in circumference]), is determined as being 125 feet from the Earth. That is what is implied by the doctrine of "the sacred soul who alone discovered the truth"![83]

353: As for the height of the Moon, what could one even say? For if the Sun is 125 feet distant from us, and far lower than mountains, how far distant from the Earth must the Moon be when [by *(14)*] its circuit is, by the minimum calculation, $^1/_{13}$ of the solar circuit?[84]

357: But even if Epicurus could pay no attention to these [calculations], nor uncover them in an inquiry that was beyond a fellow who valued pleasure, he should at least have paid attention to the actual power of the Sun, and to have reflected [on the following]: *(a)* that the Sun illuminates the whole sky, which is almost immeasurably large; *(b)* that it heats the Earth so that some parts of it are uninhabitable because of extreme heat;[85] *(c)* that through its considerable power it provides an Earth that is alive[86] so that it produces crops and sustains animal life; and that it alone causes animals to subsist, and also crops to be nourished, grow, and come to fruition; *(d)* that it alone is what causes not only the daytimes and nighttimes, but also summer, winter, and the other seasons; *(e)* that it alone is the cause of people being black, white, and yellow, and

82. At I.7.123–124 a maximum height of 15 stades is given for mountains.

83. Epicurus was frequently lauded in extravagant terms by members of his school; see further Goulet 215 n. 286 and Pease (1955) on Cic. *De nat. deor.* I.43 and 72, and also lines 461–462, 467 and 487–488 below.

84. Cleomedes could have argued that if the Earth is a point in relation to the Moon (implied by *[8]* above; cf. n. 67 above), and if the Moon is 1 foot wide (its apparent size), then for the Epicureans it should also be at the *same* distance from the Earth as the Sun.

85. Cleomedes accepted that the torrid zone was uninhabitable (see I.1.88–89, 210–211, 266–267; and I.2.73–76), and criticized Posidonius' arguments for its habitability (I.4.90–131).

86. *empnous:* literally "endowed with pneuma," i.e., with the force that is, in effect, the soul of the cosmos.

differing in other visible aspects, depending on how it sends out its rays to the latitudes of the Earth; *(f)* that the power of the Sun, and it alone, renders some places on the Earth well-watered and teeming with rivers, others dry or lacking in water; some barren, others adequate for crop production; some acrid and foul-smelling (like those of the Fish-Eaters),[87] others fragrant and aromatic (like places in Arabia); and different places capable of producing different kinds of crops.

376: The Sun is, in general, the cause of virtually all the variety found among things on the Earth, since the Earth shows considerable contrast at some latitudes. We can, for example, learn of the variety of things reported in Libya, in the territory of Scythia, and in Lake Maiotis,[88] where the crops, animals and, in a word, everything are subject to major transformations, including the temperatures of the air, and its varying states. Then there are the differences observed throughout the whole of Asia and Europe[89] in springs, crops, animals, metals, and hot springs, and in every type of air—very cold, very torrid, temperate, light, dense, moist, dry—as finally in all the other differences and peculiarities observed at each latitude. Of all these the power of the Sun is the cause.

387: The Sun, furthermore, has such a great superfluity of power that the Moon too receives its light from it,[90] and so has this as the exclusive cause of all its power in its different phases, since the Moon not only fashions enormous changes in the air by controlling it and thereby fashioning innumerable purposeful results, but is also the cause of the flowing and ebbing of the Ocean.[91]

393: The Sun's power has also a further observable property: that

87. The "Fish Eaters" *(Ikhthuophagoi)* are often associated with the Arabian Gulf; see Hdt. 3.19 and Strabo 16.4.4.

88. This is the ancient name for the Sea of Azov. See Strabo 2.5.23, and also 2.1.16 on its climate.

89. Asia, Europe, and Libya (see lines 378–379 above) were the continents of the inhabited world known to the ancients; see, for example, Hdt. 4.42 and 198.

90. For the details see II.4.21–32.

91. On lunar power see also II.3.61–67.

while fire cannot be extracted by reflection from ordinary fire, we contrive to extract fire by reflection from the rays of the Sun, despite its being such a vast distance away from the Earth.

396: Also, as it goes through the zodiacal circle (that is, as it effects this type of course), the Sun by itself harmonizes the cosmos, and so, by being the exclusive cause of continuing stability in the comprehensive ordering of the whole cosmos, it provides the whole cosmos with an administration that is fully concordant.[92] And if the Sun changes its position, either by abandoning its own place, or by disappearing completely, not a single thing will then be born or grow—in fact nothing will "subsist" at all, but everything that exists and is visible will be dissolved together and so be destroyed![93]

404: Epicurus, then, should have attended to all this, and reflected on whether a fire that was 1 foot wide could have a power that was so extensive, so great, and so prodigious.[94] But in astronomy, in the area of sense presentations, and in every investigation generally, he was the same as in his treatment of the first principles of the cosmos, the theory of the goal of life, and in ethics generally[95]—a man far blinder than a bat![96] No wonder, since pleasure-loving fellows certainly cannot uncover the truth

92. Despite the "administrative" activity of the Sun, it is the heavens in their totality that play this role; see I.2.1–4. An earlier Stoic, Cleanthes (ca. 331–230 B.C.), had represented the Sun as the "controlling organ" *(hēgemonikon)* of the cosmos (*SVF* 1.499), but Posidonius did not follow him; see Kidd *Comm.* 145.

93. See also Cic. *De nat. deor.* 2.91 for this argument.

94. Lucr. 5.592–613 argues that a Sun of extremely small size can cause major terrestrial effects, but concedes the Stoic point by envisaging (at 610–613) an adjacent band of invisible heat that augments solar power.

95. For Epicurus' treatise on the goal of life see Usener *Epicur.* 119.13–123.17 and Diog. Laert. 10.27, and for that on sense presentations Diog. Laert. 10.28. The areas of philosophy identified here could also correspond to the categories of Cleomedes' own program of teaching.

96. The English idiom is used here. The Greek refers to *spalakes* (blind rats or moles), which were thought to have eyes under their skin; see Arist. *De an.* 425a10–11.

in what exists. That is for *men* who are naturally disposed to virtue and value nothing ahead of it, not for lovers of a "tranquil condition of flesh" and the "confident expectation regarding it."[97]

414: In an earlier generation they drummed the Epicureans and their scriptures out of communities in the belief that doctrines that had reached such a level of blind perversion offered people offense and corruption.[98] Today, by contrast, because people are, I think, undone by effeminate luxury, they esteem the members of the sect and their actual treatises so highly that they seem to have a stronger desire for Epicurus and the members of his school to speak the truth than for the gods and Providence to exist in the cosmos. In fact some would even pray for Providence to be destroyed rather than have Epicurus convicted of false statements.[99] That is the wretched state that they are in—so reduced by pleasure that they revere its advocate above everything in life!

426: In addition to all the absurdities mentioned, the Epicureans also claimed that heavenly bodies were kindled on rising, and extinguished on setting.[100] That is just like someone saying that people exist while they are being seen, but die when they are not seen—or his applying like reasoning to every other visible thing! So Epicurus was such a clever and inspired man that it did not even occur to him that, because the Earth has a spherical shape, each [heavenly body] sets and rises at different times in different [regions], and so by his doctrine would have to be extinguished

97. Both expressions were used in Epicurus' "On the Goal of Life" *(Peri telous);* see Usener *Epicur.* 121.34–122.3

98. For corroboratory evidence see Goulet 214 n. 277.

99. On alleged Epicurean impiety see, for example, *SVF* 2.1115–1116, Posid. F22a–bEK, and Usener *Epicur.* 246.20–248.23.

100. This was in fact just one of a set of multiple explanations for the alternation of daytimes and nighttimes, as Algra (2000) 183 notes; see Epicur. *Pyth.* 92, and Lucr. 5.650–662 and 758–761. Ptol. *Alm.* 1.3, 11.24–12.18 refuted it, as did the earlier Platonist Dercyllides (see I.2 n. 9), according to Theon *Expos.* 199.21–22. Posidonius dismisses the audible quenching of the setting Sun at F119.3–6 EK. On the illogic of multiple explanations see Wasserstein (1978) 490–492.

and kindled simultaneously. At every alteration in horizons, that is, there would have to be, in repeated progressions, incalculably numerous cases of the destruction of bodies—bodies that were both destroyed and rekindled, since this would happen at every alteration in horizon.[101]

438: We can learn about the alterations in horizons from countless other [phenomena], but preeminently from reports of [the lengths of] daytimes and nighttimes at the solstices among different races.[102] Thus the [lengths of] nighttime at the summer solstice are reported [as follows for these places]:

Meroe in Ethiopia	11 hours
Alexandria	10 hours
The Hellespont	9 hours
Rome	under 9 hours
Marseilles	8½ hours
Among the Celts	8 hours
Lake Maiotis	7 hours
Britain	6 hours

It is obvious from these [reports] that the Sun sets and rises at different times in different [regions]. This happens also for people below the same parallel circles (that is, with identical seasons), whether they are located further east and encounter the Sun's onset sooner, or in the west and do so later.[103] So if there are countless alterations in horizons (there being a different one at every latitude of the Earth), heavenly bodies will have to be extinguished and kindled incalculably many times. Anything more

101. If the Epicureans thought that the Earth was flat (cf. I.5 n. 6), then horizons would *not* alter (I.5.30–44), but their theory would still have other problems.

102. On Ptol. *Alm.* 2.6, the major ancient evidence on variations in lengths of daytimes, see Neugebauer (1975) 44. For more elementary accounts see Gem. *Isag.* 6.7–8, Mart. Cap. 8.877 (cf. 6.595), Plin. *NH* 2.186, and Strabo 2.5.38–42.

103. These are the *perioikoi;* see I.1.236–251 and cf. I.5.37–39.

unthinkable in its display of every kind of reckless ignorance could not be even conceived!

452: Certainly not even the illuminations of the Moon, despite being very vivid, restrain the Epicureans from such ridiculous claims.[104] I mean, how could the Moon be illuminated and shine throughout the night, if the Sun is extinguished on setting? Or how is the Moon eclipsed on falling into the shadow of the Earth, if it is not even illuminated at all? Or how does it exit from the shadow and become illuminated again when there is no Sun below the Earth? Or how does the Sun itself, if it is extinguished, reach its point of rising again? Epicurus, in fact, believed in an old wives' tale, like the Iberians' report that the Sun on falling into the Ocean makes a noise when it is extinguished like red-hot iron in water. That is how "the first and only man to discover the truth" arrived at this doctrine! And he did not even understand that every part of the heavens is at an equal distance from the Earth, but believed instead that the Sun sank into the sea and rose again from the eastern sea—kindled by water in the east, but extinguished by it in the west![105]

467: That is what the "sacred wisdom" of Epicurus discovered! But, by Zeus, it occurs to me to compare him to Homer's Thersites. For Thersites was the worst man in the Achaean army, as indeed the Poet himself says and portrays Odysseus as saying. His own words are "He was the ugliest man to come to Troy" and so on, while he depicts Odysseus saying to Thersites: "There is no other mortal man, I vow, worse than you."[106] But despite being like this, he still did not keep his peace, but

104. The option of lunar illumination by the Sun is in fact included in Epicurean multiple explanations; see Epicur. *Pyth.* 95–96 and Lucr. 5.705–714.

105. There is no evidence of the Epicureans causally linking solar kindling and extinction with a circumambient ocean. (Ptol. *Alm.* 1.3, 11.24–26 just refers to the Sun falling "to the Earth.") Cleomedes did not mention this additional absurdity earlier, but (cf. lines 13–19 and n. 7 above) only a less bizarre theory of solar reconstruction ("kindling") through interaction with the air.

106. *Iliad* 2.216 and 248–249. On the general topic of the Stoic reading of Homer see Long (1992).

first wrangled boastfully with the kings as though he too had status, then dared rank himself among even the leaders by mentioning "[the women] whom we Achaeans give you first whenever we take a stronghold . . . [the man] whom I or another Achaean might bring in in chains."[107] In this vein Epicurus too boasts of being important, given that he tries to include himself among the philosophers, and not just that, but also affirms the right to take first prize, and thereby reveals himself as more thirstily ambitious than even Thersites. The latter, after all, boasts only of being a prince and an equal to the kings, but does not also assign himself first prize, whereas Epicurus claims that he alone has found the truth through his vast wisdom and knowledge, and so thinks it right that he should also take first prize.

489: That is why I would believe it to be quite wrong for someone to say to *him:* "Babbling Thersites, clear orator though you are, hold off!"[108] For I would not also call *this* Thersites "clear," as Odysseus does the Homeric one, when on top of everything else his mode of expression is also elaborately corrupt.[109] He speaks of "tranquil conditions of flesh" and "the confident expectations regarding it," and describes a tear as a "glistening of the eyes," and speaks of "sacred ululations" and "titillations of the body" and "debaucheries" and other such dreadful horrors. Some of these expressions might be said to have brothels as their source, others to resemble the language of women celebrating the rites of Demeter at the Thesmophoria, still others to come from the heart of the synagogue and its suppliants—debased Jew talk, far lower than the reptiles!

503: But despite being like this in discourse and doctrines, he still does not blush to rank himself with Pythagoras, Heraclitus, and Socrates,[110] even asserting the right to occupy the first place among them, just like

107. *Iliad* 2.227–228 and 231.

108. *Iliad* 2.246–247.

109. For ancient attacks on Epicurus' style, and especially his use of neologisms, see Usener *Epicur.* 88–90, and Pease (1955) on Cic. *De nat. deor.* 1.86.

110. Posidonius respected Pythagoras (T91 and T95EK); for Heraclitus see I.8.96–97; on Socrates, the early Stoics, and the Epicureans see Long (1988).

temple robbers trying to rank themselves with priests and hierophants by asserting the right to hold first place among them! Or imagine Sardanapalus trying to square off against Hercules in an endurance test, and grabbing Hercules' club and lion skin and telling him "I have more right to these!"[111]

511: Will you not be off, evil degenerate, to your saffron-robed whores, with whom you will dally on couches, whether combing purple wool, or wreathed in crowns, or with your eyes painted, or even entertained by the *aulos*[112] in excessive and unseemly drunkenness, and then coming to the final act like a worm wallowing in utterly vile and excremental slime? So will you not be off, "most brazen and shameless soul," routed from Philosophy, to Leontion, Philainis, and the other whores, and to your "sacred ululations" with Mindyrides, Sardanapalus and all your boon companions?[113] Do you not see that Philosophy summons Hercules and Herculean men, certainly not perverts and their pleasures? Indeed, it is evident, I think, to cultivated people that Epicurus has nothing to do with astronomy, let alone with philosophy.[114]

111. The contrast between this hedonistic Assyrian monarch (seventh cent. B.C.) and Hercules was standard; see, for example, Juvenal 10.361–362. On Hercules' club and skin as symbols of strength see Cornut. *Theol.* 63.12–21 (at *SVF* 1.514).

112. This was a reed instrument and the principal wind instrument of Greek music; see Michaelides (1978) 42–46.

113. Leontion was an associate of Epicurus; see Usener *Epicur.* 411 and Pease (1955) on Cic. *De nat. deor.* 1.93. The other names are conventional symbols of hedonism; see *SVF* 3, pp. 198–200, and Goulet 216 nn. 298 and 299.

114. Astronomy, this sentence implies (see Goulet 39 n. 31 on its construction), is subordinate to philosophy, as claimed by Posidonius F18EK (see Appendix).

CHAPTER TWO

1: We have demonstrated that the Sun is not 1 foot wide, and so certainly not the size that it appears to be. Next we shall try and establish that it is larger than the Earth. This has already been in effect demonstrated, yet something else was being primarily established in the earlier discussion.[1] Here, however, we shall speak directly about this [thesis], starting out from the phenomena alone.[2]

7: In the first lecture course we demonstrated that the Earth, through having the ratio of a point [to the size of the cosmos], conceals none of the [celestial sphere's] 360 degrees, indeed not even a small fraction of a degree, since, as is demonstrated by the equinoxes, precisely 180 degrees always show above the Earth, along with the 6 zodiacal signs, and half of the equinoctial circle.[3] So since the Earth does not conceal even a small fraction of a degree, whereas the Sun occupies a magnitude of almost ½ degrees,[4] the Sun is larger than the Earth.

1. In the calculations of the size of the Sun at II.1.286–312 the relative size of the Earth was mentioned only at II.1.294–296 in connection with Eratosthenes' measurement.
2. For reasoning similarly based "directly" on the phenomena see also I.5.104–113.
3. Cf. I.8.37–43 with I.8 n. 11.
4. See II.1.329–330.

13: Now if we also hypothesize something equal in size to the Earth rising or setting [like the Sun], it will not spend any time at the horizon.[5] That is because just as the Earth, given its position in the exact center [of the cosmos], does not conceal even a small fraction of 1 degree, so too something equal in size to the Earth will not spend any time at the horizon when it rises or sets. Yet the Sun both rises and sets over an extended interval of time. It is therefore larger than the Earth.[6]

19:[7] Also, when one spherical body is illuminated by another, then if they are equal [in size] to one another, the shadow of the illuminated body is sent out in a cylindrical form. But where the illuminated body is the larger, the shadow is funnel-like,[8] with its [outer] ends being continually further widened and its forward progress being without limit. But if the body that causes illumination is larger, it is necessary that the shadow of the body that is illuminated be configured in the shape of a cone. Now since both the Sun and the Earth are spherical bodies, and the former causes illumination while the latter is illuminated, it is necessary that the shadow of the Earth be sent out with a shape that is either funnel-like, cylindrical, or conical. But it is neither cylindrical, nor funnel-like.

5. Taking the ratio of the Earth's diameter to that of the Sun as approximately 80,000:520,000 stades (or 1:6.5) (cf. II.1.295–296 and 311–312), then, adapting the example at II.1.145–148, in which a horse ran 10 stades while the Sun rose, the Earth (or a body of equivalent size), located at the same distance from the Earth as the Sun, would appear to rise for a terrestrial observer in the time taken by such a horse to run just over 1.5 stades.

6. For a more elaborate version of this argument by the Stoic Dionysius of Cyrene (a near contemporary of Posidonius; cf. II.1 n. 26) see Philodem. *De signis* cols. 10–11 (sect. 15 De Lacy [1978]), and Barnes (1990) 2661–2662.

7. See Figure 21 for the shapes proposed in this paragraph.

8. The adjective so translated *(kalathoeidēs)* literally means "basket-like," where the container *(kalathos)* has a base significantly narrower than its opening; i.e., it is shaped somewhat like a modern filter funnel, or, in relation to the present context, like an inverted cone, or what Pliny *NH* 2.51 calls a *turbo rectus* (an upright spinning-top).

Therefore it is conical,[9] and, if that is so, the Earth has as the cause of its illumination something larger than it—the Sun.[10] In the discussion concerning the Moon[11] we shall demonstrate that the Earth's shadow is neither funnel-like, nor cylindrical. That, then, is enough on the size of the Sun.

9. This argument (also at Heraclit. *Allegr.* 46, Plin. *NH* 2.51, and Theon *Expos.* 195.5–197.7) implicitly relies on the Stoic "fifth undemonstrated argument" (see I.5.20–29 and I.6.1–8), in which all but one of a set of disjuncts are eliminated.

10. Posidonius (F9EK) argues that the Sun is larger than the Earth because the Earth's shadow is conical in a lunar eclipse.

11. At II.6.79–108.

CHAPTER THREE

1: The notion that the Moon too is not the size that it appears to be can also be formed from what was said above about the Sun (that is, most of what was said there can also be applied to the Moon),[1] but it is the eclipse of the Sun that primarily demonstrates this. That is because the Sun is eclipsed only when the Moon passes under it, and obstructs our line of sight; a solar eclipse, in other words, is a condition affecting not the Sun, but our line of sight.[2] So whenever the Moon passes under the Sun such that it is in conjunction with the Sun, and at that conjunction is located on the circle through the middle [of the zodiacal band],[3] it necessarily sends out to the Earth a conical shadow, reportedly extending over more than 4,000 stades (the Moon's shadow equals the total area in which the Sun is invisible when the Moon moves below it). So if its conical shape is extended over this much of the Earth, or even more still, the base of the cone (also equal to its diameter)[4] is clearly many times larger.

1. On the apparent size of the Moon see II.1.114–128 (with II.1 n. 27 on the Epicureans).
2. For this definition see *SVF* 2.650 and Posid. F125EK. See also II.4.127–131.
3. That is, it is located on the zodiacal circle.
4. See also II.1.67.

15: Also, an observation of the following kind occurred at a solar eclipse.[5] On an occasion when the Sun was totally eclipsed at the Hellespont, it was observed at Alexandria as eclipsed beyond $^1/_5$ of its own diameter,[6] which is just over 2 digits in appearance. (The apparent size of the Sun, that is, and similarly of the Moon, is held to be 12 digits.)[7] So from this it is clear that the 2-digit appearance of the size of the Moon and Sun is coextensive with a distance on the Earth equivalent to that between Alexandria and the Hellespont, both of which are located below the same meridian.[8] So if by hypothesis the [conditions] of this eclipse remain fixed, then for people initiating a journey from Alexandria to the Hellespont the 2-digit appearance[9] of the Sun seen at Alexandria would become proportionately less. As there are 5,000 stades from Alexandria to Rhodes, and another 5,000 from there to the Hellespont, the appear-

5. The eclipse is that of March 14, 189 B.C.; see Neugebauer (1975) 316 n. 9. It was recorded by Hipparchus; see Pappus *In Ptol.* 5.11 (68.5–9), and cf. Préaux (1973) 255–256.

6. In other words, the Sun was obscured to $^4/_5$ of its diameter.

7. Cf. II.4.117, and see Figure 22. By Ptolemy's time (cf. *Alm.* 6.7), it was the practice to define the maximum obscuration or magnitude of an eclipse in terms of digits, where 1 digit *(daktulos)* is $^1/_{12}$ of the diameter of the eclipsed body. The eclipse in question thus had a magnitude of 12 digits in the Hellespont and almost 10 digits in Alexandria. Such digits are not the same as those digits of angular measure which are found in Babylonian astronomical texts, and amount to 5' of arc each (cf. Toomer [1984] 322 n. 5). Nor are they the digits, or finger's-breadths, that were $^1/_{16}$ of a foot *(pous)*, or $^1/_{12}$ of a hand's span *(spithamē)*.

8. There is in fact no such coextension (or alignment). As Neugebauer ([1975] 964) notes, the "obscuration of one sphere by another does not vary linearly with the displacement of an observer on a third sphere."

9. *phasis:* the term translated "appearance" here is being used, as it often is, as a synonym for the commoner term for appearance, *phantasia;* cf. line 20 above for "the 2-digit appearance *(phantasia).*" In astronomy, however, it acquires the more technical sense of "appearance at a significant configuration with the Sun" (cf. Toomer [1984] 22). Thus the fixed stars and planets are said to have phases, e.g., the first visibility of a fixed star at sunset, and at II.5.70–71 Cleomedes refers to "the phase of [lunar] illumination."

ance of the Sun seen at Rhodes will necessarily be 1 digit.[10] Then as they go from Rhodes to the Hellespont, this appearance too will be proportionately diminished, and will finally be out of sight when they reach the Hellespont. So clearly, if the 2-digit appearance of the size of the Moon and Sun is coextensive with so great a quantity of the Earth, it is necessary that their whole bodies be coextensive with 6 times such a quantity of the Earth.[11]

34: From this procedure the notion can be formed that the [remaining] heavenly bodies too are of enormous size (but certainly not the size they appear!), and particularly the fixed stars, which are the farthest away.[12] For while the difference in their sizes is observed to be large, none appears less than 1 digit. Venus in fact sends out the appearance of 2 digits, making its diameter 1/6 of the Sun's diameter, assuming that they are the same distance from the Earth, but otherwise in proportion [to their true distances].[13] The size of the bodies that appear 1 digit in diameter is 1/12 of the Sun's diameter, if they are assumed to be at the same height as the Sun, but since they are at a greater height, the proportion of their [true] distance will be taken into account.[14]

10. Strabo 2.5.40 gives the approximate distance from Alexandria to Rhodes as 3,600 stades, and that from Rhodes to Alexandria in the Troad as 3,400.

11. Here again (cf. I.7 nn. 9 and 21, and II.1 nn. 58 and 73) a calculation is implicitly based the principle that two ratios (or spatial coextensions) are the same: i.e., 2 digits:10,000 stades :: 12 digits:60,000 stades. The implicit conclusion, then, is that the Sun must have a diameter of at least 60,000 stades, sufficient to show, as a preliminary "notion" (cf. *ennoein* at line 34 below, and I.1 n. 40), that the Sun is not the size it appears to be.

12. For the Epicurean claim that the fixed stars are as large as they appear see Epicur. *Pyth.* 91, and Lucr. 5.585–591.

13. All supralunary heavenly bodies are "fiery" in proportion to their distance into the fire-sphere of the aether (II.1.335–336; cf. II.5.4), yet inherent luminance is not a factor in this analysis, or in that offered in the next paragraph.

14. Here this formula is used to admit that a celestial distance *is* incorrect, just because it is a pure hypothesis. Cf. I.7.46–47 (with I.7 n. 11) where it is a safeguard against the plausible measurement of a terrestrial distance being incorrect.

43: Thus the question of whether some heavenly bodies are also equal to the Sun's size, or even exceed it in size, should not be abandoned. If, for example, one of them were elevated so far that the Sun, if also imagined elevated just as far, will be seen possessing the size of a star, then it will be equal [in size] to the Sun. But if elevated farther, it will be larger in proportion to its height. So since the fixed stars at the outermost circumference of the heavens are very distant, though none is less than 1 digit in appearance, they will all be larger than the Sun.[15] (Furthermore,[16] the Earth, being a point in relation to the height of the Sun, would either not be seen at all by a human being when seen from the height of the Sun, or else would be seen to have the size of an extremely small star, whereas from the sphere of the fixed stars it would not even be seen at all, <not even if assumed to be equally as bright as the Sun>.[17]) It is evident, then, that all the stars seen at this height from the Earth are larger than the Earth, as of course is the Sun itself too, to which many fixed stars are also probably equal in size, or even exceed it in size. That, then, is our discussion concerning this topic.

61: As for the size of the Moon (specifically its not being 1 foot wide) evidence can also be derived from its power, since it not only illuminates the whole sky, fashions major changes in the air, and has many things on the Earth in sympathy with it, but is also the exclusive cause of the ebbing and flowing of the Ocean.[18]

65: The preceding [discussion] is an adequate argument that neither the Sun, the Moon, nor any other heavenly body, is the size it appears to be.

15. This visibility is caused by a luminance that increases in proportion to the distance from the Earth; see n. 13 above.

16. This parenthesis (lines 51–55) repeats I.8.21–26, where the radius of the Earth is shown to be negligible in most celestial observations.

17. The clause in angle brackets is at lines 52–53 in the manuscripts, where it disrupts the reasoning; here it parallels the argument at I.8.25–26.

18. On the Moon's "power" see II.1.387–392. On the causation of tides see Posid. F138EK, and cf. F106EK.

68: Now while none of the other heavenly bodies (at least those visible to us) is held to be smaller than the Earth, astronomers claim that the Moon is smaller than the Earth,[19] offering as their primary evidence the fact that its diameter measures out the Earth's shadow twice.[20] Again, at solar eclipses, as we have already said,[21] the Sun has been observed partially eclipsed at Alexandria during a total eclipse at the Hellespont. This would not happen unless the Earth had a significant size relative to the Moon:[22] in other words, if there is this much difference over a distance of 10,000 stades, the Moon evidently does not cast a shadow on much of the Earth. But if the Moon were equal to, or larger than, the Earth, it would cast a shadow on a considerable area of it during its courses below the Sun. But in fact there will even be areas of the Earth where the Sun is totally visible, while it is being totally eclipsed elsewhere.

81: The Moon does appear large, in fact equal in size to the Earth, and larger than the other heavenly bodies, when in reality it is smaller than they, since it is closest to the Earth of all the heavenly bodies, and thought to be located right at the junction of the air and the aether.[23] That it *is* the closest of all [the heavenly bodies] to the Earth is demonstrated from [the following considerations].[24] *(a)* For those who view the

19. The Stoics (*SVF* 2.666) and Posidonius (F122EK) are both reported as claiming that the Moon was larger than the Earth. Theiler (2.179) and Kidd (*Comm.* 472) are reluctant to attribute this view to Posidonius, and argue that he agreed with Aristarchus that the diameter of the Moon is half that of the Earth (cf. II.1.286–288). See also Pease (1955) on Cic. *De nat. deor.* 2.103.

20. See II.1.286–288 on this being determined at total lunar eclipses.

21. At lines 15–33 above.

22. I.e., if it was not observationally insignificant (or "a point") in relation to the Moon; see I.8.106–112 and cf. I.8 n. 2.

23. See also I.2.37–38.

24. Statements *(a)*–*(e)* all record observations. The physical theory introduced in *(b)* is not essential to the argument from observation. Thus the Moon's proximity to the Earth, like the cause of its eclipses (see II.6.35–36; cf. II.6.56 and 194–196), is directly demonstrable from the phenomena.

Moon with special care it is demonstrated by sight alone,[25] since no other heavenly body goes under it, whereas it is seen to pass under all the planets. From this it is demonstrated that they are more distant than it. *(b)* Its own body is mixed with air, and is rather murky [in appearance],[26] because it is not in the unadulterated [part] of the aether, like the rest of the planets, but, as we have said, at the junction of the two elements [the air and the aether]. *(c)* The Moon alone falls into the Earth's shadow, but none of the other heavenly bodies do; otherwise they would at different times appear brighter and fainter, since every body that is composed of fire appears brighter in a shadow, but fainter under the rays of the Sun.[27] *(d)* In contrast with the other heavenly bodies the Moon has a unique sympathy with bodies on the Earth, just because of its greater proximity to the Earth.[28] *(e)* It goes through its own circuit in 27½ days,[29] whereas no other planet has a period of less than 1 year.

100: It is evident from these [phenomena] that the Moon is the closest to the Earth of all the heavenly bodies.

25. "Sight alone" would not, of course, be an adequate criterion for establishing the nature of something unobservable; cf. I.5.1–9.

26. See II.5.1–4, and II.5 n. 3.

27. See II.6.101–105, and cf. Sext. Emp. *PH* 1.119. Also cf. II.4 n.10 below on the special case of the Moon.

28. See lines 61–67 above.

29. See I.2.41–42.

CHAPTER FOUR

1: There have been several theories concerning the illumination of the Moon.[1] Berossus actually claimed that the Moon was "half fire," and that it moved with a plurality of motions.[2] First is the one in longitude;[3] second the one in latitude (that is, in height and depth [relative to the zodiacal circle]), which is also seen occurring in the case of the five planets;

1. For the three theories considered in this chapter see Apuleius *De deo Socratis* 117–119, with Donini and Gianotti (1982). For the association of the third theory (lines 21–78) with Posidonius see nn. 8 and 19 below.

2. Jacoby *FGrH* 3.C.1, no. 680, at 395–397 distinguished this Berossus from Berossus the Babylonian, a historian of the third century B.C. (Lines 1–9 here = Jacoby Fr. 18; lines 1–17 = Schnabel [1923] Fr. 18.) But his view is still being debated; see, for example, Burstein (1978) 31–32 and Verbrugghe and Wickersham (1999) 13–15 for arguments against the distinction, and Kuhrt (1987) 36–44 for an able defense. On Berossus' lunar theory see Vitruv. *De arch.* 9.2.1 and Lucr. 5.720–730. For a reconstruction see Toulmin (1967).

3. Here we have emended Todd's text at II.4.3 by deleting a clause that states that this motion in longitude is one "which occurs together with this cosmos." The problem with such a qualification is that it would identify Berossus' first motion as the Moon's diurnal motion, which is not longitudinal, whereas its longitudinal motion is sidereal, i.e., in the opposite direction to the cosmos (as identified collectively for the planets at the I.2.8–11). In this way we address the problem of the omission of sidereal motion raised by Goulet 220 n. 336. We

and third is the one around its own center. Berossus believes that the Moon waxes and wanes as it rotates with this third motion, that is, as it turns different parts of itself toward us at different times, and that this rotation occurs in a time equal to its reaching conjunction with the Sun.

10: His doctrine is easily refuted. First, since the Moon exists in the aether, it cannot be "half fire" rather than being completely the same in its substance like the rest of the heavenly bodies.[4] Second, what happens in an eclipse also conspicuously disconfirms this theory. Berossus, that is, cannot demonstrate how, when the Moon falls into the Earth's shadow, its light, all of which is facing in our direction at that time, disappears from sight.[5] If the Moon were constituted as he claims,[6] it would have to become *more* luminous on falling into the Earth's shadow[7] rather than disappear from sight!

18: Others say that while the Moon is illuminated by the Sun, it illuminates the air by reflection, as is seen happening also with mirrors, bright silver objects, and the like.

21: A third option claims that the Moon's light is mixed both from its own <body>[8] and from the Sun's light, and that such a [state] comes about

---

regard the text at II.4.2–3 as having been contaminated by the reference to the motion that "occurs together with the cosmos," which was probably originally a gloss that was mistakenly inserted into the text. For planetary latitudinal motion at lines 4–5 see I.2.64–69.

4. The Moon, that is, cannot have two radically distinct parts, if it is located in the aether, since while the aether may be less dense at greater distances from the Earth (cf. II.1.336), it must be equally dense at any specific distance. On the other hand, it may be difficult to differentiate the Moon if it shares in all the physical properties of its medium; see Todd (2001).

5. See lines 82–94 below for an account of how lunar eclipses occur by the theory advocated in this chapter.

6. That is, if it were inherently luminous by being "half fire."

7. For this principle see II.3.91–95 and II.6.103–105; also Plut. *De fac.* 933D.

8. This supplement, confirmed by lines 80–81 below, rules out the interpretation of this third theory as involving the admixture of two kinds of light: solar light and an inherent lunar light. Goulet 221 n. 346 rightly rejects this proposal, made by Cherniss (1957) 123 n. c, who is followed by Kidd *Comm.* 474–475. If

through its not remaining unaffected. That is, unlike solid bodies that give off light, it does not have solar rays rebounding from it and illuminating the air by reflection in a process of reception that involves its reciprocating the rays and thereby sending them back in our direction. Instead, the Moon is altered by the light of the Sun, and through such a blending possesses its own light not intrinsically, but derivatively[9] (just as fully heated iron possesses light derivatively), since it is not unaffected, but is transformed by the Sun. This option is sounder than the one claiming that the Moon causes illumination by reflection because rays rebound from it, as is seen happening where bodies that are solid give off light.

33: The impossibility of the Moon sending out light by reflection might be best summarized by the following [arguments].

34: *(a)* It is not impossible for reflection to occur from relatively solid bodies (reflections are also seen occurring from water, since even water is to some extent compact), but it *is* impossible for reflection to occur from rarefied bodies. After all, how could reflection occur from air or fire,[10] when such bodies are naturally disposed to absorb light rays, yet are not illuminated by them just on the surface, but fully absorb

---

nothing else, inherent lunar light is excluded by the criticism of Berossus' theory above. Also, *pace* Kidd *Comm.* 476, this theory of admixed lunar light could be Posidonian (cf. n. 19 below). The fact that at Posid. F127EK the Moon is called "luminous and fiery," and at F122.1–2 EK is said to be mixed from air and fire, does not mean that it also has an inherent and visible lunar light. Any such luminance can be totally lost (see lines 82–94), while any inherent light is visible only under special circumstances (see n. 10 below).

9. *kata metokhēn:* literally, "by participation."

10. Air and fire mingle at the point of their conjunction (I.2.37–38; II.3.83–84) in the lower part of the aether, where the Moon is located (see line 11). The Moon is therefore "mixed with air" (II.3.88; II.5.2 and 6), but too low in the aetherial fire-sphere to be inherently luminous. Hence its igneous component furnishes a "murky" appearance, clearly visible only when the Moon is darkened during a lunar eclipse; see I.2.38, II.3.89, and II.5.2–4 with II.5 n. 3.

anything that impinges on them, just as sponges invariably absorb water?[11]

42: *(b)* The light from bodies that illuminate by reflection is sent out a short distance, but the Moon not only sends out its luminance as far as the Earth, but also illuminates the whole sky. Yet bodies that illuminate by reflection do not send out luminance even 2 stades, as can be seen with mirrors as well as with every single body that illuminates by reflection.

48: *(c)* If anyone suggests that in illuminating by reflection the Moon sends its light farther than the [solid] bodies just mentioned because it is extremely large, we shall respond that both small and very large bodies that illuminate by reflection are subject to the same proportional progression:[12] that is, a larger area will be illuminated by large bodies in respect of length, yet light will certainly not be sent out a greater distance forward. Instead, whether the body that illuminates by reflection is 1 foot or 1 stade wide, it will send out its light over a distance that is equal in respect of its depth.

56: *(d)* It is in addition quite evident that if the Moon caused illumination by reflection, it would not illuminate the Earth at either the crescent or the quarter.[13] That is because objects that illuminate by reflection send out their light at right angles, and so, since the Moon is spherical, its light would be sent out to the west in the phases just men-

11. On lunar density see the problems and solutions at lines 81–94 and 95–107 below. Since the Moon's "sponginess" does not allow it to be totally permeated by its absorbed light (lines 101–102 below), *di' holōn* (line 40 here) must be translated "fully" rather than "totally." See also n. 24 below.

12. That is, the larger the reflecting surface, the larger the volume of air illuminated two-dimensionally ("in respect of length").

13. For this argument see also Plut. *De fac.* 929F-930A. The ancient Greeks used *dikhotomos* (literally "cut in half") to describe the Moon when the Sun's light illuminates exactly half of its face. This happens at what we call the first and last quarters.

tioned, and thus straight toward the Sun.[14] Indeed, not even when full would it cause illumination with the whole of its circle. It would do so if its shape were flat,[15] but since it is a sphere and thus has the extremities of the circle that is visible to us sloping round,[16] illumination from these sloping [extremities] will be sent out at equal and right angles, with the result that only the very middle of the Moon will illuminate the Earth, not its whole circle. In other words, the light from the very middle of the Moon can be sent toward the Earth at right angles, but the light from its sloping [extremities], which do not face the Earth, cannot. So if the Moon caused illumination by reflection, its whole circle would not illuminate the Earth. But it is evident that the Earth *is* illuminated from its whole circle.[17] In fact, as soon as its outer rim rises above the horizon, it illuminates the Earth, although its parts that slope round face the heavens, certainly not the Earth. So since the Moon illuminates not only with its middle, which faces the Earth, but also with its sloping [extremities], which do not face the Earth, clearly it does not send out its light by reflection.[18] Instead, it is because it is illuminated through-

14. The argument here is compressed. Cleomedes posits that spherical mirrors send out their light at right angles to their surfaces, i.e., radially (cf. Goulet 222 n. 350). His point is that the Moon, when it is at the crescent or quarter, would thus not reflect any light to the Earth. See Figure 23. Stephen Menn has raised in a private communication the intriguing possibility that Cleomedes has garbled a somewhat better argument, originally framed in terms of reflection at *equal* angles, by casting it as one about reflection at right angles.

15. Cf. II.5.37–40 where a Moon with a flat shape is also said to be incapable of undergoing phases.

16. The adjective *periklinēs* ("sloping round"), used frequently in this context, refers to the bulbous nature of a hemispherical body. It is used, for example, of the dome of a building.

17. Cf. Plut. *De fac.* 930E on lunar illumination geometrically demonstrated as occurring by reflection from a curved surface.

18. This conclusion is reached without taking into account the size of the Moon, and its distance from the source of its illumination. When noted at lines 102–103 below, they are used to defend the view that the Moon cannot be totally penetrated by the Sun's light.

out by the rays of the Sun (that is, has its light in a blended form) that it illuminates the air.

79:[19] Since the Moon causes illumination in this way rather than by reflection, obviously its light is blended both from its own body and from the rays of the Sun. Yet there are thought to be the following problems [with this theory].

82: *How does the Moon's light disappear as soon as it falls into the shadow of the Earth?* Conversely: *How is the same [light] visible in it as soon as it leaves [that shadow]?* But there is no need to raise this problem and puzzle over it when something similar is seen when the *air* is illuminated.[20] If, for example, a light is brought into a darkened room, the internal air is immediately illuminated; and if the light that illuminates it is extinguished, the air is darkened at exactly the same time as the extinction.[21] This is also seen occurring in the case of the Sun when at its rising the air is immediately illuminated, while it is darkened at the same time as the Sun is concealed by the horizon.[22] (Even if by hypothesis the Sun is extinguished by falling into the Ocean,[23] not only would the air be dark-

19. The three problems that follow (lines 79–126) are sufficiently interconnected to have had a common source, which was probably Posidonius, even though he is only mentioned (line 98) in connection with the second. (Theiler F291 stops, without good reason, at line 107.) Kidd's hesitation on this point (*Comm.* 475–476) is based on his interpretation of the earlier theory of lunar illumination as the blend of lunar and solar light. But, as we have seen (n. 8 above), that theory is one of admixed light, to which Posidonius could have subscribed; in that case, he could have defended it by all three of the arguments presented here.

20. Since the Moon is a separate body, and not just any random volume of air, the analogy here is loose.

21. Alex. Aphr. *De an libr. mant.* 139.17–19 argues that light cannot be suddenly extinguished in a confined space, if it is, as his Stoic opponents claim, a body.

22. But the "northern lights" (I.4.196–207), and refractively caused solar illumination (II.6.168–177 and 187–191), can both occur after sunset.

23. This is the Epicurean theory, criticized at II.1.459–466.

ened when this happened, but it would also get dark at exactly the same time as the extinction.) It is not, I think, at all puzzling to have a similar result in the case of the Moon, too, whenever it falls into the shadow of the Earth. Such is the nature of bodies that are rarefied.

95: Another problem raised in this area is: *Why in solar eclipses do the rays of the Sun not send out light by completely penetrating the Moon, as they do clouds, which are denser than the Moon?* Posidonius duly responds that not only is the surface of the Moon illuminated by the Sun (as in the case of solid bodies that have only their surface illuminated), but that the Moon, as a rarefied body, has rays from the Sun penetrating it to a very great distance, yet not totally.[24] The reason is that the Moon has a considerable volume because of its very large diameter, and because the Sun is no small distance from it. Cloudy air,[25] by contrast, inasmuch as it has no volume,[26] takes in rays that easily penetrate it. (It may be relevant to mention that the Moon's compactness, through which the rays of the Sun cannot escape, also has a unique physical quality.)[27]

108: A further problem raised is: *How does the Moon, as the smaller [of the two], conceal*[28] *the Sun by obstructing its whole body, that is, by being coextensive with its whole diameter?* Now some predecessors believed that in

24. Since the Moon is like a sponge, though not one fully permeated by water (see n. 11 above), lunar illumination does not directly exemplify the "total blending" involved in Stoic cosmology, where pneuma pervades the cosmos totally; cf. I.1.72–73, and see Todd (1976) 29–73.

25. That is, rarefied air. For clouds dense and voluminous enough to reflect sunlight see II.6.171–173, with II.6 n. 22.

26. "Volume" here, and in the preceding sentence, translates *bathos*. Cherniss (1957) 103 n. c argues that, as at Plut. *De fac.* 929D, this term refers to density. But the reference to the Sun's distance implies that when its rays reach the Moon they are weak enough for the latter's volume to block them. Density is introduced as an afterthought; see next note.

27. Kidd *Comm.* 457 and 478 is probably right to see this parenthesis as Cleomedes' own comment. It certainly conveys some reservations about volume alone being able to explain the Moon's absorption of solar rays.

28. "Conceal" translates the verb *episkotein*, although it literally means "cast darkness on," a sense applicable to lunar eclipses (see line 133 below).

total [solar] eclipses when the centers of the deities are in a straight line, the rim of the Sun is observed encircling the Moon by protruding in all directions.[29] But this is not part of what we detect; if it were, then, since the Sun is much larger than the Moon, the protrusions would be seen by us as extremely bright rather than as revealing a minimal extension. So it must be said that although the Moon is smaller than the Sun, nothing prevents its concealing the whole of the Sun, since it is equal, at least in appearance, to the Sun.[30] That it is equal in appearance is also evident just from a [solar] eclipse, but is best proven from the following procedure:[31] when a body is positioned at an appropriate distance, and conceals the whole diameter of the Moon by being coextended with its total size, it also conceals the Sun. And, in general, there is nothing to prevent larger bodies being concealed by smaller ones, and this can be caused in several ways. After all, even in our ordinary experience extremely small bodies conceal mountains and whole seas, and at all events whatever conceals something does not have to be either larger than what is concealed, or even equal to it.[32]

29. Cleomedes is effectively reporting the view that all total solar eclipses are annular; cf. P. Par. 1 col. 19.16–17, a papyrus document dating from the second century B.C.

30. This equality can be calculated by water clocks (II.1.184–191 and 297–299), or expressed through "digital" measurement (II.3.15–43). Cleomedes would appear to be siding with Ptolemy here in the view that there are no annular eclipses of the Sun, i.e., that all total eclipses of the Sun entail complete obscuration. According to Ptol. *Alm.* 5.14, 417.1–11, while the apparent diameter of the Sun is constant, the apparent diameter of the Moon is the same as that of the Sun only when the Moon is at apogee, i.e., at its farthest, and hence smallest; see Neugebauer (1975) 106.

31. It is a "procedure" *(ephodos)* because an axiomatic principle derived from optical theory (see n. 32 below) is applied to observations.

32. The principle involved here (cf. also II.1 n. 50) can be seen as a corollary and extension of Eucl. *Opt.* prop. 5 (8.6–7) ("objects of equal size unequally distant appear unequal," tr. Burton [1945] 338). Thus two objects of unequal size (the Moon and Sun) are unequally distant, yet appear equal when they are aligned

127: The Sun, then, is eclipsed through being obstructed by the Moon; certainly this happens only at their conjunction.[33] Also, a solar eclipse is a condition affecting not the deity itself, but our line of sight: that is, when the Moon comes between us and the Sun, our line of sight, since it is obstructed by the Moon, cannot impinge on the Sun. A lunar eclipse, by contrast, is a condition affecting the deity itself, since the Moon, whenever it falls into the Earth's shadow, is deprived of the Sun's light and plunged into darkness. This happens whenever the Sun, the Earth, and the Moon are in a straight line. That the Moon is eclipsed only by falling into the shadow of the Earth will be demonstrated once we conduct our discussion concerning the wanings and waxings of the Moon.[34]

---

at "an appropriate distance" (*summetron diastēma*, line 119) so that the nearer and smaller object obscures the more distant and larger one.

33. Knowledge of solar, as well as lunar, eclipses was assumed earlier at I.5.39–44, I.8.106–112, II.1.455–457, II.3.4–33, and lines 12–17 above.

34. That is, in II.6.

CHAPTER FIVE

1:[1] The Moon, as has been demonstrated,[2] exists in closest proximity to the Earth of all the heavenly bodies. It therefore has its body mixed with air and somewhat murky, and this becomes particularly striking in its total eclipses.[3] Now just as the Sun also naturally illuminates every other body that is not totally composed of fire, so too it casts its rays on, and illuminates, the Moon, which is both compact and mixed with air. Accordingly, the part of the Moon that is turned to the Sun is illuminated.

8: Now if the Moon always maintained the same relation to the Sun, then a single part of it would always be illuminated. But since, in accordance with its motion based on choice, it approaches the Sun at one time and withdraws from it at another, as it goes from conjunction to full Moon and from full Moon to conjunction, the light from the Sun therefore goes

1. Cf. the account of lunar phases given up to line 40 below with Gem. *Isag.* 9.
2. At II.3.81–99.
3. Since the Moon has no inherent light (see II.4 n. 8), the murkiness evident in total eclipses can only result from its inherent, but relatively limited, heat. That is, it is located at the edge of a fire-sphere, the aether (II.4.11), but acquires part of its substance from the adjacent element air (I.2.37–38; II.3.83–84; lines 2 and 6 here). Its heat is then notably visible in the darkness of an eclipse, since bodies composed of fire are always more luminous under such conditions (II.3.93–94).

round the whole Moon in its circuit of it.[4] By moving relative to its illumination from the Sun, the Moon is, in other words, affected in just the same way as is the Earth through being stationary. The Earth, that is, always has an equal amount of light from the Sun, yet, in the course of the Sun's period, has different parts illuminated by it at different times. This is because both the Sun's luminance and the Earth's shadow complete a circular course along with[5] the Sun, and the tip of the Earth's shadow is directly opposite the center of the Sun. In this way the Moon too always has the same [amount of] light from the Sun (it is certainly not illuminated in differing [amounts] at different times!), yet different parts of it get illuminated at different times, as it approaches the Sun and again withdraws from it, and in this way it has the light from the Sun encircling the whole of its body.

24: Thus at conjunction it is the hemisphere of the Moon facing the heavens that is illuminated, since that is the part of it facing the Sun at that time. But as it passes beyond the Sun, and in proportion to its withdrawal turns its hemisphere that is facing the Earth toward the Sun, it first causes a crescent shape on being illuminated from the side, then a half shape[6] as it increasingly revolves toward the Sun, then a gibbous shape, and after that a full shape when it is in opposition to the Sun. So in the course of reaching opposition from conjunction, the Sun's light goes down from the hemisphere of the Moon facing the heavens to the one facing us, and in this way the Moon is said "to wax" up until full Moon. But when, after being in opposition, it passes beyond opposition, it, by contrast, wanes as the light is carried round from the hemisphere of the Moon facing us to the one that is facing the heavens, right up until conjunction. So if the Moon's shape were flat, it would be full as soon as it

4. In effect, then, the Moon makes one revolution on its axis in a synodic month; cf. II.4.5–9.

5. *sumperinostein;* see I.8 n. 20.

6. *dikhotomos:* i.e., the shape of the Moon at the first quarter (cf. II.4 n. 13). See also lines 73, 88 and 90 below.

passed by the Sun after conjunction, and would remain full until [the next] conjunction. But since it in fact has its shape in the form of a sphere, it produces the types of its shapes in the way described.

41: The cause of the Moon's having differences in its shapes could be more effectively summarized if we used the following procedure to learn what happens to it.[7] Two circles are conceived of in the Moon: *A*, the one by which its dark part is separated from its illuminated part; *B*, the one by which the part visible to us is separated from the part that is invisible. Each of these circles is smaller than *C*, the circle that can divide the Moon into two equal parts, that is, its great circle. Because the Sun is larger than the Moon, it illuminates more than half of it, and thus *A* (the circle that separates the dark from the illuminated part) is smaller than *C* (the great circle of the Moon). *B* (the circle in our line of sight) is, by the same token, necessarily smaller than *C* (its great circle), since we see less than half of the Moon. The reason is that when a spherical body is seen by two eyes, and the distance between them is less than the diameter of the [sphere] that is being seen, the part [of the sphere] that is seen is less than half.[8] So since *B* divides the Moon not into equal, but into unequal, parts, it too is smaller than *C*, the great circle.

56: Both *A* and *B*, however, appear as great circles relative to our perception, and while they always have the same size, they still do not maintain the same fixed position, but cause numerous interchanges and configurations relative to one another as at different times they coincide with one another, or slope to intersection at an oblique angle.[9] Most such intersections are minimal interchanges, but, as is the case with a genus, all are of two kinds: a right-angled [intersection], and one in which they

7. Lines 44–64 involve a "procedure" (*ephodos;* see Introduction n. 38), since their reasoning relies on independently identifiable geometrical and optical principles.

8. This is a verbatim statement of Eucl. *Opt.* prop. 27 (44.14–15); cf. also Plut. *De fac.* 931C.

9. See Figure 24 for the cycle of the phases of the Moon and their correlation with the two circles described in lines 56–80.

intersect obliquely with one another.[10] There are also only two coincidences: when they coincide at conjunction, and at full Moon.

65: Now when the Moon passes by the Sun after conjunction, circles *A* and *B* distance themselves from one another, and slope to intersection at an oblique angle, so that all that is left illuminated, at least in relation to us, is the small [area] between the circumferences of both. This type of transition, from the coincidence of the circles to their intersection, completes the Moon's crescent shape, since as the circles continually move toward intersecting one another at right angles, they also increase the phase of illumination, since the [area] between the intersection of the circles is always illuminated in such a progression.

72: When the figure of intersection reaches right angles, the Moon is seen at the [first] quarter. But when the circles proceed from this figure to obtuse angles, they cause the deity's gibbous shape, while they cause full Moon by again being fully coincident at opposition. Then by proceeding again from this coincidence to yet another, and by completing the same shapes as they wane, they proceed to the point at which all the luminance disappears when the circles *A* and *B* exactly coincide with the part of the Moon that faces the heavens. That is essentially our discussion concerning the waxings and wanings of the Moon.[11]

81: The earliest natural philosophers and astronomers also realized that the Moon acquires its light from the Sun, as is clear just from the etymology of the word—the name of the Moon *(selēnē)* is derived from its always-having-new-light *(selasaeineon)*[12]—as also from the passing on

10. The text of this final part of the sentence is uncertain, and the translation follows a Byzantine paraphrase (see *Caelestia* ed. Todd at II.5.61–62.) which maintains the required meaning.

11. This account of the phases of the Moon completed here ignores the effect of the Moon's motion in latitude, as well as the subtleties introduced at lines 41–56 above.

12. For this etymology see Pl. *Crat.* 409b12. On Posidonius' interest in etymology, which continued an earlier Stoic tradition, see Kidd *Comm.* 77 and 699.

of torches to people entering the festival of Artemis (symbolic of the Moon acquiring its light externally).[13]

87: Earlier [thinkers] claimed that the Moon had three shapes: the crescent, the quarter, and the full (hence the custom of making Artemis also three-faced).[14] More recent ones added to this trio the shape now called "gibbous," larger than the quarter, but smaller than the full, Moon.

92: *Mēn* is applied in four significations.[15] *(a)* The [lunar] goddess is called *Mēn* when she is crescent-shaped,[16] as is *(b)* the actual condition of the air between conjunctions (as we regularly say: "the month *(mēn)* has been humid or temperate"). Also called *mēn* are *(c)* the interval of time between conjunctions, and, finally, *(d)* the interval of 30 days (as in our saying that we have been out of town, or in town, "for a month," without meaning in any way "from conjunction to conjunction," but just the sum total of 30 days). The first two [entities][17] (the crescent-shaped goddess, and the condition of the air) are bodies, whereas the next two are incorporeal, since time itself is also incorporeal.[18]

13. On Artemis and the Moon see *SVF* 2.748, and the further references at Goulet 223 n. 369.

14. For these three shapes linked with Hecate see Cornut. *Theol.* 72.7–13 and cf. Plut. *De fac.* 937F. For the four configurations see Posid. F122EK with Kidd *Comm.* 473.

15. For *(a)* and *(c)* see *SVF* 2.677, p. 199.30–34. For *(c)*, the only astronomically significant sense (as lines 102–141 here show) see Gem. *Isag.* 8.1. For supplementary semantics see lines 148–149 below.

16. *Mēn* was originally a male Anatolian deity (Mannes), represented with a crescent Moon behind his shoulders. Similarity to -μην, the root of the word for "month" (μείς), seems to have led to the form *mēn* being applied to a temporal period.

17. In Stoic metaphysics they are termed "somethings" *(tina)*; see, for example, *SVF* 2.331.

18. A "signification" *(sēmainomenon)* for the Stoics is by definition the incorporeal meaning (the *lekton;* cf. I.1 n. 48), in contrast with a corporeal speech-act or the object spoken about; see *SVF* 2.166. It is being used in an extended fashion here to identify the reference of the word *mēn* in *(a)* and *(b)*, as well as the

102: The conjunctions of the Moon with the Sun do not always maintain an equal time interval for the following reason.[19] The Sun, as already stated,[20] gets both closer to the Earth and higher in accordance with its course based on choice. So when it is lower, it necessarily goes through the zodiacal sign more quickly, but when higher does so more slowly.[21] For when it is lower it goes through a shorter arc, but through a longer one when higher.

107: We might learn this too from what happens with respect to the sections of cones:[22] that those near the bases are wider, those closer to the vertex narrower. Now the cones flowing out from the eye to the heavens have the [point] right at the pupils as their vertex, and have the object of vision on which they impinge as their base, and since the Earth is the center [of the cosmos], the bases of the cones flowing out from it to all the zodiacal signs will be equal.

114: Now if it so happened that the Sun moved at neither a greater nor a lesser height, but always kept the same height from the Earth, then it would go through the zodiacal signs in equal periods of time,

---

meanings in *(c)* and *(d)*. Although the time intervals identified in *(c)* and *(d)* are called bodies in one Chrysippean quotation (*SVF* 2.665), the incorporeality of time can transfer to the time intervals as intervals, without reference to the bodies that determine their character; see Brunschwig (1988) 106. The point made in this text also applies to the instances of time intervals at II.1.150–151 and 182–183, and II.2.17–18. It shows the extent to which astronomy is being approached from the perspective of Stoic philosophy.

19. The analysis that follows is a commentary on sense *(c)* from the preceding paragraph; it also elaborates the brief reference to the lunar month at I.2.42.

20. At I.4.57–71. See Figure 7.

21. The varying speeds attributed to the Sun here, and then later to the Moon and the other planets, are all apparent (see Theon *Expos.* 135.6–11). The issue of their real motion, which is a function of their density (cf. II.1.334–338 and II.1 n. 79), is ignored in this context.

22. The three propositions introduced in this paragraph are the assumptions for an *ephodos* at lines 114–141.

and in that way its conjunctions with the Moon would also maintain an equal interval of time.[23] But since this is not the case, but the Sun is instead observed moving at its greatest height in Gemini, and at its lowest in Sagittarius, then in Gemini it will go more slowly through the section of the visual cone[24] (a wider one because it is closer to the base), whereas it will go through the section in Sagittarius more quickly, since here, by contrast, the section of the cone is narrower (that is, closer to the vertex).[25]

123:[26] So when conjunction occurs at the start of Gemini, the month will necessarily be abbreviated, since the Moon is moving closer to the Earth there, while the Sun is at its greatest height. That is because the Moon will overtake the Sun while the Sun is still in Gemini, a sign through which it goes in 32 days. But if conjunction occurs near the start of Sagittarius, the Moon will not catch up to the Sun while the Sun is still in this sign, since the Sun takes 28 days to go through it.[27] This month will, therefore, be the longest of all, since the Moon passes through Sagittarius more slowly, and the Sun does so quickly, and so the Moon catches up with it slowly. The result will be proportionate in the intervening signs.[28]

133: In the same way it is also proven that all the planets have high

23. This would be true only if the Moon's circuit was also concentric about the Earth.

24. Literally "the cone that forms the line of sight" *(kōnos tēs opseōs).* That is, every line of sight has visual pneuma in such a shape; see II.1.57–65 and 252–255, and cf. II.6 n. 27.

25. Cf. I.4.62–71.

26. On the duration of lunar eclipses see II.6.68–78, which presupposes the present analysis, as would any account of the duration of solar eclipses. Cherniss (1957) 126 *a* is, however, wrong to claim that the present text applies to solar eclipses.

27. Geminus in his calendar has the Sun taking 32 days to traverse Gemini (108.1 Aujac [1975]), but 29 (rather than 28) to traverse Sagittarius (103.3).

28. Cleomedes seems to suggest here that the Moon's circle has a fixed perigee and apogee (see Figure 25), but this would not be correct.

points and low points [relative to the Earth][29] in each of the zodiacal signs. For given that all the zodiacal signs are divided into 30 degrees, then when planets go through some of them more quickly, and others more slowly, they obviously go more quickly through the sections of the [visual] cones that they encounter when lower, whereas when the sections of the [visual] cones are wider, their passage is slower because of their height. Since all the planets are heightened and lowered, all of their circuits are comparably eccentric, since because of the variation in their heights they are not equidistant from the Earth in every direction.

141: So since the Moon's [circuit] is also like this, it is spread below the zodiacal band at an oblique angle to the whole of it. Specifically, it touches the northern [circle of the zodiacal band][30] to the extent that the Moon itself invariably approaches the northern [regions], and [touches] the southern [circle] in the same way.[31] So, given this, it necessarily intersects with the circle through the middle [of the zodiacal constellations] at two points,[32] which are variously termed "points of contact" *(sunaphai)* or "nodes" *(sundesmoi)*.

29. The terms that describe these distances, *hupsos* and *tapeinōma* (literally "highness" and "lowness"), are also used analogously in astrology to identify degrees of planetary influence; see I.2 n. 22. At I.2.64–69 they refer only to positioning in the zodiac.

30. Like Goulet 223 n. 373, we reject the suggestion in Neugebauer (1975) 962 that the circles approached by the Moon are the arctic and antarctic circles. On the inclination of the Moon's orbit to the ecliptic cf. II.7.1–2.

31. Cleomedes does not commit himself here to identifying the northernmost circles of the zodiacal band with the northernmost and southernmost parallel circles reached by the Moon.

32. We have followed most critics in deleting a relative clause at lines 145–146 that glosses the phrase describing this circle as "that which is called 'heliacal' and 'ecliptic' *(ekleiptikos)*." This clause is awkwardly positioned in the sentence relative to its antecedent, and so is probably a marginal comment added to the text later, given also that the term *ekleiptikos* (sc. *kuklos*) ("ecliptic circle") is attested in only two sources for the period to which Cleomedes' treatise is datable: P. Oxy. 4138a ii.121 (second century A.D. in Jones [1999]) and Ach. *Isag.* 53.9–10 (mid-second to mid-third century A.D.; see Mansfeld and Runia [1997] 300).

148: Just as "the Sun" is used in two senses—both in the sense of itself and that of its light—so we also standardly use "the Moon" in the same two senses.[33]

150: We shall next conduct our discussion concerning the eclipse of the Moon, our object being to avoid sharing with old hags the belief that at eclipses witches drag the Moon down![34]

33. This sentence may have been displaced from a more logical location after line 86 above.

34. On this folk belief see Mugler (1959), Hill (1973), and Bicknell (1984). Cf. II.1.459–461 for a similar ridicule of folk astronomy.

CHAPTER SIX

1: The Moon is eclipsed by falling[1] into the shadow of the Earth whenever the three bodies—Sun, Earth, and Moon—are in a single straight line, with the Earth in the middle. This can happen only at full Moon.[2] And the Moon falls into the Earth's shadow in the following way. The Sun moves, as already stated,[3] with its own circuit located below the circle that is exactly at the middle of the zodiacal band. Accordingly, the Earth, when illuminated by the Sun, necessarily sends out a shadow, as do all other solid bodies that are illuminated. Now since this shadow has a conical shape, it does not occupy the whole zodiacal band, and so is not aligned with its total breadth, since it terminates in a vertex. But this shadow, since it is necessarily directly opposite the center of the Sun right at the exact center of its vertex, is itself also located below the exact middle of the zodiacal band. Now this shadow far exceeds the distance of the

1. "Falling" here and elsewhere translates a compound verb *(peri-piptein)*, the prefix of which refers to the Moon's circular orbit "around" *(peri)* the Earth.

2. For this as a Stoic definition see *SVF* 2.678 (also Posid. F126EK). For other elementary accounts of lunar eclipses see Gem. *Isag.* 11, and Theon *Expos.* 193.23–198.8.

3. At I.2.53–59 and I.4.52–53.

Moon, though without going up as far as the remaining heavenly bodies. So when the Moon is detected as being in opposition to the Sun, and either to the right (that is, north) of the zodiacal circle, or on the opposite side, it evades the shadow of the Earth, and for this reason the Moon is not eclipsed at every full Moon. But when in opposition to the Sun the Moon is detected as being so situated that a single straight line can be extended through the centers of the Sun, Earth, and Moon, then, by falling right into the shadow of the Earth, it is fully eclipsed. The shadow of the Earth, in other words, moves in direct opposition to the Sun, and is, as it were, "dragged" by it, just as Homer says: *The shining light of the Sun fell in the Ocean, dragging black night over the fertile land.*[4]

24: Since the shadow moves along with the Sun in this way, and at its very tip is directly opposite the Sun's center, then the Moon, as it proceeds in accordance with its motion based on choice, meets the shadow moving from east to west as it itself moves west to east.[5] And by falling into the [Earth's shadow] in this way, [the Moon] is deprived of rays from the Sun (just as we too are when someone stands in our way when we are in sunlight). But it is not always the case that the Moon as a whole is darkened by the Earth (that is, totally concealed by its shadow), but on occasion [the Moon is darkened] just partially. This hap-

4. *Iliad* 8.485–486.

5. Cf. lines 38–42 below, and Plut. *De fac.* 932F-933A. The Earth's shadow and the Sun both have a proper motion eastward, that is, in the direction opposite to the daily rotation, while the Moon similarly has a proper motion eastward, although a much faster one. This means that the Moon overtakes the Earth's shadow each month from the west, with an eclipse occurring if it enters this shadow. Cleomedes misleads the reader when he states (in lines 26–27) that the Moon, while moving east, meets the Earth's shadow as this shadow is moving only westward (cf. lines 41–42). For, while it is true that both the Moon and the Earth's shadow, because of the difference in their proper motions eastward, share to different degrees in the diurnal motion from east to west, it is wrong to say that the Moon and the Earth's shadow meet because they are moving in opposite directions.

pens when through being in opposition to the Sun it touches the [circle] through the middle [of the zodiacal constellations], yet is not detected as having its center at the exact middle [of the zodiacal band]; for this is how a specific part, but not the whole, of it falls into the [Earth's] shadow.

35: That the Moon is eclipsed by falling into the Earth's shadow, and only in that way, can be seen from the phenomena alone.[6] *(a)* It is eclipsed only at full Moon (the only time in fact that it can fall into the Earth's shadow while in opposition to the Sun). *(b)* In any total eclipse the parts of the Moon facing the east are seen to be the first to disappear, because as it sets out for the east in a motion opposite to the heavens, it meets the Earth's shadow, which always moves from east to west. But when it starts to emerge after the eclipse, it has those of its [parts] that face the east emerging first. That is, it is absolutely necessary that as the Moon meets the shadow, the first of its parts to encounter the Earth's shadow and be concealed are again the first to emerge after being concealed. *(c)*[7] Whenever the Moon is partially eclipsed, and is affected in this way as it goes down from north to south, then the parts of it that face south necessarily disappear, since in the downward course they take the lead in falling into the shadow and are in this way concealed, whereas the parts that face north escape the shadow. But when the Moon goes up from the south to the north and effects a partial eclipse by being in opposition to the Sun with its center not yet at the exact middle of the zodiacal band (that is, in line with the center of the Sun), then the parts of it facing north are eclipsed, since they take the lead in falling into the Earth's shadow, whereas the parts facing south are visible.

56: So all these [observations] establish for us by essentially visual means that the sole cause for the Moon's eclipse is the process by which on falling into the shadow of the Earth (that is, being darkened by it) it

6. Since "the phenomena alone" *(auta ta phainomena)* here represent "clear" visual evidence (cf. lines 56 and 195), they can serve as a criterion (cf. I.5.4, and II.3 nn. 24 and 25).

7. For lunar motion in relation to the zodiacal circle see II.5.141–147.

is deprived of the impact derived from the rays of the Sun that illuminate the part of the Moon that is always turned toward the Sun.

60: Furthermore, the segments of the Moon that are illuminated at an eclipse are seen to be curved.[8] This too happens of necessity since the Moon, which is spherical, falls into the conical shape of the Earth's shadow, and so its illuminated segments are also seen to be curved. In other words: When a spherical shape encounters a conical shape and has the part that is in contact with the conical shape always disappearing from sight, necessarily the remaining part, which has not yet disappeared from sight, has its shape curved along the segment (that is, is crescent-shaped).[9]

68: The following too has been observed in the case of the lunar eclipse: the Moon effects a total eclipse at a very great height, when very close to the Earth, and at an intermediary distance.[10] When eclipsed at a very great height it emerges more rapidly, but does so slowly when very low, and when in between it also has an intermediary duration for its eclipse in between the [extremes] mentioned. This [variation] clearly reveals that the Moon is eclipsed in no other way than by falling into the shadow of the Earth. That is, when it is eclipsed at a very great height, it emerges more rapidly through encountering the narrower part of the shadow; but when very close to the Earth, it has to go through a wider extent of the shadow, and thus the duration of its eclipse is greater. But when on occupying an intermediary height there is a proportionate outcome, it also has an intermediary duration for its eclipse.

79: This [evidence of varying eclipses] proves that the Earth's shadow also has a conical shape; in fact, these [phenomena], given the way they

8. Because of the general principle introduced at lines 64–67 (cf. also Arist. *De caelo* 297b25–30 and Plut. *De fac.* 932E-F), this proof is an *ephodos* (see Introduction n. 38). On a traditionally related claim, that the curved shadows appearing on the Moon during lunar eclipses demonstrate the Earth's sphericity, see Neugebauer (1975) 1093–1094, who describes it as "mathematically inconclusive."

9. This curve is in three dimensions, not two.

10. At II.5.102–132 the Moon's eccentric orbit was analyzed with reference to the case of conjunction.

are, are proven by one another. That is, a lunar eclipse is demonstrated to occur only by the Moon's falling into the shadow of the Earth; conversely, variations in eclipses of the Moon demonstrate that the Earth's shadow is conical, since the Moon spends a longer time in eclipses that are closer to the Earth, yet emerges more rapidly in eclipses that are at a greater distance from the Earth, whereas in eclipses at intermediary distances the duration of the eclipse is also intermediary. Partial eclipses too show that the Earth's shadow is conical, since the Moon then has segments illuminated in such a way that its shape becomes crescent. This would not occur unless it fell into a shadow with a conical shape.

90: But it is from the following [argument] that the conical shape of the Earth's shadow might be most effectively demonstrated.[11] If its shadow were in fact cylindrical or funnel-like (that is, if the body that illuminates it—the Sun—were equal to, or smaller than, the Earth), then the shadow that was funnel-like would occupy most of the heavens by terminating in a broad span, with the result that not only would the Moon be eclipsed every month, but it would also remain in the [Earth's] shadow all through the night. But if the shadow were cylindrical, it would occupy the whole breadth of the zodiacal band, since it would not terminate in a vertex, and the Moon would be duly eclipsed each month by falling into the shadow. But because the Earth's shadow is actually conical, and so terminates in a narrow vertex, the Moon evades it when it is detected occupying the northern or southern parts of the zodiacal band at full Moon. (If the shadow were cylindrical or funnel-like, it would also advance as far as the [fixed] stars. As a result the stars would at different times have a brighter or fainter appearance: brighter when in shadow, since every body that is composed of fire is brighter in the darkness of a shadow, but fainter when in the rays of the Sun.[12]) As none of this is observed among the phenomena, it is clear

11. At II.2.19–30 this argument was used to demonstrate that the Sun was larger than the Earth. See Figure 21. On the translation "funnel-like" see II.2 n. 8.

12. Cf. II.3.91–95.

that the Earth's shadow is necessarily conical.[13] If so, it is evident that the body that illuminates it—the Sun—is larger than it.[14]

109: Lunar eclipses are such as we have demonstrated. But statements made about paradoxical eclipses seem to contradict the theory that establishes that the Moon is eclipsed by falling into the shadow of the Earth. For some say that a lunar eclipse occurs even when both luminaries are observed above the horizon,[15] and that this indicates that the Moon is not eclipsed by falling into the Earth's shadow, but in some other way. For if an eclipse does occur when both the Sun and the Moon are seen above the horizon, the Moon cannot at that time be eclipsed by falling into the Earth's shadow. Furthermore, if both [Sun and Moon] are visible above the horizon, and if the Earth's shadow can no longer be at the place where the Moon is seen to be eclipsed, then the place where the Moon is located is illuminated by the Sun! If this [theory] is correct, we shall have to lay claim to a different cause for the Moon's eclipse.

122: Earlier scientists confronted with such statements tried to solve this problem as follows. They said that the Moon could fall into the Earth's shadow (that is, be precisely opposite the Sun), even though both the luminaries were above the horizon. This could not happen if the Earth's shape were flat (i.e., a plane), but, because its shape is spherical, both divinities' bodies could be observed above the horizon directly opposite one another. To explain. Because of the protrusions of the Earth's curvatures the [two divinities] will not actually face one another in direct opposition. Even so, those who are standing on the Earth could not be prevented from seeing both [divinities] because it is the Earth's curvatures on which they are standing. These curvatures do not im-

13. The logical necessity here is again (cf. II.2.19–30, with II.2 n. 9) implicitly based on the Stoic fifth undemonstrated argument.

14. Cf. II.2.27–28.

15. For such an eclipse reportedly observed in the west by Hipparchus see Plin. *NH* 2.57.

pede people standing on them from seeing both the bodies above the horizon, although they will obstruct the divinities when they are in direct opposition to one another. Thus while [the Sun and Moon] will not face one another, *we* will not be prevented from seeing both of them because we are standing on the Earth's curvatures.[16] The latter obstruct those bodies that are in low areas at the horizon, whereas the curvatures on which we are stationed are more elevated.[17]

139: That was how earlier scientists solved the problem adduced here, but their position may not be sound. This is because while our line of sight might be affected in this way at an elevation, since the horizon becomes conical when we are elevated into the air far above the Earth,[18] this is no longer the case when we are located *on* the Earth. For despite the existence of curvatures on which we are located, our line of sight is eliminated by the size of the Earth.[19] So it must not be stated, or believed to be at all possible, that a lunar eclipse occurs when both [Sun and Moon] are observed above the horizon by us while we are standing on the Earth.

149: Instead, we must confront [proponents of paradoxical eclipses] initially by claiming that this theory is fabricated by certain people who wish to impose a problem on the astronomers and philosophers who are engaged with these matters. For numerous lunar eclipses have occurred, both total and partial, and have all been recorded, yet nobody is reported

16. Such a sight line would mean an Earth no longer discountable ("a point") in relation to the distance of the Sun (as argued in I.8). Either the cosmos would be smaller, or the Earth larger. See I.8 n. 34.

17. This observer is at ground level, but has a "conical" horizon (line 142), i.e., one of more than 180°. The Sun and Moon are thus in the "low areas" beneath this *enlarged* horizon, and the Earth affects observation as though it were a mound over which we look at bodies on either side of it.

18. Curvatures are inherent to a spherical Earth (I.1.247 and 250; I.3.17 and 30; I.5.116 and 120), but visible only from an elevated position (I.8.138–139) due to the Earth's size.

19. For a ground-level observer the sight line literally "disappears into" *(enaphanizetai)* the distance, and the Earth appears to be flat (see I.5.11–13); i.e., we have a horizon of 180°, even though the Earth is spherical.

as having recorded this type of eclipse, at least up to our lifetime—no Chaldean, Egyptian, or other scientist or philosopher. The claim is just a fabrication. Second, if the Moon were eclipsed in any other way than by falling into the shadow of the Earth, it could also be eclipsed when *not* at full Moon, that is, when it advances a large or small distance away from the Sun, and eclipsed again after full Moon when it approaches the Sun and is waning. But in fact, although it undergoes numerous eclipses (eclipses being frequent enough), it has never been eclipsed except at full Moon, that is, when in opposition to the Sun—in fact only when it is possible for it to encounter the shadow of the Earth. Certainly all its eclipses are predicted by people who construct astronomical tables, because they know that, whenever coincidence occurs, it is at full Moon that the Moon is detected either totally or partially below the exact middle of the zodiacal band, and that in this way it effects either partial or total eclipses. It is therefore impossible for a lunar eclipse to occur when *both* luminaries are seen above the horizon.

168: Since there are by nature a wide variety of conditions that affect the air, it would not be impossible for us to encounter an image of the Sun[20] as not yet having set after it had already set (that is, after it was below the horizon). The cause[21] could be a rather dense cloud present in the west that is illuminated by the rays of the Sun and sends out an image of the Sun to us.[22] Or it could be the occurrence of a counter-Sun,

20. This "image" *(phantasia)* is a genuine illusion, unlike cases of visible objects projecting illusory appearances (e.g., I.8.21 and 159–160), or the cases of the Sun's appearance cited in II.1.

21. The multiple explanations that follow, which may be Posidonian, are incompatible with one another, like some sets of explanations in Epicurean physics (see Wasserstein [1978]). Cf. I.4.90–109 where, by contrast, Posidonius is reported as entertaining multiple, though complementary, explanations.

22. On this phenomenon (known as parhelion, or mock Sun) see Kidd *Comm.* 467–470 (on Posid. F121EK) and D. Kidd (1997) 476–477 (on Arat. *Phaen.* 881). Clouds can be permeated by the Sun's rays (II.4.95–97 and 103–104), but also acquire moisture from the atmosphere at the horizon (line 175 below; cf. II.1.29–30), and so can reflect light (see II.4.35–36).

since many things like that appear in the air, and particularly around Pontus.[23] But the ray that flows out from the eyes could also be refracted on encountering air that is damp and moist, and encounter the Sun after it is already concealed below the horizon.[24]

178: Something similar to the latter is also observed happening in our ordinary experience. For example,[25] if a gold ring is placed in a cup, or in some other vessel, then if the vessel is empty, the object placed within is not visible at an appropriate distance[26] because the visual pneuma runs unimpeded down from the brim of the vessel in a straight line. But when the vessel is filled with enough water to become level with the brim, then from the same distance the ring lying within the vessel is visible. This is because the visual pneuma no longer runs [straight] down from the brim, but comes into contact with the water that has filled [the vessel],[27] and is thus refracted, goes to the bottom of the vessel, and encounters the ring.[28] Something similar could, then, occur with damp and sodden air too, so that when the ray from the eye is refracted and bends below the horizon,

23. For a counter-Sun (anthelion), sometimes confused with parhelion, occurring in the eastern sky, in the area of Pontus, see Anaxagoras at DK 59A86.

24. On atmospheric conditions affecting astronomical observations, with reference to this passage, see Lloyd (1982) 134–135, and cf. Ptol. *Opt.* 5.23–26 (tr. Smith [1996] 238–240). Cf. also II.1 n. 11.

25. For this example see also Archim. at Olymp. *In meteor.* 211.18–23 (= Archim. Fr. 18; cf. Fr. 17), Sen. *NQ* 1.6.5, Ptol. *Opt.* 5.5 (tr. Smith [1996] 230–231), and Damian. *Opt.* 14.3–6 (ed. Schöne [1897]). Cf. Eucl. *Catoptr.* 286.17–19 for the general principle involved.

26. This distance is "appropriate" (*summetron;* cf. II.4 n. 32) in the sense that eye and object are at a distance and angle that ensures the object's visibility. For this same general sense see Alex. Aphr. *De an.* 41.17–19, Sext. Emp. *AM* 7.188 and 438, and cf. Sedley (1976) 49 with n. 90.

27. "[Level] to the brim" (*κατὰ τὰ χείλη*, line 185) is the manuscript reading printed in Todd's edition. It is omitted here as a gratuitous iteration of the same phrase in the preceding line. Since the vessel is already said to be full to the brim (lines 182–183), no additional reference to this fact is needed.

28. The visual cone (see II.1.57–70 and II.5.107–141), also relevant to refraction, is not mentioned.

it encounters a Sun that has already set, so that an image of it is engendered as still being above the horizon. Perhaps something else much like this could also on occasion produce an image in us of the two [heavenly] bodies being above the horizon after the Sun has already set. Still, that the Moon is eclipsed *only* by falling into the shadow of the Earth is a cognitively reliable[29] [conclusion] derived from the phenomena. That, then, is enough on the eclipse [of the Moon].

29. *enarges:* that is, not just visually but cognitively reliable, and applied to a process of reasoning. See Introduction 10, and II.1 n. 61.

CHAPTER SEVEN[1]

1: The Moon is said to move a greater distance than do the other planets toward each [side] of the circle through the middle of the zodiacal constellations; next in order is Venus, which goes 5 degrees to each [side] in its chosen motion, then Mercury (up to 4 degrees),[2] Mars and Jupiter (up to 2½ degrees), [and] Saturn (up to 1 degree on each side).[3]

5: Mercury and Venus do not move to every [angular] distance from the Sun, but Mercury [moves] 20 degrees at most, Venus 50 degrees at most; the remaining three, just like the Moon, move to every [angular] distance from the Sun.[4]

8: Mercury effects [superior] conjunction with the Sun in 116 days,

1. Since this chapter offers no rationale for the data presented, and since lines 11–14 could follow II.6 without any interruption in thought, lines 1–10 may be an interpolation.

2. In the list of planets at I.2.20–42, Mercury, not Venus, is located just above the Moon, another reason perhaps (see preceding note) for regarding this material as interpolated.

3. Neugebauer (1975) 1014–1016 analyzes the evidence in Ptolemy's *Handy Tables* for extremal latitudes, and compares it to the values given in the *Almagest*.

4. Neugebauer (1975) 804–805 catalogues ancient evidence on the maximum elongation of the inner planets.

when the latter comes in between it [and the Earth]; Venus resumes the same position in relation to the Sun in 584 days, Mars in 780 days, Jupiter in 398, Saturn in 378.[5]

11: That will be as far as our discussion of these [matters] will go, at least for the present.[6] These [two] lecture courses[7] do not comprise the writer's actual doctrines, but have been amassed from certain treatises, earlier as well as more recent ones, with most of the statements taken from Posidonius' [works].[8]

5. The interval in which a planet returns to the same position in relation to the Sun (e.g., conjunction) is known as its synodic period. Neugebauer (1975) 782–785 catalogues ancient evidence on planetary periods, and notes (785) that "beyond the Babylonian parameters only the synodic periods listed in Cleomedes II.7 are astronomically meaningful." See also Neugebauer (1975) 965.

6. For evidence of a forthcoming course see I.1.94–95, 173–174 and 191–192.

7. The "lecture courses" *(skholai)* are the sets of lectures that seem to have comprised the two books of the *Caelestia;* cf. II.2.7 where "the first of the *skholika*" refers to the whole of Book I.

8. This typical disclaimer (cf. also I.8.161–162 above; see Whittaker [1987] 102 with n. 77) should not be deleted as a gloss; see *Caelestia* ed. Todd *Praef.* xviii.

Figures

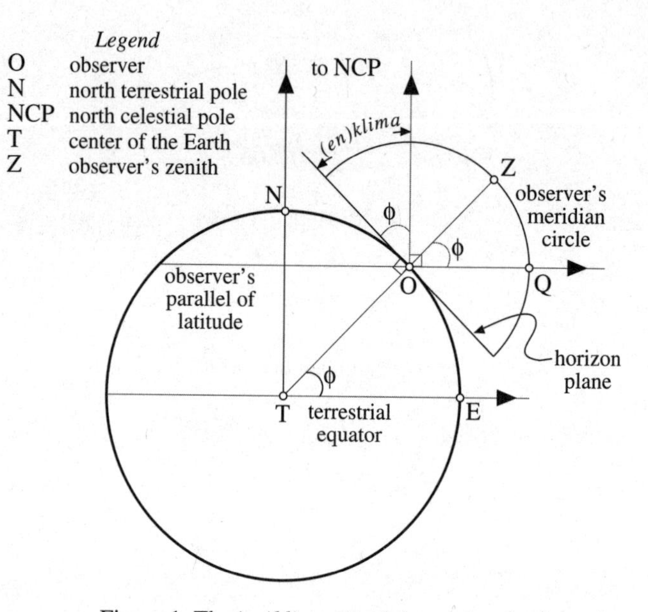

Figure 1. The *(en)klima* (ϕ) of the north celestial pole for an observer in the northern hemisphere (I.1.176)

ϕ is equal to angle ZOQ which is in turn equal to angle OTE, the observer's latitude.

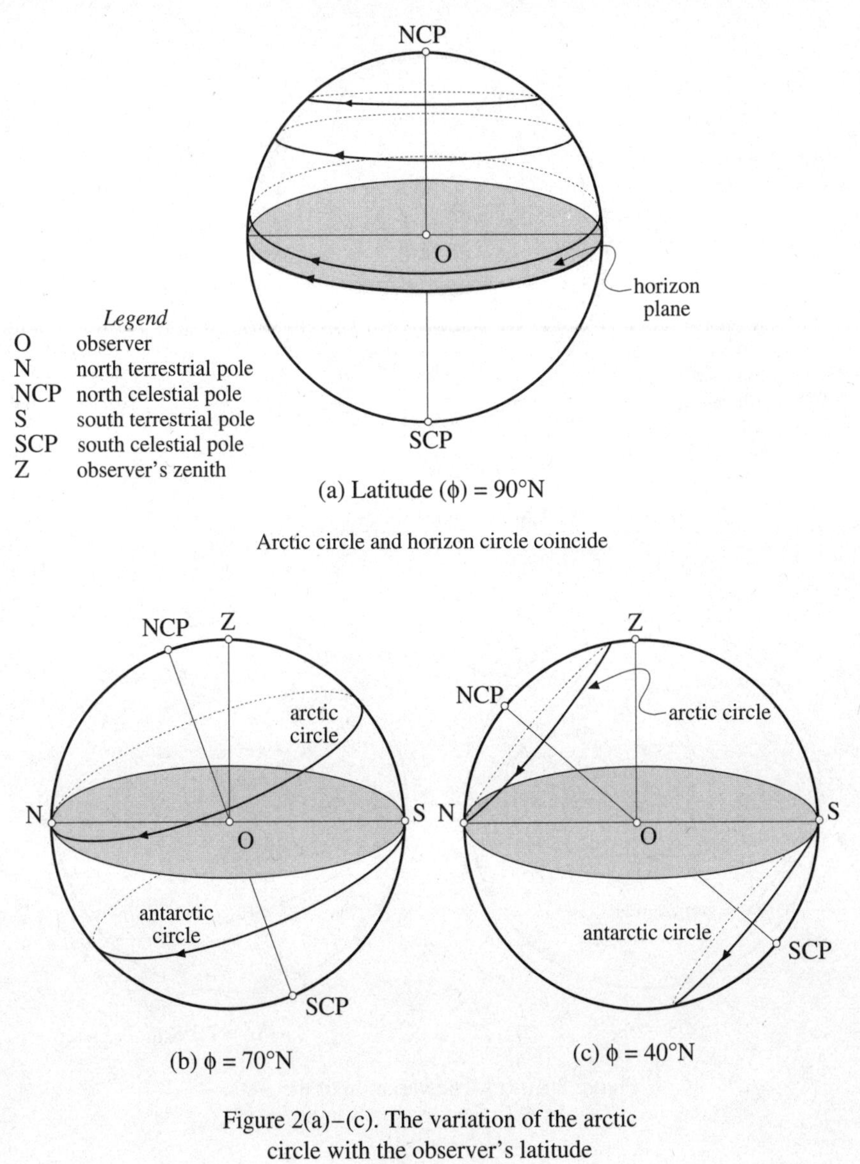

Figure 2(a)–(c). The variation of the arctic circle with the observer's latitude (I.1.193–201)

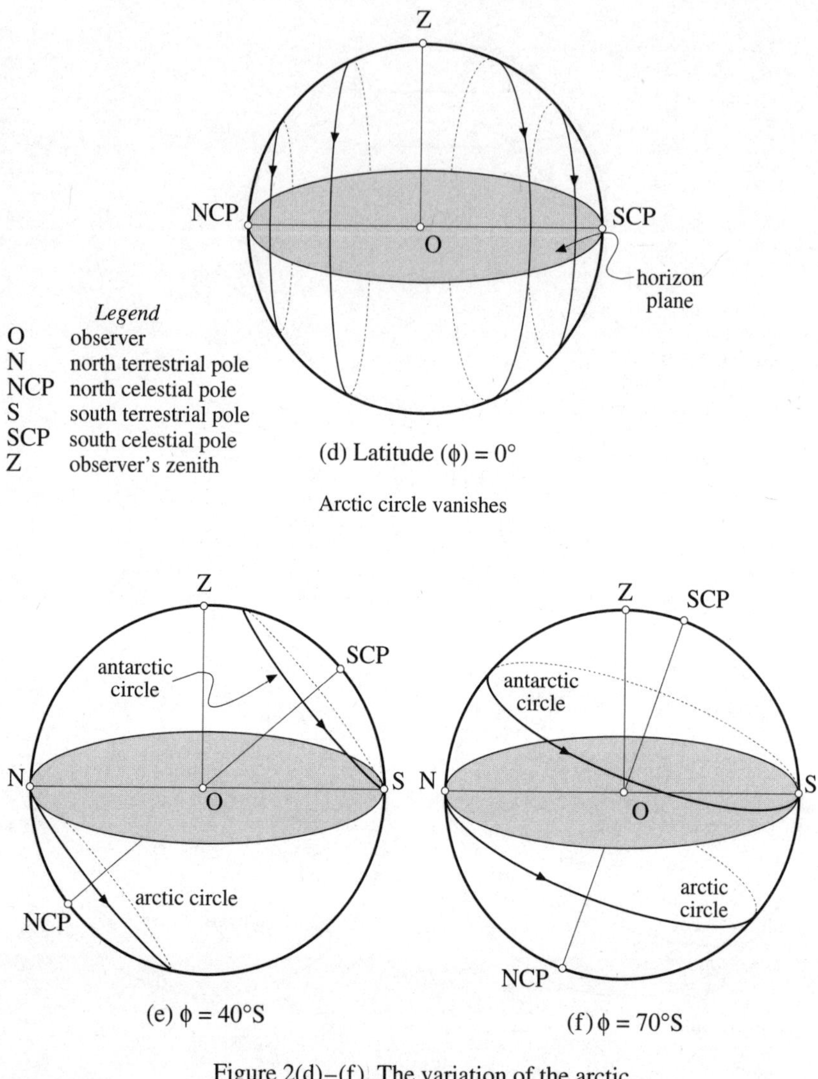

Figure 2(d)–(f). The variation of the arctic circle with the observer's latitude (I.1.193–201)

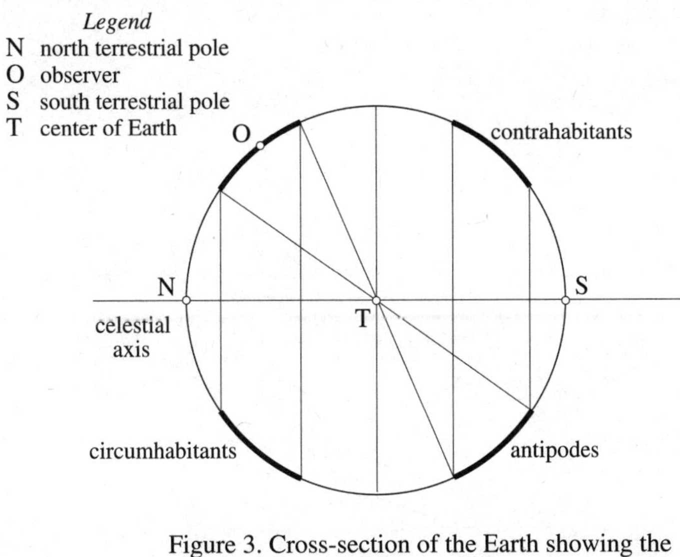

Figure 3. Cross-section of the Earth showing the locations of the inhabitants of the temperate zones (I.1.209–234)

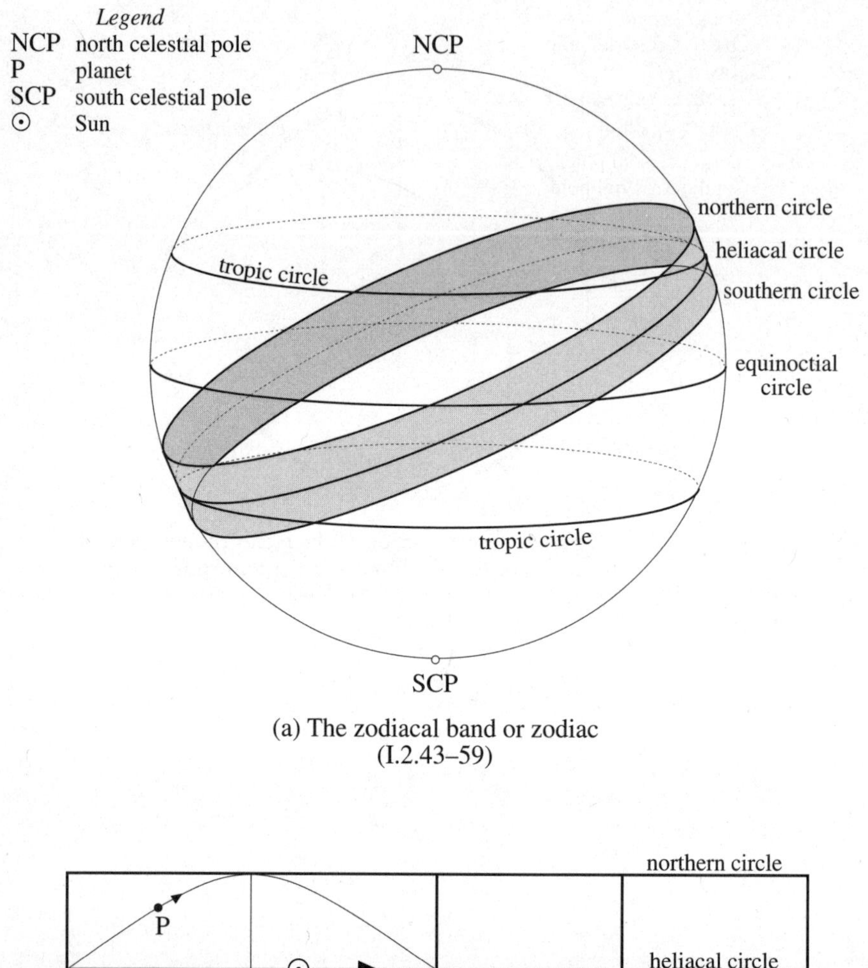

(a) The zodiacal band or zodiac
(I.2.43–59)

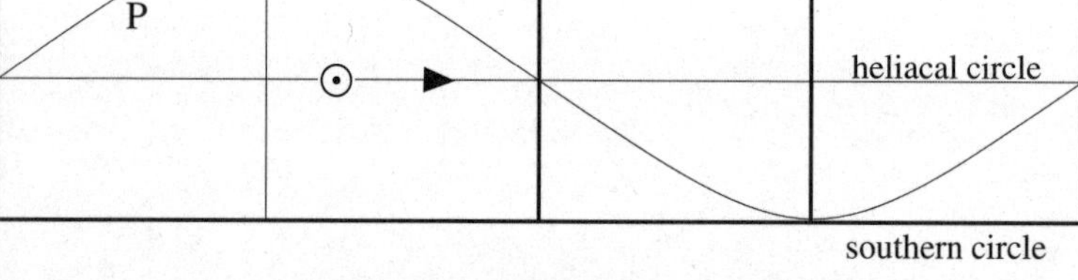

(b) Planetary motion in the zodiacal band (shown laid out flat)
(I.2.60–72)

Figure 4

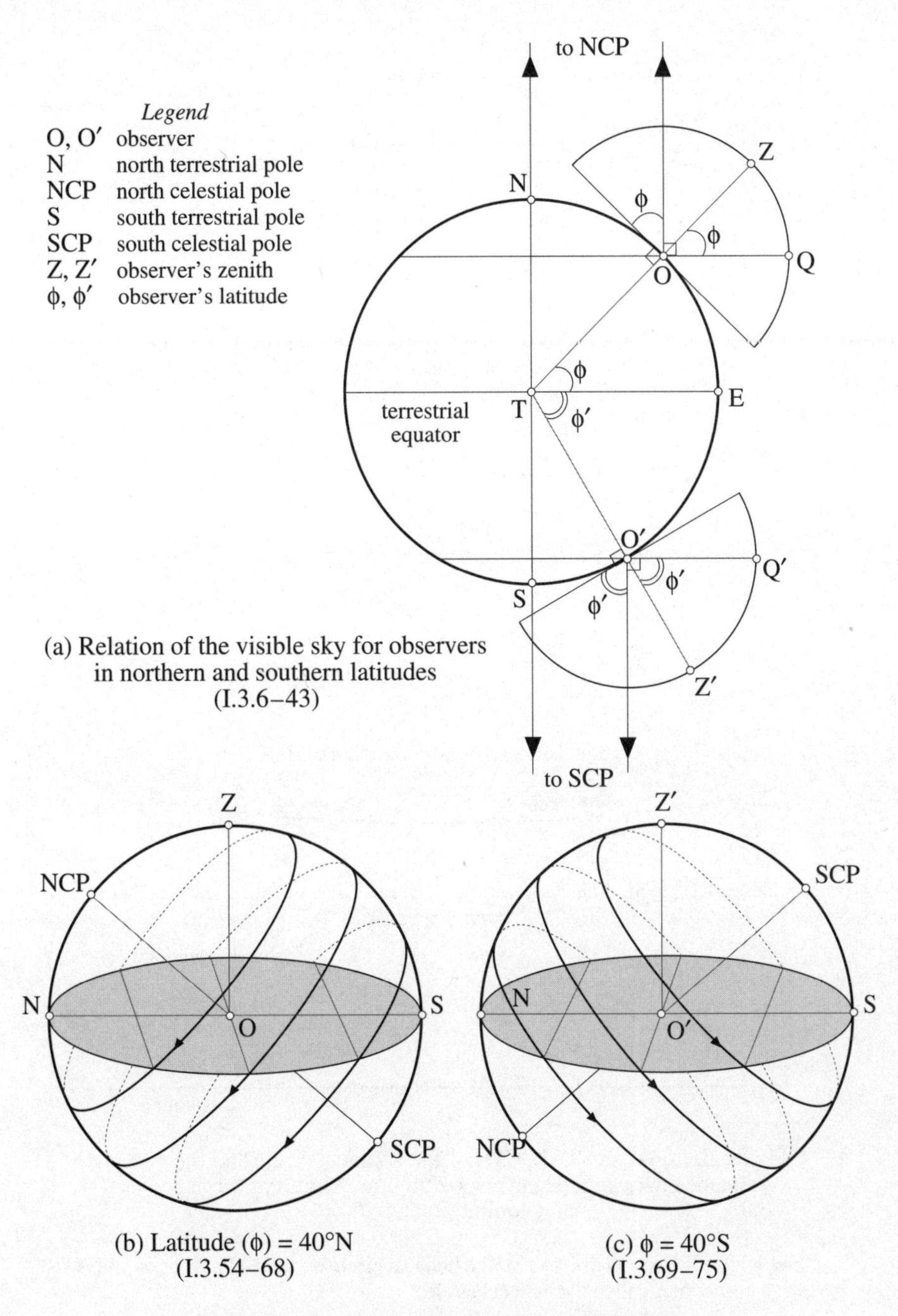

(a) Relation of the visible sky for observers in northern and southern latitudes (I.3.6–43)

(b) Latitude (φ) = 40°N (I.3.54–68)

(c) φ = 40°S (I.3.69–75)

Figure 5. The interrelation of the equinoctial and solstitial circles for observers at northern and southern latitudes

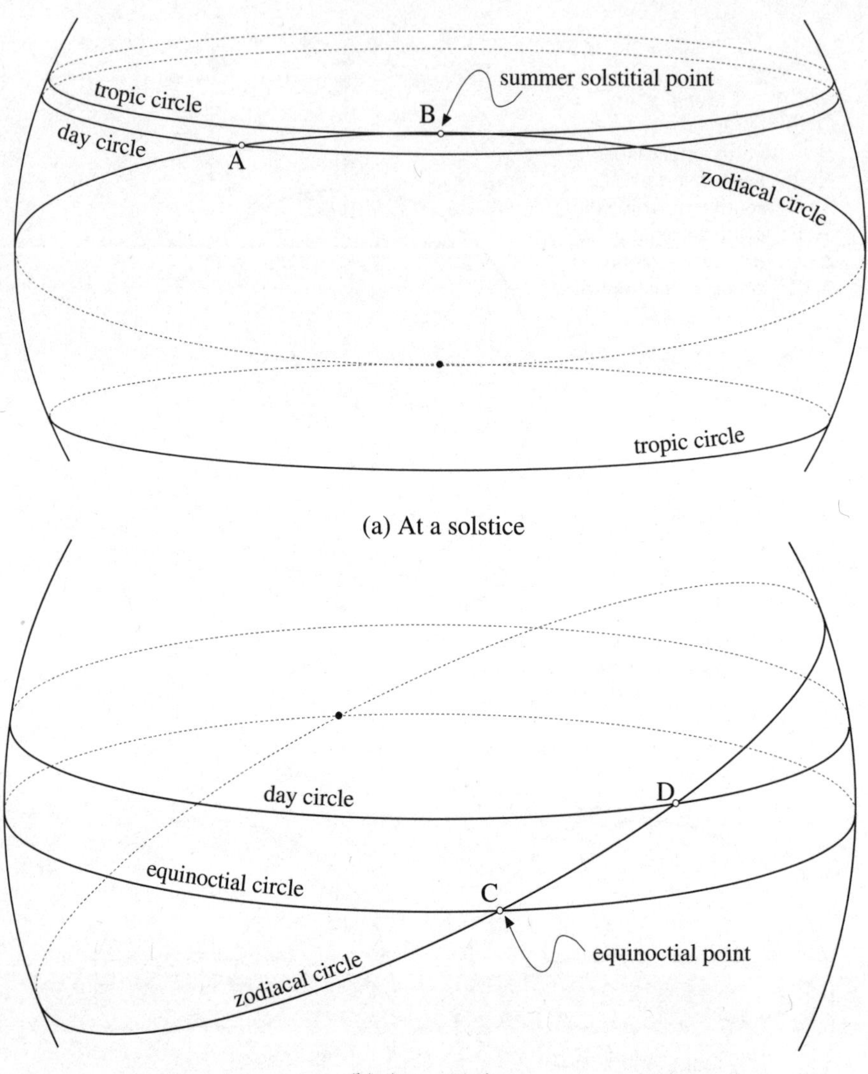

Figure 6. Equal arcs (AB, CD) of the zodiacal circle and the varying distance between the corresponding day circles at the solstices and equinoxes (I.4.30–43)

The zodiacal circle is the trace of the heliacal circle on the zodiacal band and is itself sometimes called the heliacal circle.

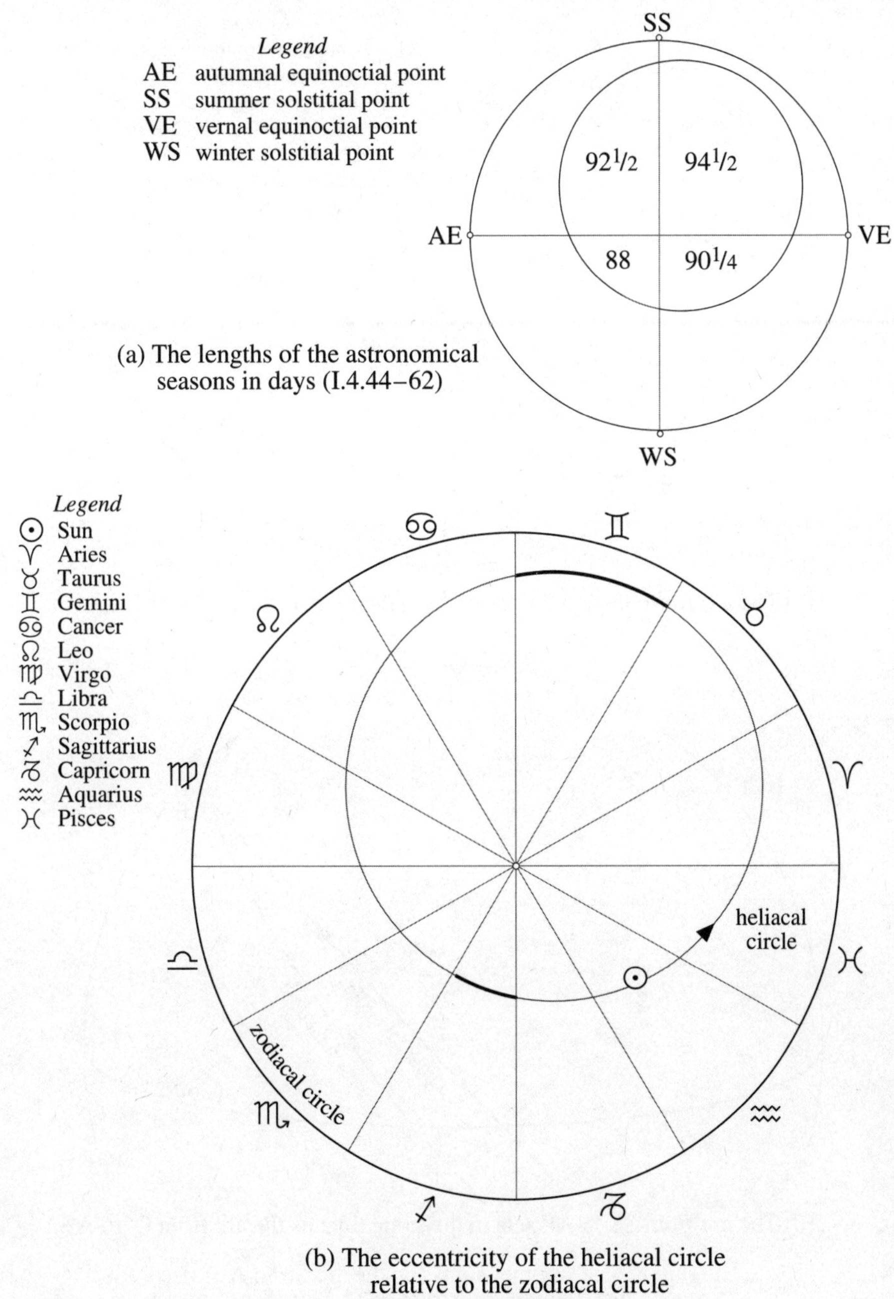

(a) The lengths of the astronomical seasons in days (I.4.44–62)

(b) The eccentricity of the heliacal circle relative to the zodiacal circle (I.4.62–71)

Figure 7

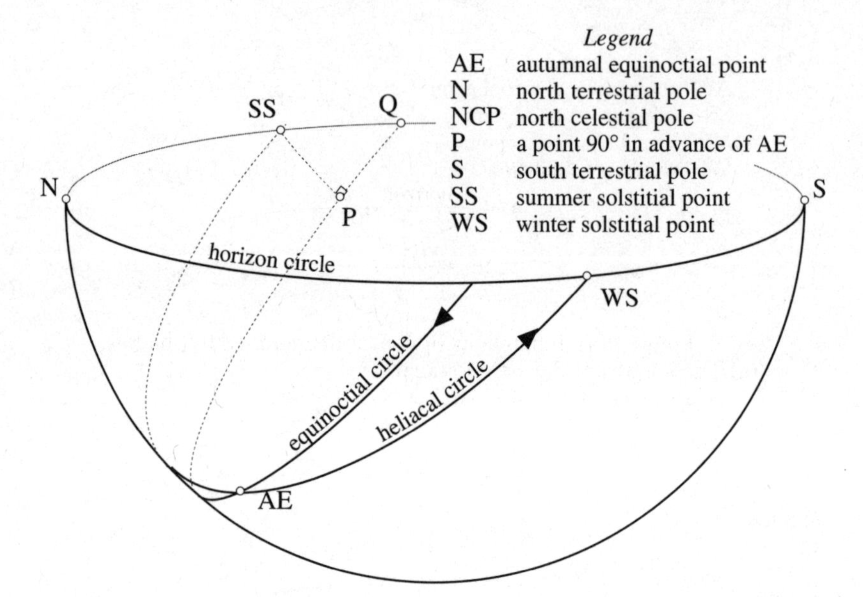

(a) The arc from SS to AE rises in the same time as the arc from Q to AE.

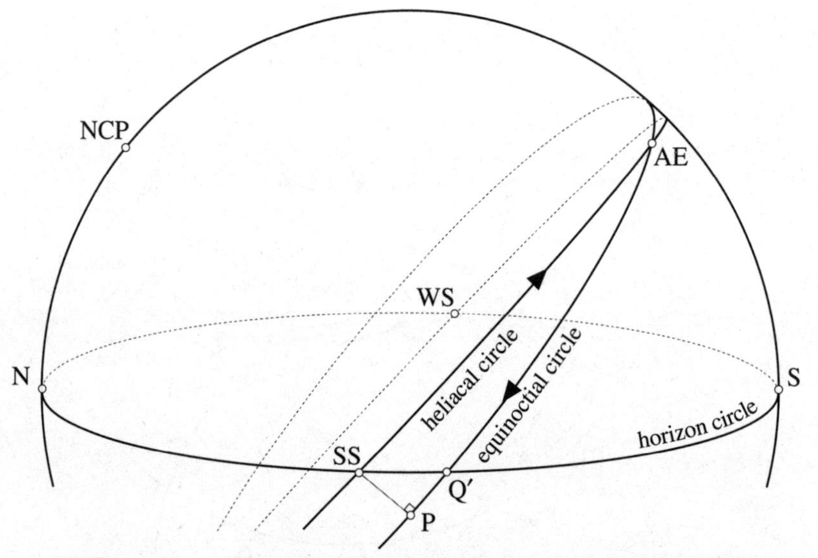

(b) The arc from SS to AE sets in the same time as the arc from Q′ to AE.

Figure 8. The summer signs (on the arc SS to AE) rise more slowly than they set (I.4.80–86)

The arc from Q to AE is longer than the arc from Q′ to AE.

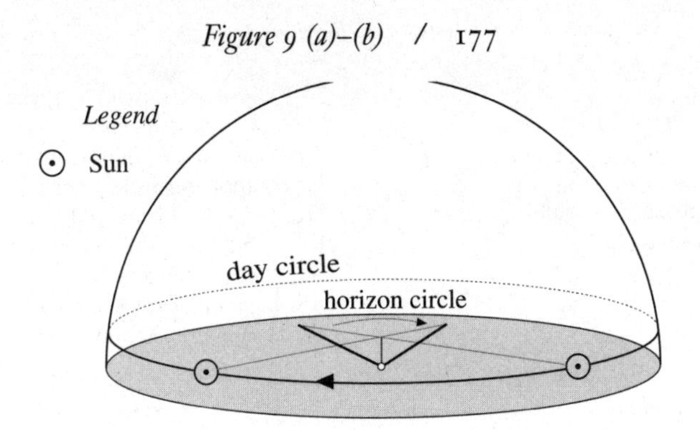

Figure 9(a). Inhabitants of the Earth encircled by shadow
(I.4.133–139)

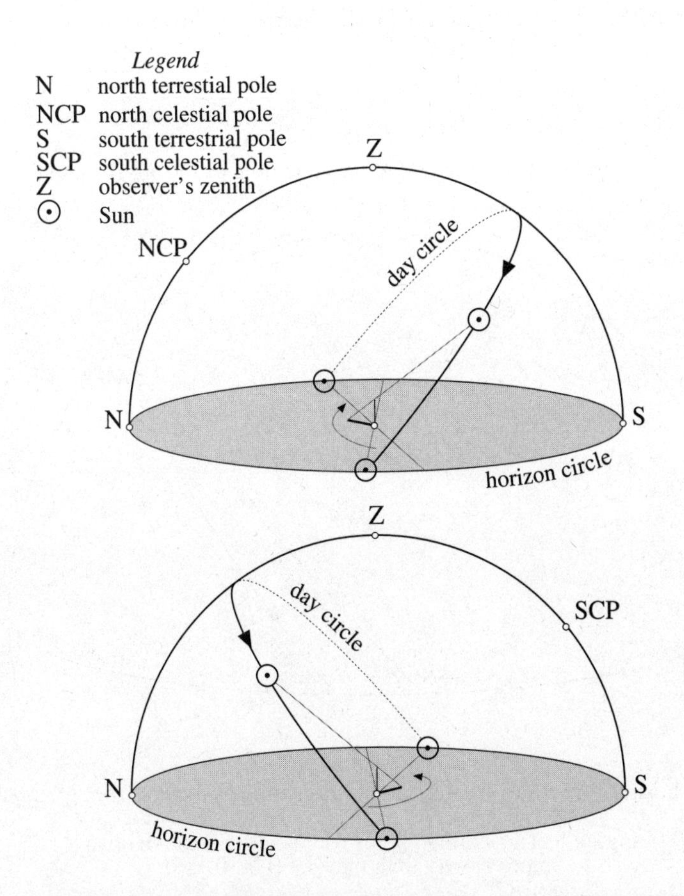

Figure 9(b). Inhabitants of the Earth shadowed unidirectionally
(I.4.139–143)

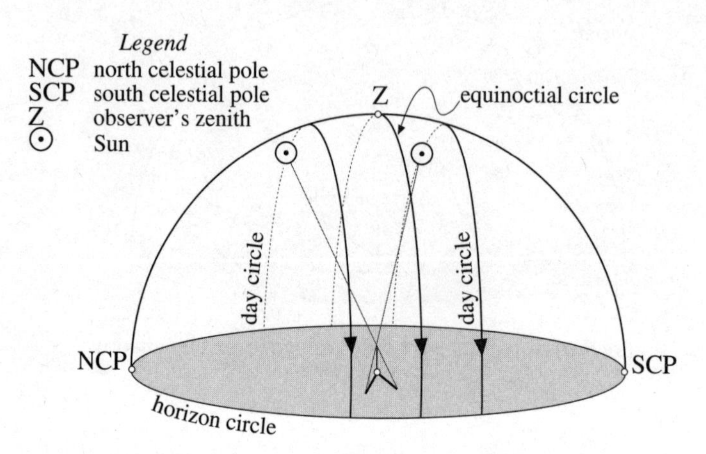

Figure 9(c). Inhabitants of the Earth shadowed bidirectionally (I.4.143–146)

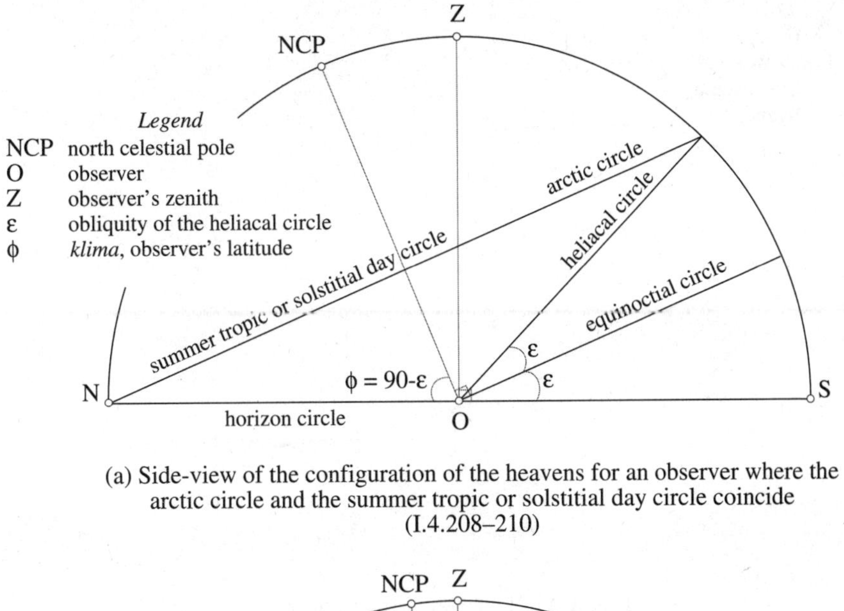

(a) Side-view of the configuration of the heavens for an observer where the arctic circle and the summer tropic or solstitial day circle coincide (I.4.208–210)

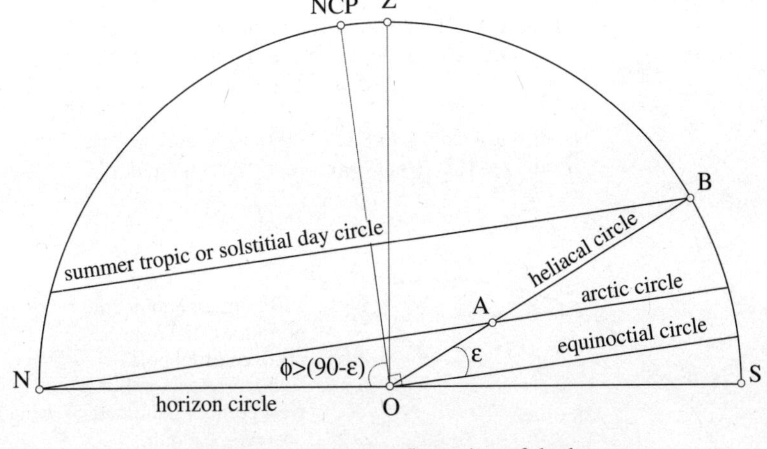

(b) Side-view of the configuration of the heavens for an observer to the north of Thule (I.4.218–219)

There will be continuous daylight for twice as long as it takes the Sun to travel from A to B.

Figure 10

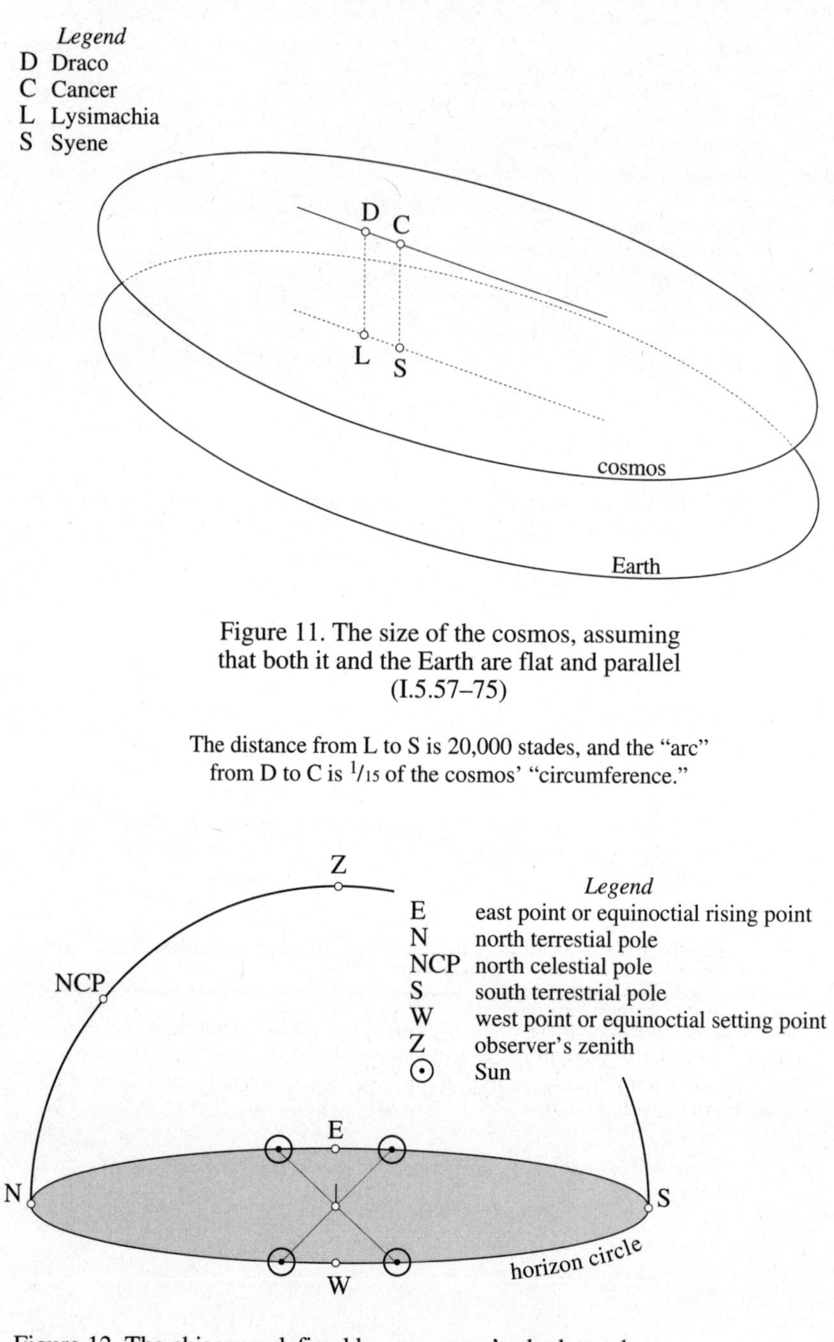

Figure 11. The size of the cosmos, assuming that both it and the Earth are flat and parallel (I.5.57–75)

The distance from L to S is 20,000 stades, and the "arc" from D to C is 1/15 of the cosmos' "circumference."

Figure 12. The chiasmus defined by a gnomon's shadow when the Sun is at the summer and winter solstitial points (I.6.26–31)

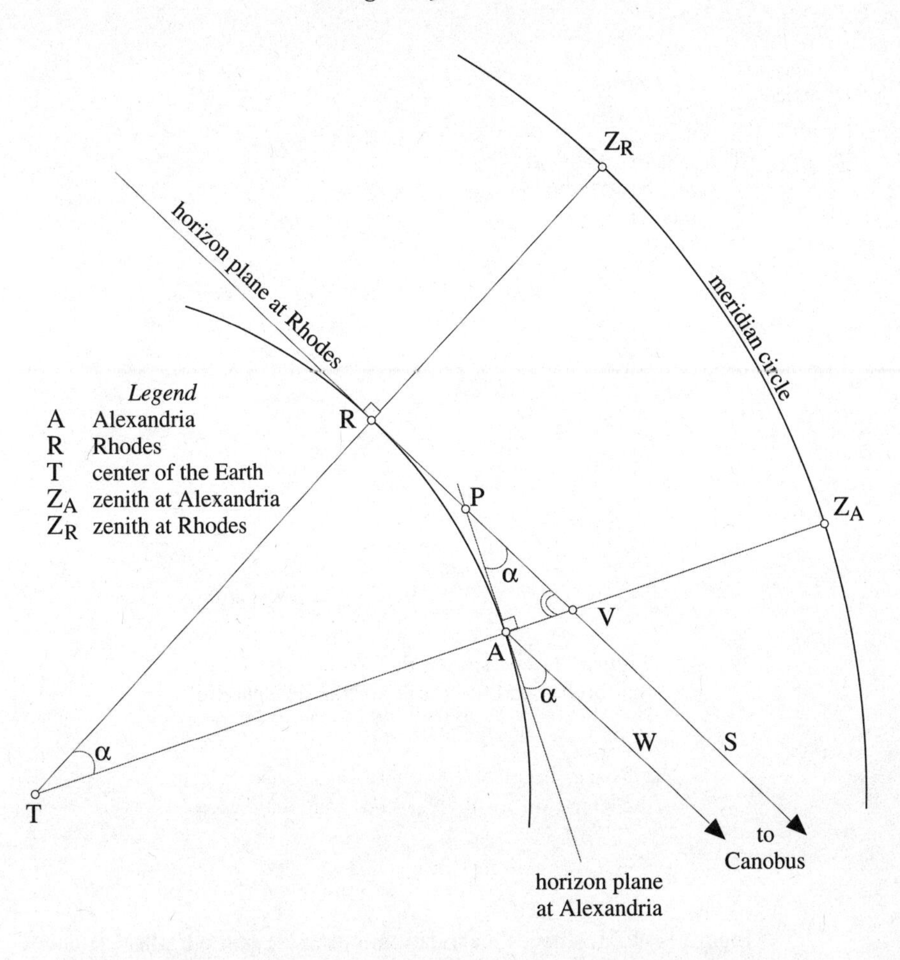

Figure 13. Posidonius' calculation of the size of the Earth (I.7.8–47)

The elevation of Canobus above the horizon at Alexandria (angle α) is 1/48 of a circle and equal to the arc from the zenith at Rhodes to the zenith at Alexandria. Since the arc from Rhodes to Alexandria is 5,000 stades long, the circumference of the Earth is 240,000 stades.

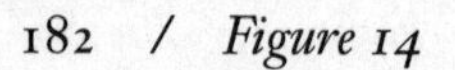

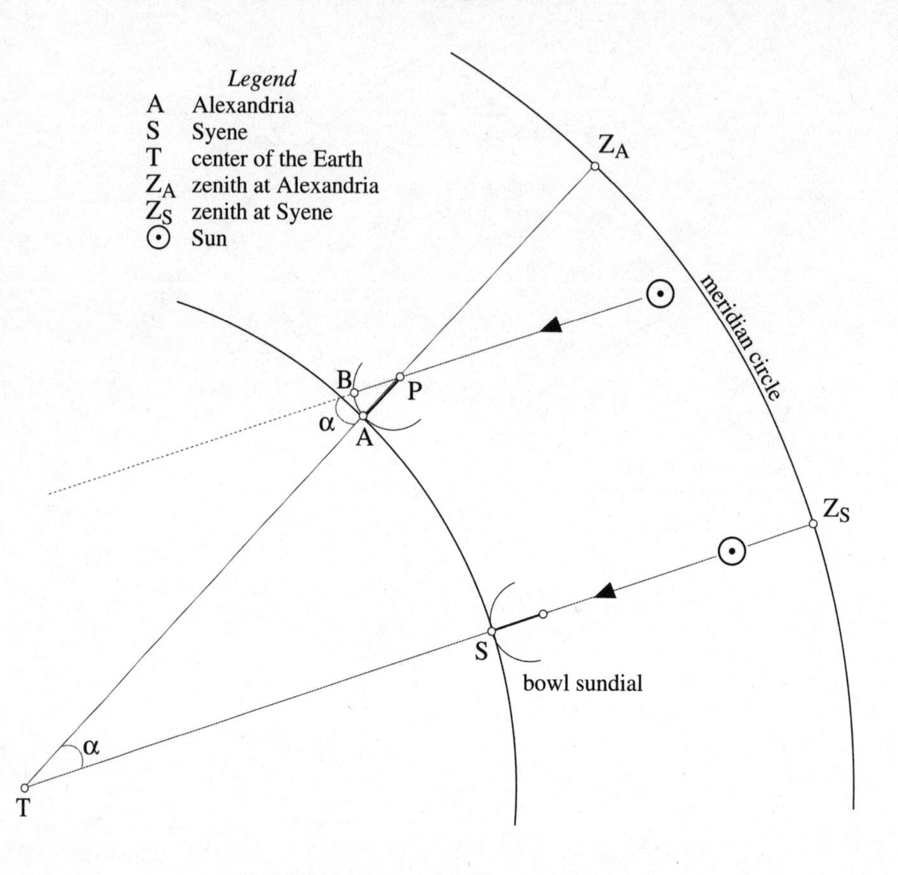

Figure 14. Eratosthenes' calculation of the size of the Earth (I.7.49–110)

Since angle BPA equals angle ATS and arc BA is $^1/_{50}$ of a circle, arc AS is also $^1/_{50}$ of a circle. Thus, given that the arc from Alexandria to Syene is 5,000 stades long, the circumference of the Earth is 250,000 stades.

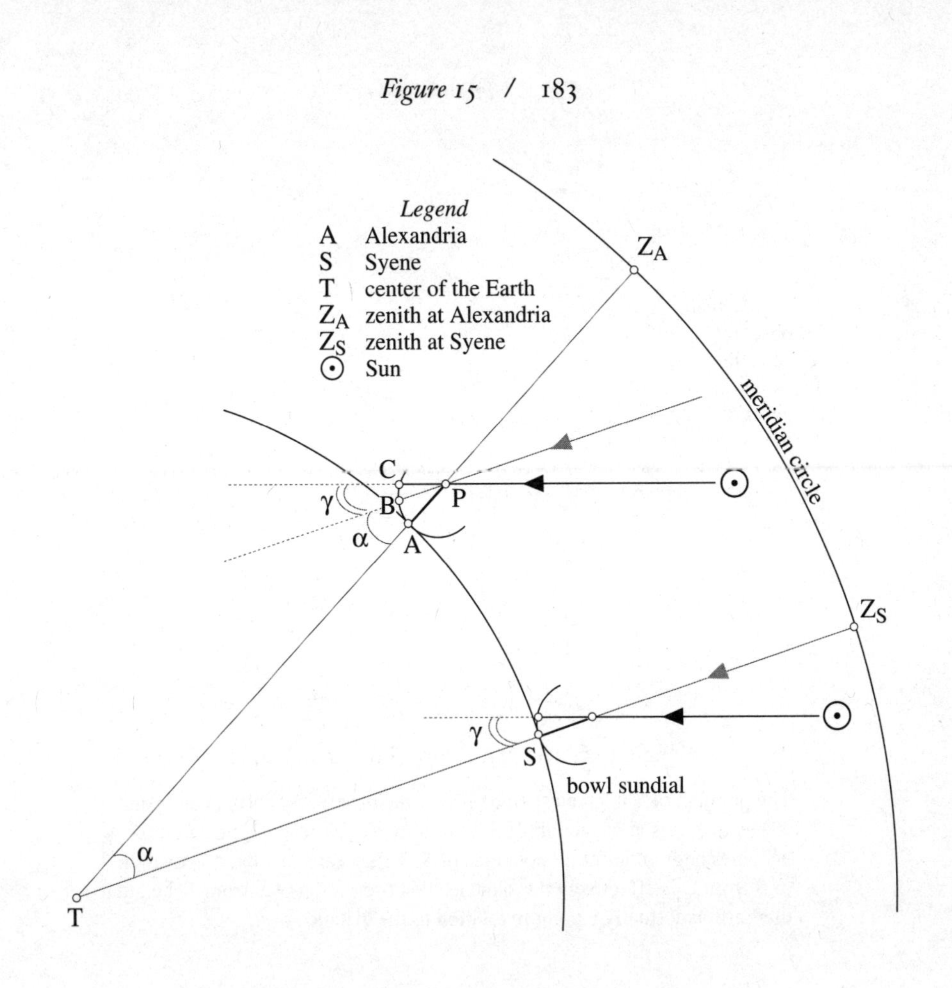

Figure 15. Eratosthenes' calculation of the size of the Earth adapted for the Sun at winter solstice (I.7.111–118)

Angle CPB = γ. Angle CPA – angle CPB = angle BPA. But angle BPA = angle ATS. So, since arc BA (= arc CA – arc CB) is 1/50 of a circle, arc AS is also 1/50 of a circle. Thus, given that the arc from Alexandria to Syene is 5,000 stades long, the circumference of the Earth is 250,000 stades.

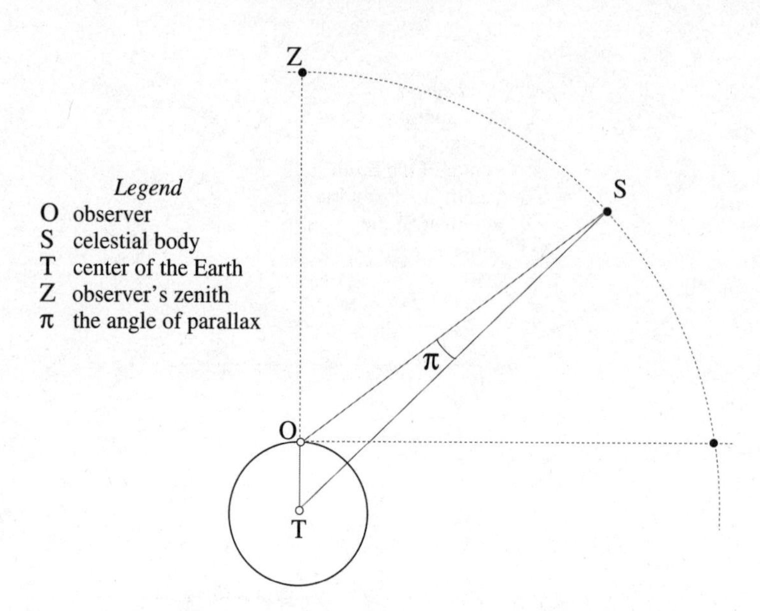

Figure 16. The parallax of S for an observer at O (I.8)

The parallax of S is greatest when it is at the observer's horizon and vanishes when it is at his zenith. If S is suitably distant from O, its parallax becomes negligible for all positions of S. This means that the observation of S from O is effectively the same as the observation of S from T, i.e., that the Earth is virtually a point in relation to the distance of S.

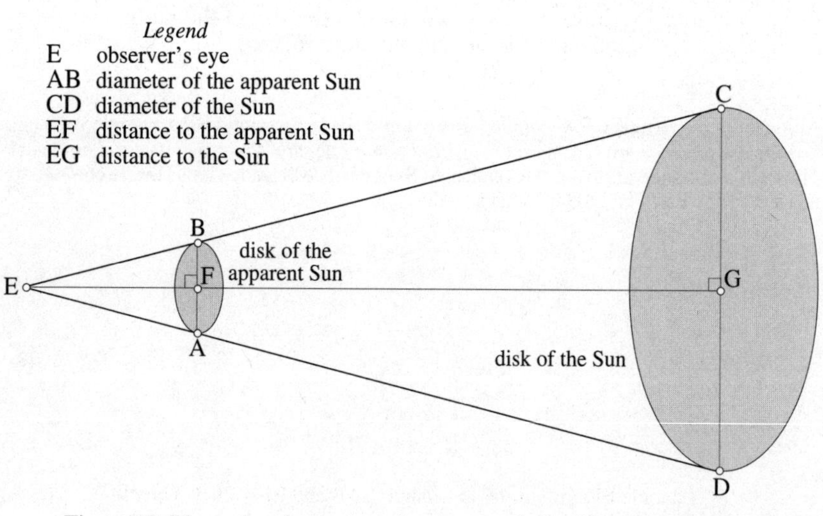

Figure 17. The real and apparent visual cones defined by the Sun (II.1.57–75)

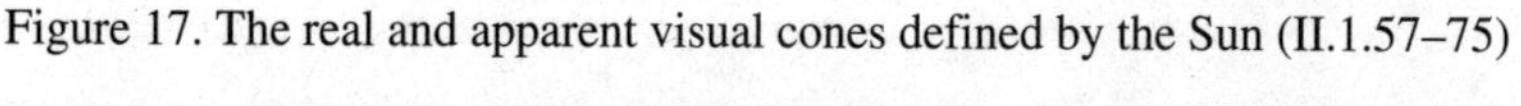
Since triangle AEB is similar to triangle DEC, then EF:EG :: AB:DC

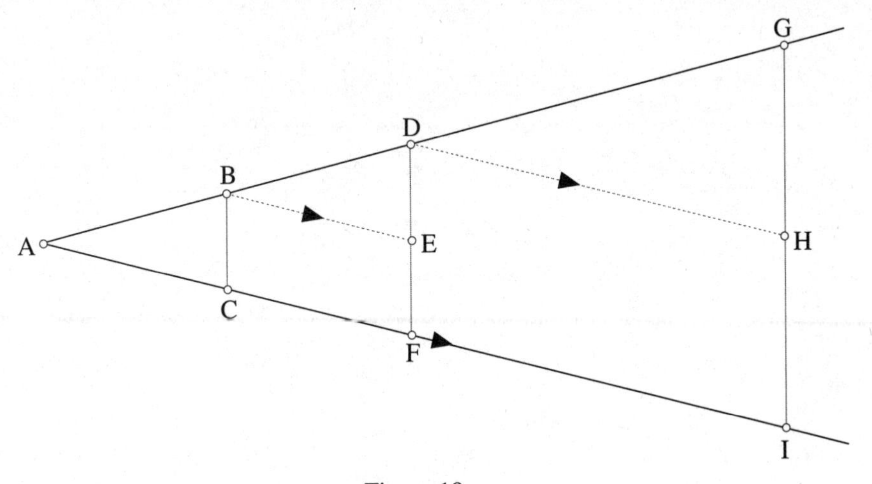

Figure 18
(II.1.243–249)

Isosceles triangle ADF is similar to isosceles triangle ABC and AD = 2AB. Therefore, AD:AB :: DF:BC :: 2:1. Similarly, isoceles triangle AGI is similar to isoceles triangle ABC and AG = 2AD = 4AB. Therefore, AG:AB :: GI:BC :: 4:1.

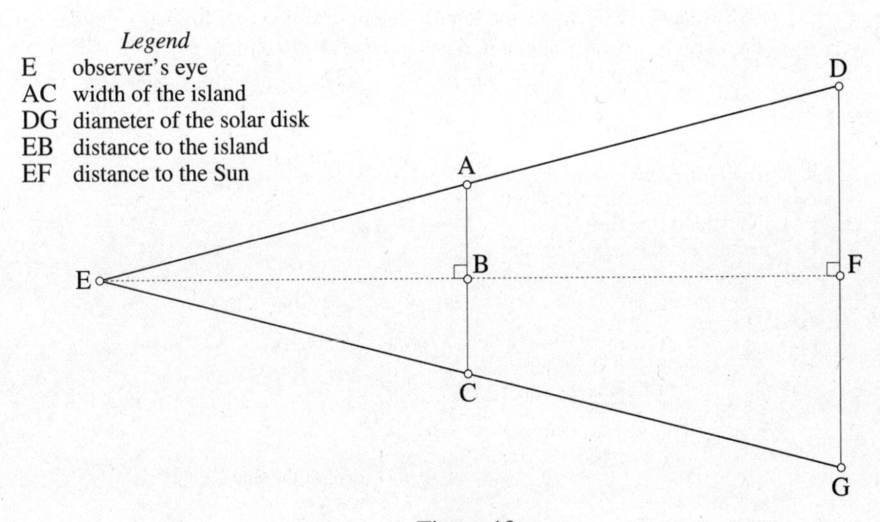

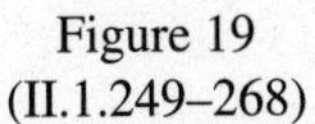

Figure 19
(II.1.249–268)

Triangle EDF is similar to triangle EAB and EF = 2EB. Therefore, EF:EB :: DF:AB :: 2:1. Since DF = FG and AB = BC, DG = 2AC.

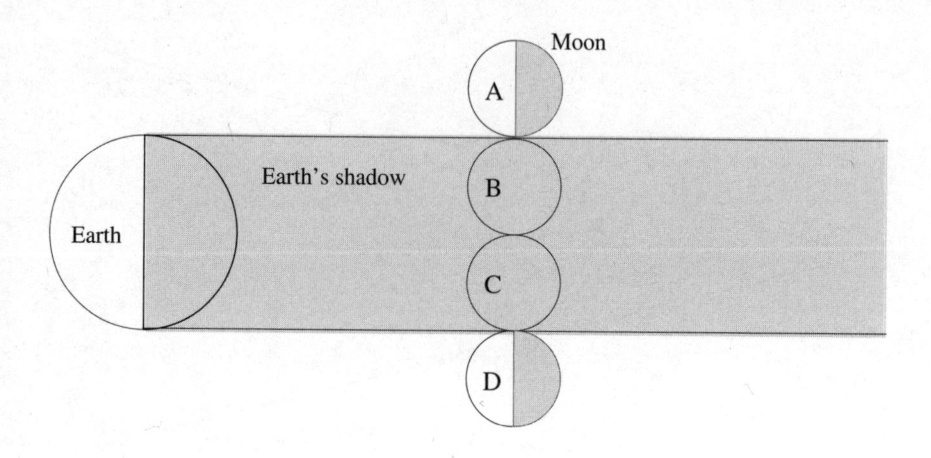

Figure 20
(II.1.286–294)

The time it takes the Moon to go from position A to position B is the same as the time it takes the Moon to go from position B to position C. So the Moon measures the Earth's shadow twice. If the Earth's shadow is cylindrical, then the diameter of the Moon is half of the Earth's diameter.

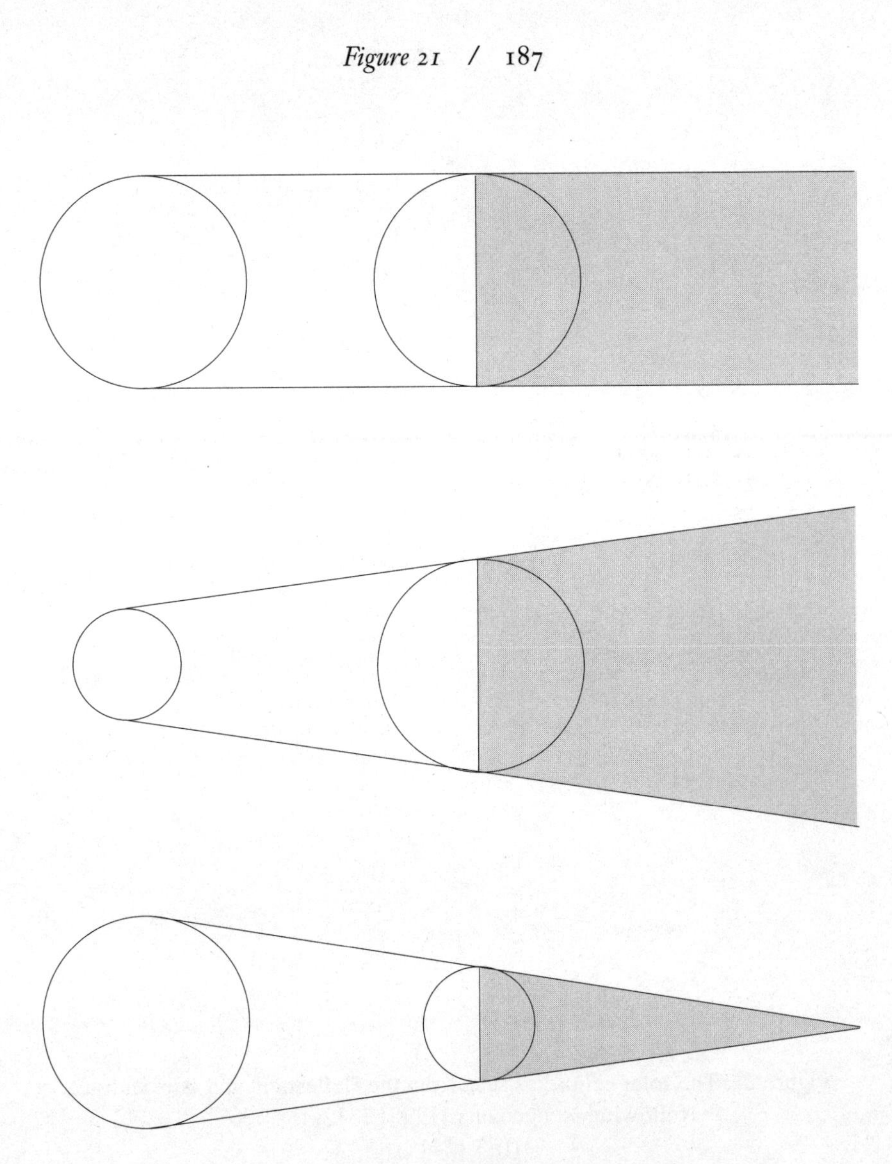

Figure 21. Types of shadow cast by a spherical body when illuminated by a spherical body (II.2.19–24)

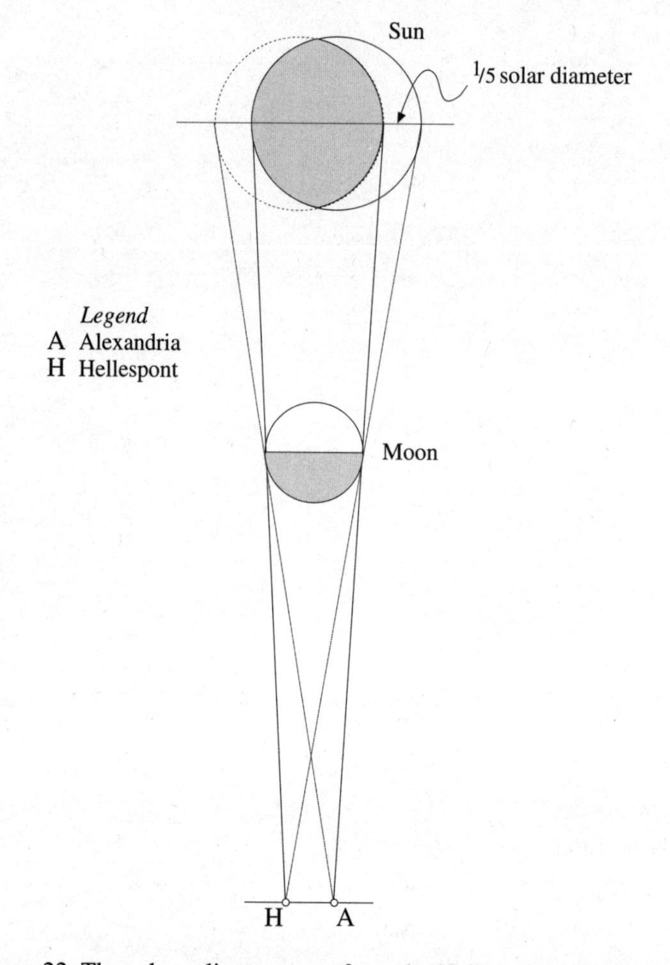

Figure 22. The solar eclipse as seen from the Hellespont and Alexandria (following Neugebauer [1975] 1312, fig. 290) (II.3.15–19)

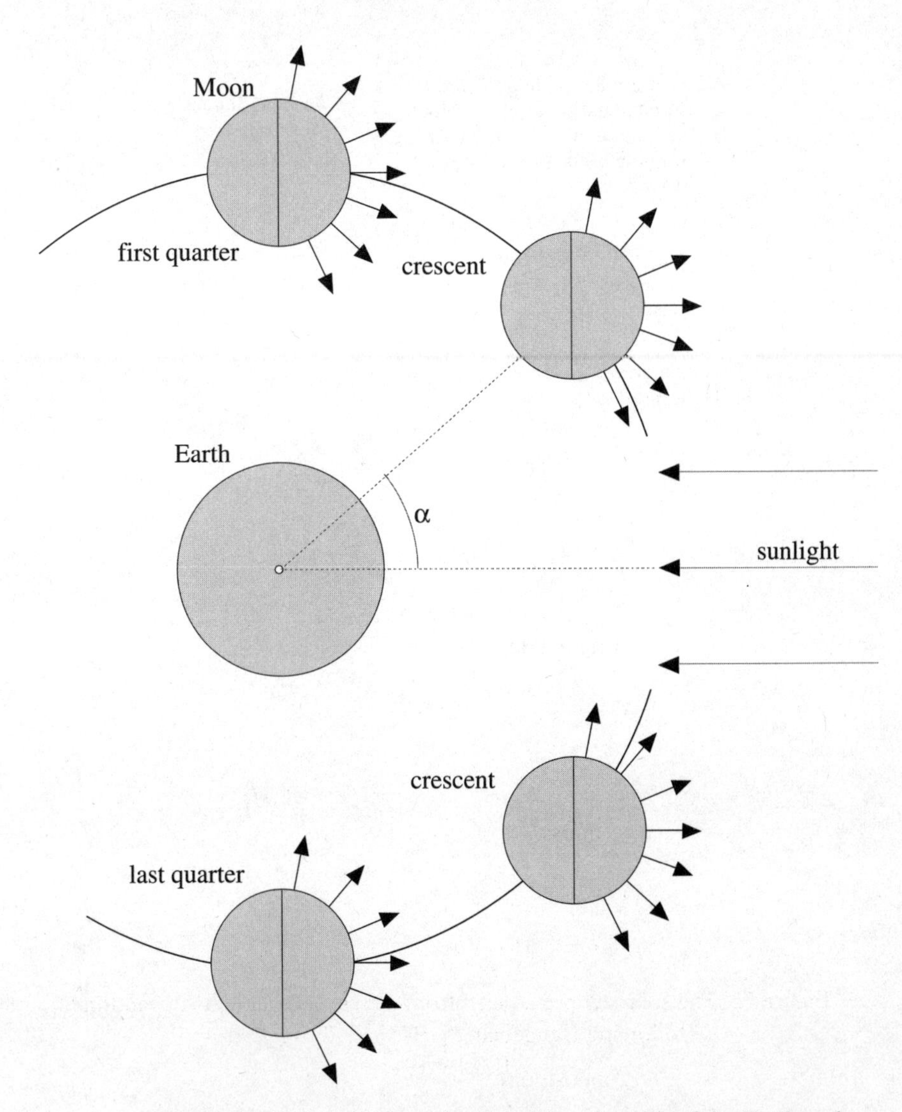

Figure 23. View from the pole of the Moon's orbit of the Earth and Moon in cross-section, showing the Moon at the crescent and the quarter as it reflects the Sun's light at "right angles," that is, radially (II.4.56–61)

According to Cleomedes, when angle α defined by the Moon, the Earth, and the Sun is 90° or less, no reflected light from the Moon reaches the Earth.

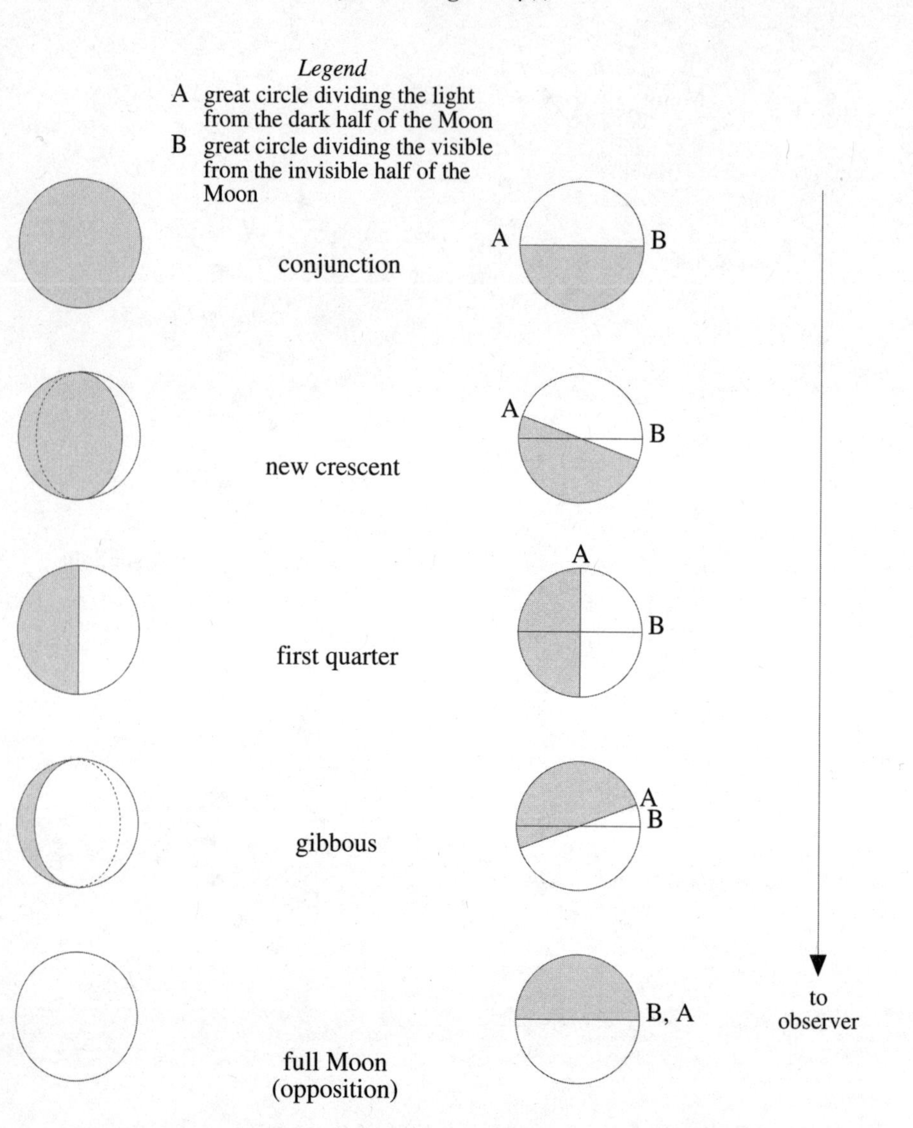

Figure 24(a). The phases of the Moon from conjunction to full Moon (II.5.41–80)

The great circles, A and B, are viewed from above the pole of the Moon.

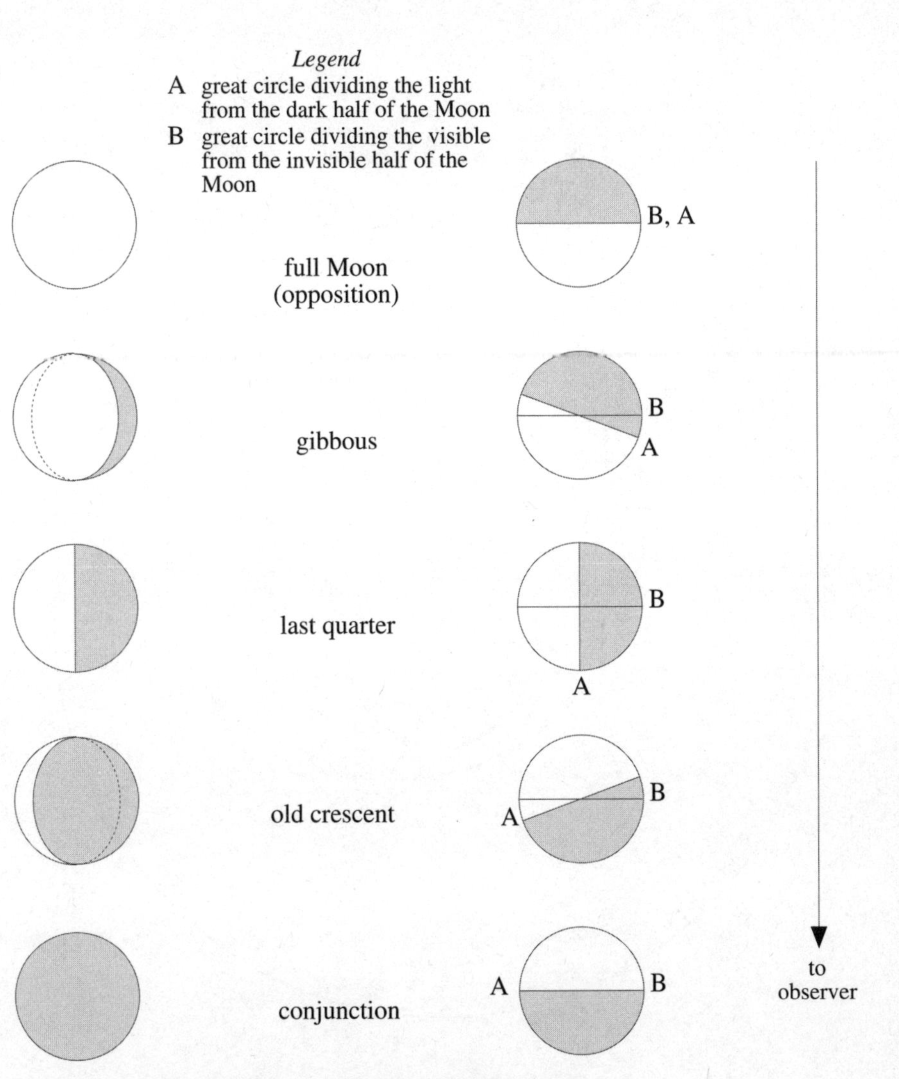

Figure 24(b). The phases of the Moon from full Moon to conjunction (II.5.41–80)

The great circles, A and B, are viewed from above the pole of the Moon.

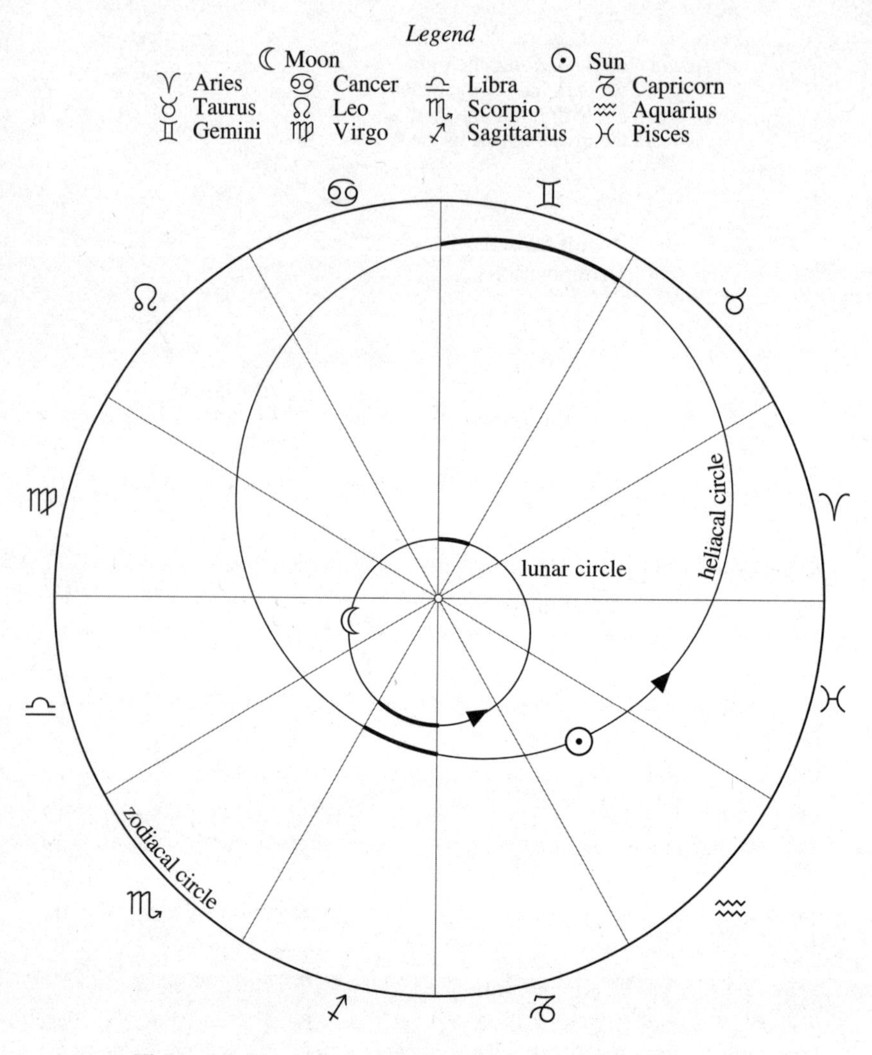

Figure 25. The variation in the length of the synodic month (II.5.102–132)

APPENDIX: POSIDONIUS ON PHYSICS AND ASTRONOMY
(Fragment 18EK[1]*)*

In the Introduction we highlighted Posidonius F18EK to clarify Cleomedes' dependence on Posidonian principles both for the general structure of his treatise, and for its methodology. Our case is principally based on lines 5–32 and 39–49 of the text translated below. They develop a programmatic distinction between physical theory and astronomy with reference to their differing approaches to the same subject matter. Physical theory is founded on the intrinsic qualities of the material cosmos, while astronomy concerns itself with incidental properties accessible through observation, and makes calculations about them. This descriptive distinction is reinforced by the prescription (lines 30–32 with 39–49) that in the crucial area of the motion of the heavenly bodies, astronomy must derive from physical theory its account of what in the cosmos really is in motion and what is stationary.

Now F18EK is not a fragment, in the sense of an authentic segment of an original work. It is a report by the sixth-century A.D. Neoplatonist commentator Simplicius of a quotation by a late second-century A.D. Peripatetic commentator, Alexander, from an epitome of a Posidonian treatise made around two

1. The text, unless otherwise noted, is that in EK, the line numbers of which are also followed. For some modern translations see Heath (1913) 275–276, Edelstein (1936) 319–320, and Aujac (1975) 110–113; for commentary see Aujac (1975) 161–162 and Kidd *Comm.* 129–136.

centuries earlier by Geminus.[2] Material may therefore have been omitted and rearranged by the original epitomizer, and further adapted for report by the later writers. This process of transmission has undoubtedly contributed to the elliptical character of this text, and our notes to the translation will try to supply a context for its argument. One section of F18EK, however, needs some preliminary discussion, even though it does not bear directly on any Cleomedean material. This is a passage concerning planetary motion at lines 32–39, which we shall argue is not integral to the argument of this text, although it is an important commentary on some of its content. If its status is properly understood, then it may no longer be used, as it has previously been, as evidence for Posidonius' general position on the role of astronomy in relation to physical theory.

F18EK, we suggest, offers a coherent argument without lines 32–39, as can be seen from the link between the two sections that surround them. Thus lines 30–32 state that astronomers will introduce hypotheses (without reference to any specific subject matter) to save the phenomena, while at lines 39–42 we learn that such a dependence on hypotheses is why[3] Heraclides of Pontus (380s–ca. 310 B.C.) tried to explain the "unsmooth" *(anōmalos)*[4] motion of the Sun by claiming that the Earth moves while the Sun is stationary. The details of Heraclides' theory do not interest Posidonius, only the fact that it exemplifies the procedure of explaining phenomena by saving them through a hypothesis (see n. 35 below). That general procedure is recapitulated at lines 42–45 with reference to celestial motion, and a final comment (46–49) is made about the need for astronomers to derive from natural philosophers the principles that define the motions that are possible for heavenly bodies. In this sequence of reasoning *any given astronomer* (cf. n. 19 below) is envisaged as formulating a *single* hypothesis to explain the motion of the heavenly bodies on the basis of the phenomena, and then, as in Heraclides' case, becoming liable to having that hypothesis corrected as the price of not having started with principles supplied by natural philosophy.

2. This must be Geminus of Rhodes (first century B.C.), whose introduction to astronomy has been frequently cited in this study. Proclus *In Eucl. El. I* 38.1–42.8 (at Aujac [1975] 114–117) records a taxonomy of the sciences consistent with that of F18EK; see in particular 41.29–42.6 on astronomy. The view that the author of the introduction to astronomy was an earlier Geminus (Reinhardt [1921] 178–183) has not won wide acceptance.

3. Thus at line 39 below, διό ("that is why") has to introduce a specific instance of the general claim made at lines 30–32.

4. See n. 29 below.

Now lines 32–39 begin with "for example" (*οἷον*), yet offer no example of the use of a hypothesis to save the phenomena. Instead, as the following formatting of our translation indicates, they pose a question and offer a complex clarification of its content.

> [*Question*]: Why do the Sun, the Moon, and the planets appear to move unsmoothly?
>
> [*Clarification*]: (i) After all, whether we hypothesize that their circuits are eccentric, or that the heavenly bodies go round along epicycles, the apparent unsmoothness of their motion will be saved, and (ii) [we] will have to go through *all* the modes according to which the phenomena[5] can be caused, so that (iii) our systematic treatment of the planets will resemble a theory of causes [set out] according to each possible mode [of explanation].

The question here is more general in nature than anything found in the preceding discussion, since it mentions the planets in toto, whereas F18EK had previously largely confined itself to issues involving the Sun, Moon, and Earth. But, more important, why is the question posed at all? If lines 32–39 are to be consistent with the basic thesis of F18EK, this question can only imply the need for a *physical* account of planetary motion. In that case, the purpose of the clarification that follows is to define the situation that will arise in the absence of such an account, *not* for astronomers but for the "we" of this passage. This collectivity must consist of Stoic philosophers who are committed to explaining planetary motion through physical theory. At least, if Posidonius is the author of lines 32–39, he could hardly be addressing astronomers as "we" after distinguishing them from philosophers in the preceding part of F18EK.

If our proposal here is correct, lines 32–39 express in broader terms than the rest of the passage why Heraclides' hypothesis (at 39–42) has to be rejected by a Stoic philosopher. That is, Stoic physical theory can demonstrate that the Earth is stationary while the Sun moves, whereas lines 32–39 reveal what the explanation of planetary motion would be like if there were no such physical theory to fall back on.

Thus part (i) mentions two hypotheses (those of eccentric and epicyclic motion) as familiar to "us" as ways of saving the phenomena. Whereas lines 30–32 had suggested that different astronomers could formulate different hypotheses, part (ii) now claims that if "we" admit these two hypotheses, "we" will be required to state every possible hypothesis that would explain unsmooth planetary

5. On this phrase see n. 32 below.

motion,[6] and part (iii) sums up by describing such an inventory as only "resembling" a theory of causes (an *aitiologia*). That is, it is not a real theory of causes, because it does not depend on physical theory, which, as we have seen, prescribes a *single* cause. Anyone saddled with this consequence might be able to reconcile different hypotheses *mathematically*, but would have to admit that different hypotheses had different, and irreconcilable, *physical* implications (see n. 30 below). There could be no better argument for requiring here, as elsewhere in F18EK, a single physical theory as the presupposition of astronomy.

Now a practicing astronomer of Posidonius' time did not, and would not, want to entertain a multiplicity of hypotheses, and it is implausible to suggest, as some scholars have, that lines 32–39 are actually licensing an astronomical program in which multiple hypotheses will be generated as a natural part of planetary theory, when that is pursued as an activity parallel, rather than strictly subordinate, to physical theory.[7] Instead these lines are best understood as a Stoic philosopher's sophisticated reaction to the problem raised by the historical fact that different

6. Presumably this would entail developing each hypothesis for each planetary body, if the hypotheses are to be quantified appropriately. This would amount to 21 hypotheses if to epicyclic and eccentric motion we add homocentric motion and attempt to explain the motion of each planet by each of these hypotheses. The writer may, of course, have had more than these three kinds of hypotheses in mind; after all, there were Babylonian arithmetical schemes for the planets as well as Greek versions of these schemes in play during the relevant time period; see Bowen and Goldstein (1991), and Jones (1999), especially at 1:5–34. Geminus for one uses a Babylonian scheme at *Isag.* ch. 18 to account for the Moon's unsmooth motion; see Bowen and Goldstein (1996) 167–171.

7. For this position see Kidd (1978a) and *Comm.* 132–136, and cf. Aujac (1975) 162 n. 6. Kidd sees the subordination of science to philosophy claimed by Posidonius as compatible with science's inherent employment of what he (Kidd) calls a "hypothetical method" (cf. n. 33 below). Kidd, however, fails to distinguish between the historical fact of different astronomers having *different* hypotheses regarding planetary motion, and the unique, and undesirable, situation envisaged at F18.32–39 EK, in which a *single* physical theorist will be obliged to adopt *multiple* hypotheses. The upshot of those lines is that Posidonius must be presented as deprecating hypotheses entirely rather than as wanting philosophy to "arbitrate" between them (Kidd *Comm.* 136). For further criticism of Kidd's conception of hypotheses in Posidonian texts see on *Caelestia* I.7 at nn. 4 and 11; also, on Posidonius and multiple explanations see on *Caelestia* I.4 at n. 25 and II.6 n. 21.

astronomers favored different hypotheses concerning planetary motion.[8] As such, lines 32–39 can be compared with an argument used by the Sceptic Aenesidemus (first century B.C.): that single causal explanations are unacceptable when, given the variety of evidence available, many modes of explanation are possible.[9] Aenesidemus sees such multiple explanations as unavoidable, and as undermining, by their very multiplicity, the whole program of causal explanation. The author of F18.32–39 EK, on the other hand, argues that multiple explanations will arise for anyone (philosophers as well as astronomers) when there is no criterion, in the form of an irrefutable physical theory, that can be used to decide which explanation is true.

F18.32–39 EK, then, does not in its present location in the text form a continuous part of the argument of this "fragment." It is at best tangential to it, both in being a query about how one should explain the apparent unsmooth motion of the planets in general, and in being a reflection on what follows for philosophers if they do not have a proper physical account rather than an inventory of astronomical practice. It then (lines 46–49) leads into a prescription directed to astronomers to temper their hypothesizing by starting from a physical account. Posidonius himself could certainly have employed such reasoning, both as a justification for physical theory, and as a further way of warning astronomers that,

8. Thus we reject the paraphrase of this particular text by Lloyd (1978) 213: "it is [the astronomer's] business to say in how many ways it is possible to save the phenomena." For Posidonius it is *nobody's* "business" to multiply hypotheses; instead, by being formulated with reference to physical theory, they lose any hypothetical status, and are necessarily not multiple. Lloyd (1978) 213–214 attributes an incoherent position to Posidonius in claiming that astronomy "presupposes" physics, while the astronomers' "business" is to state multiple hypotheses. He manages this by failing (at 214) to take full account of F18. 46–49. Here the astronomers are said *(a)* to take *"first principles,"* not "presuppositions," from the natural philosophers, and *(b)* to demonstrate "through" those principles only two types of motion (parallel and oblique) by the heavenly bodies. Lloyd quotes *(a)*, and ignores *(b)*, and thus seems to think that astronomers derive from physics only more remote principles (his "presuppositions")—simple, smooth and orderly motion—that are compatible with their positing multiple explanations. Instead, the format of their explanations is rigidly prescribed in *(b)*.

9. At Sext. Emp. *PH* 1.181. See Barnes (1983) 167–169 and (1990) 2665–2666 on this argument and its analogues in Quine's concept of data "underdetermining" theory, and cf. Introduction n. 25.

if they rely on hypotheses, then a single hypothesis may not really account for the phenomena. But there is no suggestion in lines 32–39 that Posidonius also wanted astronomers to entertain multiple hypotheses[10] among which philosophers would then arbitrate. On the contrary, he could only have held out the specter of multiple hypotheses as an additional way of demonstrating the need to constrain astronomical hypotheses by physical theory. That warning, however, is not directly conveyed by the language and reasoning used in lines 32–39.[11] The author of those lines is, as we have suggested, drawing the attention of fellow Stoics to a problem closely related both to the general thesis of F18EK, and to the specific case of Heraclides' treatment of "apparent unsmooth motion." He is, in other words, offering philosophical reflection that can be seen as enriching the simpler programmatic claims in the rest of F18EK.

F18.32–39 EK, then, clearly reflects some imperfect epitomizing and selection in the evolution of this whole text.[12] Yet these lines are still compatible with the mantra of F18EK: that astronomers "take first principles from natural philosophers." Astronomers who fail to do so risk having their hypotheses corrected by philosophers, an embarrassing result, yet nothing to compare with the

10. We thus question Kidd's attempt, however tentative, at *Comm.* 136 to put F18EK into some theoretical relationship with a well-known text at Simplicius *In de caelo* 32.29–32 that shows tolerance toward astronomers who develop different hypotheses to save the phenomena. Simplicius could have based his position on F18EK, which, of course, he quotes (see n. 13 below), but in doing so he would be misunderstanding Posidonius, as surely as Kidd does in supposing the Stoic philosopher to be defining science by its use of "hypothetical method" (see n. 7 above).

11. Kidd *Comm.* 132–133 paraphrases 32–39 by initially saying that "we" entertain different hypotheses for planetary motion, but then switches to talk of "*their* study of the planets" leading to a list of explanations. "They" are Kidd's "scientists" who work with "possible hypotheses"; but the text of F18EK clearly shows that it is the "we" of lines 32–39 who will have to cope with such hypotheses, and there is no reason to identify this group as astronomers (cf. n. 7 above).

12. The use of *οἷον* ("for example") at line 32 to introduce this passage in itself suggests that some preceding discussion has been lost. This would almost certainly have concerned the "apparent unsmooth motion" of the Sun mentioned in the comment on Heraclides at lines 39–42, and used as the basis of the reflections at lines 32–39.

nightmare faced by philosophers who are unable to explain the nature of the cosmos and are therefore engulfed by the multiple hypotheses indicated at F18.32–39 EK.

TRANSLATION

1:[13] Alexander [of Aphrodisias] assiduously quotes a specific text of Geminus, derived from [the latter's] epitome that expounds Posidonius' *Meteorologica*[14]—[a text] that takes its starting points from Aristotle.[15] It goes as follows.

5: It is for physical theory to inquire into the substance of the heavens and of the heavenly bodies, and into their power and quality; and into their coming into existence and destruction. Through these [investigations][16] it can certainly offer demonstrations concerning size, shape, and ordering. Astronomy, on the other hand, does not attempt to speak about anything of that sort. Instead, it demonstrates the order of the heavenly bodies after declaring that the heavens really

13. The passage, quoted by Simplic. *In phys.* 291.21–292.21 as part of his exegesis of *Phys.* 193b22–35, was taken from Alexander of Aphrodisias' lost commentary on Aristotle's *Physics.* Geminus' epitome provided these commentators with historical confirmation for Aristotle's distinction at *Phys.* 193b25–35 between astronomy and natural philosophy, which he introduced as part of a larger distinction between mathematics and natural philosophy in *Phys.* 2.2. See also nn. 20 and 21 below.

14. This follows Diels' proposal (apparatus criticus for Simplic. *In phys.* 291.22; noted, but not adopted, in EK) to supply τῆς after ἐπιτομῆς, so as to leave no doubt that this is Geminus' epitome of the Posidonian treatise. Literally, it is "the epitome, <the one> that is an exposition."

15. Since the Aristotelian text supplies only "starting-points" *(aphormai)*, or "points of departure," the Aristotelian origins of Posidonius' position are of limited significance; see Todd (1988) 307–308 on Sandbach (1985) 61.

16. With Bake (1810) 60 we read νὴ Δία <διὰ> τούτων, a supplement which Kidd *Comm.* 130 found only "tempting." But the supplement is palaeographically justifiable, and by emphasizing that it is "through," or on the basis of, physical theory that the topics of shape, size and order are pursued, it is consistent with line 17 below, where astronomy is said, by contrast, to make demonstrations "through" arithmetic and geometry.

are a cosmos,[17] and speaks about the shapes, sizes, and distances of the Earth, the Sun, and the Moon; and about the eclipses and conjunctions of heavenly bodies; and the quality and quantity of their movements.

14: It follows that since astronomy deals with the theory of quantity, duration, and type of shape, it is reasonable for it to need arithmetic and geometry for this.[18] And concerning these matters, which are the only ones about which it undertakes to supply an account, it has the authority to make inferences through arithmetic and geometry.

18: Now astronomers and natural philosophers[19] will in many cases propose to demonstrate essentially the same [thesis] (e.g., that the Sun is large; that the Earth is spherical), yet they will not follow the same procedures.[20] Whereas [natural philosophers] will make each of their demonstrations on the basis of substance, or power, or "that it is better that it be thus," or [the processes] of coming into existence and change, astronomers will do so on the basis of the [prop-

17. The "real" cosmos for the astronomers is the heavens, and Cleomedes frequently uses the Greek term *kosmos* in this sense. The natural philosopher, however, is more interested in the cosmos as the totality of matter; see, for example, *Caelestia* I.1.3–10.

18. Cf. the similar definition of mathematical astronomy at Ptol. *Alm.* 1.1, 5.25–6.4. Like Posidonius (lines 15–16 below), Ptolemy (6.19–21) also associates arithmetic and geometry with astronomy, but, unlike him, rates it above physical theory, on the grounds that it deals with unchanging matter, i.e., the Aristotelian aether (6.9–11; 6.23–7.3). Ptolemy limits physics to an analysis of the qualities of sublunary matter (5.19–6.14), an unstable object (6.11–15) that can support only "guesswork" *(eikasia)*, not generate knowledge. Given the diffusion of Posidonius' ideas in the first and second centuries (see I. G. Kidd [1978a] 11), Ptolemy may even be reacting to the program summarized in F18EK, while still maintaining the primacy of physical theory as the source of the cosmology presupposed by astronomers. On the latter see *Alm.* 1.7, and cf. Lloyd (1978) 216–217, and Wolff (1988) 499 n. 31.

19. Here and subsequently we interpret the definite article in the singular with "astronomer" and "natural philosopher" as *generic* in sense (that is, it refers to any member of the class, or to the class collectively) and have translated it consistently in the plural form. See further n. 26 below on the importance of this point.

20. On the common subject matter of astronomy and natural philosophy see *Phys.* 193b29–30. On the language of "procedures" *(ephodoi)* (here represented by *hodoi*, literally "routes") in the *Caelestia* see the Introduction n. 34.

erties] incidental to shapes or magnitudes,[21] or on the basis of the quantity of the movement, and the time interval appropriate to it.

25: And natural philosophers will in many cases[22] deal with the cause by focusing on the causative power,[23] whereas astronomers, when they make their demonstrations on the basis of extrinsic incidental [properties], have no adequate insight into the cause in, for example, claiming that the Earth, or the heavenly bodies, are spherical.[24] Sometimes they do not even aim to comprehend the cause, as when they discourse on an eclipse.[25]

30: On other occasions [astronomers][26] make determinations in accordance with a hypothesis[27] by setting out some modes [of explanation], which if they are the case, the phenomena will be saved.

21. *Phys.* 193b27 and 32–33 also refer to "incidental properties" *(sumbebēkota).* But since these are observable, the phrase here, and at line 27 below, anticipates the later reference (line 32 below) to "appearances" *(phainomena).* For *kata sumbebēkos* ("incidentally") glossed as "apparent" *(phainomenē)* see Theon *Expos.* 188.22.

22. Not, however, that of the void, which "does not act at all" (οὐδὲν ποιεῖ), i.e., cause any effects; see *Caelestia* I.1.99–100.

23. *poiētikē dunamis* (for the translation see I.3 n. 17). This phrase may suggest the general notion of the Stoic *logos* (Kidd *Comm.* 132), which is, of course, "active" *(poioun),* but here it can be more specifically linked with earlier references (lines 6 and 21) to the power of *intracosmic* bodies to produce effects. In the *Caelestia* there are also references to the power of the cosmos to undergo change (I.8.92–95), of the Earth to nourish heavenly bodies (I.8.92–95), and of the Sun and Moon to produce terrestrial effects (II.1.357–404; II.3.61–67), while *Caelestia* I.2–4 emphasizes the Sun's causal role in producing variations in seasons and the lengths of daytimes and nighttimes.

24. Although Cleomedes grounds the sphericity of the Earth on observations (see I.5.104–113), but that of the cosmos (*Caelestia* I.5.126–138) on physical theory, the theory of centripetal motion (cf. I.6.41–43) still implies that the heaviest matter will necessarily arrange itself spherically at the center of the cosmos.

25. See *Caelestia* II.4.95–107 (with II.4 nn. 8 and 19) on the Posidonian physical theory underlying eclipses.

26. By interpreting the subject of the verb here, which is in the singular, as the class of astronomers, or "any astronomer" (see n. 19 above), we avoid having Posidonius claim that a given astronomer will adopt more than one hypothesis, as I. G. Kidd (see n. 7 above) would have him do.

27. At *Caelestia* II.1.310–311 and 332–333 similar language is used in connection with premises in a calculation of the size of the Sun, though there the

32: For example:[28] Why do the Sun, the Moon, and the planets appear to move unsmoothly?[29] After all, whether we hypothesize that their circuits are eccentric, or that the heavenly bodies go round along epicycles,[30] the apparent unsmoothness of their motion will be saved.[31] And [we] will have to go through *all* the modes according to which these phenomena[32] can be caused, so that our systematic treatment of the planets will resemble a theory of causes [set out] according to each possible mode [of explanation].[33]

---

hypotheses are assumptions made within the larger argument rather than the kind of foundational hypothesis that concerns Posidonius here.

28. See the preface to this Appendix, and n. 12 above, on the relation of this section to the rest of the passage.

29. *anōmalōs:* on this terminology see Bowen (1999) 289–296.

30. Hipparchus seems to have evaluated both the epicyclic and eccentric theories purely as hypotheses (Theon *Expos.* 166.4–10), and was criticized (Theon *Expos.* 188.15–24), in terms consistent with Posid. F18EK, for "not being supplied 'for the road' *(ephōdiasthai)* from natural philosophy *(phusiologia)*." That is, he did not adopt a "procedure" *(ephodos)* grounded in physical theory.

31. The eccentric and epicyclic hypotheses are treated here as independent rather than equivalent, since Stoic philosophers (the "we" of this passage; cf. nn. 7 and 11 above) are concerned with their physical consequences. Also, no writer of the first centuries B.C. and A.D. mentions this equivalence, and those that do address the issue of the planetary motions choose one hypothesis or the other, and not both. (On the project of saving the phenomena of the planetary motions before Ptolemy, see Bowen [2001].) Ptolemy actually suggests the possible equivalence of the two hypotheses at *Alm.* 3.3 (cf. Toomer [1984] 144 n. 32), but does not discuss it until *Alm.* 12.1. On the strength of the latter text some modern scholars (e.g., Neugebauer [1955] and [1959]) suppose that the proof of this equivalence goes back to Apollonius of Perge (third century B.C.). But this overlooks the fact that the single demonstration Ptolemy gives in *Alm.* 12.1 of the planetary stationary points is his own, and that, according to Ptolemy, his predecessors made their cases for each hypothesis separately; see Bowen (2001).

32. The phenomena that are "saved" are identical, whichever hypothesis is adopted.

33. The phrase κατὰ τὸν ἐνδεχόμενον τρόπον (37–38) is best taken in this distributive sense. The translation "the possible method" at Lloyd (1978) 213 and Kidd *Comm.* 133 is vague and question-begging, and for Kidd "possible" is also a synonym for "hypothetical"; see n. 7 above.

39: That is why a certain Heraclides of Pontus actually came forward to say that the apparent unsmooth motion of the Sun can be saved[34] even if the Earth somehow moves, while the Sun somehow remains stationary.[35]

42: For in general astronomers do not have knowledge of what is by nature at rest and what sort of things are moved.[36] Instead, by introducing hypotheses of some things being stationary, others in motion, they investigate which hypotheses will follow from the phenomena in the heavens.

46:[37] But astronomers have to take as first principles from natural philosophers that the motions of the heavenly bodies are simple, smooth, and orderly, and through *these* [principles] they will demonstrate that the choral

34. In fact the Earth's motion and the Sun's immobility would not explain this apparent unsmooth motion (manifested, for example, in the inequality of the seasons) without some additional assumption, such as that the Earth moves on an eccentric circle.

35. On this text as evidence for Heraclides' cosmology see Gottschalk (1980) 62–69. A derogatory use of "a certain" (τις) and "somehow" (πως) would make Posidonius' whole report vague and dismissive, and therefore not worth pursuing in any detail in the present context. Heraclides is chosen probably because the *single* possible explanation that he gave from the multiplicity available was so unusual, and, for the Stoic Posidonius, so exceptionally counterintuitive.

36. The Stoic natural philosopher will know this, and what causes it, on the basis of the theory of the centripetal motion of the elements, whereby Earth, as the densest element, is stable at the center of the cosmos; see on *Caelestia* I.6 at n. 14, and also I.1 n. 57. Dercyllides (early first century A.D.; see on *Caelestia* at I.2 n. 9), cited at Theon *Expos.* 200.4–12, argues, much like Posidonius, that astronomers must adopt this theory from the natural philosophers. His position may reflect Posidonian ideas; see I. G. Kidd (1978a) 11 and *Comm.* 135.

37. Contrast the present paragraph, and lines 30–39 above, with Simplic. *In de caelo* 488.3–25. Here Simplicius similarly recognizes the existence of a true account of planetary motion that is based on physical theory, yet denies that this account, whatever form it may take, is the basis for definitively selecting the true astronomical account from a set of divergent theories. Instead, he sees astronomers as accounting for the phenomena of planetary motion on the basis of a few unexamined hypotheses about that motion. He does not claim that their divergent accounts must be reconciled with the single true physical (i.e., philosophical) account. His tolerance contrasts sharply with the position adopted in F18EK. Cf. also n. 10 above.

dance[38] of all [those bodies] is circular, with some revolving in parallel circles, others in oblique circles.

50: That, then, is how Geminus (or rather Posidonius [cited] in Geminus) transmits the distinction between natural philosophy and astronomy, and he takes his starting points from Aristotle.

38. *khoreia:* Kidd *Comm.* 133 compares Pl. *Tim.* 40c3. The point about a choral dance is that it involves intersecting circles.

GLOSSARY OF SELECTED TERMS

AETHER: *aithēr*
AIR: *aēr*
ANTIPODES: *antipodes*
APPEAR (BE SEEN): *phainesthai, phantazesthai*
APPEARANCE: *phantasia*
ARC: *periphereia*
ASSUME: *hupotithesthai*
ASSUMPTION: *hupothesis*
BODY: *sōma*
CAUSE (N.): *aitia, aition*
CENTER: *kentron, meson*
——, EXACT: *mesaitaton*
CIRCLE (N.): *kuklos*
——, ANTARCTIC: *antarktikos*
——, ARCTIC: *arktikos*
——, EQUINOCTIAL: *isēmerinos*
——, GREAT: *megistos*
——, HELIACAL: *hēliakos*
——, NORTHERN: *boreios*
——, SOUTHERN: *notios*
——, TROPICAL: *tropikos*
CIRCUIT (OF PLANETARY MOTION): *kuklos*
CIRCUMFERENCE: *periokhē*

CIRCUMHABITANTS: *perioikoi*
COEXTENSIVE WITH, BE: *sumparekteinesthai*
COINCIDENCE: *epharmogē*
CONCEIVE OF: *epinoein*
CONCEIVING, PROCESS OF: *epinoia*
CONJUNCTION: *sunodos*
CONTRAHABITANTS: *antoikoi*
COSMOS (= UNIVERSE): *kosmos* (see "heavens")
——, WHOLE: *ta hola*
COURSE (OF PLANETARY MOTION): *poreia*
CRITERION: *kritērion*
CULMINATE: *mesouranein*
CURVATURE: *kurtōma*
DAY, DAYTIME: *hēmera*
DEMONSTRATE: *deiknunai, epideiknunai*
DEMONSTRATION: *apodeixis*
DETECT, DETERMINE (BY OBSERVATION OR CALCULATION): *heuriskein*
DIAMETER: *diametros*
DISAPPEAR: *aphanizesthai* (see "sight, out of")
DISTANCE: *apostasis, diastēma*
DOCTRINE: *doxa*
EAST: *anatolē*
ECLIPSE: *ekleipsis*
ECLIPSED, BE: *ekleipein, ekleipsin poieisthai*
EFFECT (A MOTION): *poieisthai (kinēsin)*
ENCLOSE: *periekhein, perilambanein*
EQUINOX: *isēmeria*
ESTABLISH (A THESIS): *kataskeuazein, paristanai*
FIRE: *pur*
(1) FOOT WIDE: *podiaios*
HABITATION: *oikēsis*
HEAVENLY BODIES: *astra*
HEAVENS: *kosmos, ouranos*
HEIGHT: *hupsos*
HELIACAL: *hēliakos*
HOLDING-POWER: *hexis*
HOLD TOGETHER: *sunekhein*
HORIZON: *horizōn*

HYPOTHESIS: *hupothesis*

HYPOTHESIZE: *hupotithesthai*

ILLUMINATE: *lamprunein, phōtizein*

ILLUMINATION: *phōtismos*

IMAGINE: *epinoein*

INCORPOREAL: *asōmatos*

INDEFINITELY: *eis apeiron*

INTERVAL (OF A NIGHTTIME AND A DAYTIME): *nukhthēmeron*

INTERVAL (OF SPACE OR TIME): *diastēma*

LATITUDE: *klima; para* (with the dative case)

LENGTHENING: *auxēsis*

LIGHT: *phōs*

LIMITED: *peperasmenos*

LINE, STRAIGHT: *eutheia (grammē)*

LINE OF SIGHT: *opsis*

LUMINANCE: *lampēdōn*

MERIDIAN: *mesēmbrinos (kuklos)*

MONTH: *mēn*

MOON: *selēnē*

MOTION: *kinēsis*

——, based on choice: *proairetikē kinēsis*

MOVE (INTRANS.): *kineisthai, pheresthai*

NATURE: *phusis*

NIGHT, NIGHTTIME: *nux*

NORTH: *borras*

NORTHERN: *boreios*

NOTION: *ennoia*

NOTION, FORM A: *ennoein*

OBLIQUE: *loxos*

OBSERVE: *theōrein, tērein, horān*

OCCUPY: *katekhein, katalambanein*

OPPOSITION, BE IN: *diametrein*

OPPOSITION, IN: *kata diametron*

PERCEPTION: *aisthēsis*

PERIOD: *periodos*

PHASE (OF MOON): *phasis, skhēma*

PHENOMENA, THE: *ta phainomena*

PLACE: *topos*

PLANE: *epipedon*

PLANETS: *planētai, planōmena*

POINT: *kentron sēmeion*

POINTER: *gnōmōn*

POLE: *polos*

POWER: *dunamis*

PROBLEM (RAISE A): *aporia, aporein*

PROCEDURE: *ephodos*

PROTRUSION: *exokhē*

PROVE: *elenkhein*

PROVIDE: *parekhesthai*

PROVIDENCE: *pronoia*

RAY: *aktis*

RECEIVE: *dekhesthai*

REFRACTED, BE: *kataklasthai, periklasthai*

REFRACTION: *anaklasis*

RISE: *anatellein, anerkhesthai, aniskhein, anapheresthai*

RISE (N.): *anatolē*

SEASON: *hōra*

SECTION (OF A CIRCLE): *tmēma*

SEE: *horān*

SEND OUT (LIGHT, SHADOW): *apopempein*

SENSE PRESENTATION: *phantasia*

SET: *kataduein, kataduesthai, duesthai*

SETTING: *dusis*

SHADOW: *skia*

SHORTENING: *meiōsis*

SIGHT: *opsis*

SIGHT, OUT OF: *aphanēs* (see "disappear")

SIGN (ZODIACAL): *zōidion*

SIGNIFICATION: *sēmainomenon*

SIZE: *megethos*

SOLSTICE: *tropē*

SOUTH: *mesēmbria*

SOUTHERN: *notios*

SPHERE: *sphaira*

SPHERICAL: *sphairikos*

STADE: *stadion, stadios*

STAR: *astēr, astron*

STARS: *astra*

——, ALWAYS VISIBLE: *aeiphanē*

——, FIXED: *aplanē*

——, OUT OF SIGHT: *aphanē*

SUBSIST: *huphistasthai*

SUBSISTING, STATE OF: *hupostasis*

SUBSTANCE: *ousia*

SUN: *hēlios*

SUNDIAL: *hōrologion*

SURFACE: *epiphaneia*

TROPIC: *tropikos*

——, SUMMER: *therinos*

——, WINTER: *kheimerinos*

UNLIMITED: *apeiros*

VERTEX: *koruphē*

VISIBLE, ALWAYS: *aeiphanēs*

VOID: *kenon*

VOLUME: *onkos*

WANE: *meiousthai*

WANING: *meiōsis*

WATER CLOCK: *hudrologion*

WAX: *auxesthai*

WAXING: *auxēsis*

WEST: *dusis*

YEAR: *eniautos*

ZENITH: *koruphē*

ZODIAC, ZODIACAL BAND: *zōidiakos*

ZONE: *zōnē*

——, CONTRATEMPERATE: *anteukratos*

——, FRIGID: *katepsugmenē*

——, TEMPERATE: *eukratos*

——, TORRID: *diakekaumenē*

BIBLIOGRAPHY

For earlier literature on Cleomedes see Todd ed. *Caelestia* xxiii–xxv, and Todd (2004).

I. PRIMARY SOURCES

These are identified either in the List of Abbreviations at pp. xv–xvi above, or in headings in the *Index Locorum* at pp. 0–0. For studies that contain texts and translations see Part II of this bibliography under Aujac (Geminus), Burton (Euclid, *Optica*), Cherniss (Plutarch), De Lacy (Philodemus), Goldstein (Ptolemy, *Planetary Hypotheses*), Goulet (Cleomedes), Heath (Aristarchus), Kieffer (Galen), Maass (Achilles, and *Aratea*), Pease (Cicero, *De natura deorum*), Romeo (Demetrius of Laconia), Roseman (Pytheas), Schöne (Damianus), A. M. Smith (Ptolemy, *Optics*), M. F. Smith (Diogenes of Oenoanda), and Toomer (Ptolemy, *Almagest*). Basic information on the primary sources used in this study can be conveniently obtained from D. J. Zeyl, ed., *Encyclopedia of Classical Philosophy* (Westport, Conn., 1997).

II. SECONDARY WORKS

Where items are reprinted, references in the text are to the *original* publication.

Algra, K.
(1988) "The Early Stoics on the Immobility and Coherence of the Cosmos." *Phronesis* 33: 155–180.

Algra, K.

(1995) *Concepts of Space in Greek Thought.* (Philosophia Antiqua 65). Leiden.

(2000) "The Treatise of Cleomedes and Its Critique of Epicurean Cosmology." In Erler and Bees (2000): 164–189.

Aujac, G.

(1966) *Strabon et la science de son temps.* Paris.

(1975) *Géminos: Introduction aux phénomènes.* Paris.

Ayres, L.

(1995) Ed. *The Passionate Intellect: Essays on the Transformation of Classical Traditions Presented to Professor I. G. Kidd.* (Rutgers University Studies in Classical Humanities 7). New Brunswick and London.

Bake, J.

(1810) *Posidonii Rhodii Reliquiae Doctrinae.* Leiden.

Barker, P., et al.

(2002) Eds. *Astronomy and Astrology from the Babylonians to Kepler: Essays Presented to Bernard R. Goldstein on the Occasion of His 65th Birthday.* (= *Centaurus* 44).

Barnes, J.

(1983) "Ancient Skepticism and Causation." In Burnyeat (1983): 149–203.

(1989) "The Size of the Sun in Antiquity." *Acta Classica* (Universitatis Scientiarum Debrecensis) 25: 29–41.

(1990) "Pyrrhonism, Belief and Causation. Observations on the Scepticism of Sextus Empiricus." *Aufstieg und Niedergrang der römischen Welt* 2.36.4: 2608–2695.

Barnes, J., et al.

(1982) Eds. *Science and Speculation.* Cambridge.

(1988) Eds. *Matter and Metaphysics.* (4th Symposium Hellenisticum). Naples.

Bicknell, P. J.

(1984) "The Dark Side of the Moon." In Moffatt (1984): 67–75.

Bodnár, I. M.

(1997) "Alexander of Aphrodisias on Celestial Motions." *Phronesis* 42: 190–205.

Bowen, A. C.

(1999) "The Exact Sciences in Hellenistic Times: Texts and Issues." In Furley (1999): 287–319.

(2001) "La scienza del cielo nel periodo ptolemaico." In Petruccioli (2001): 806–839.

(2002) "Cleomedes and the Measurement of the Earth: A Question of Procedures." In Barker et al. (2002).

Bowen, A. C., and B. R. Goldstein.

(1991) "Hipparchus' Treatment of Early Greek Astronomy: The Case of Eudoxus and the Length of Daytime." *Proceedings of the American Philosophical Association* 135.2: 233–254.

(1996) "Geminus and the Concept of Mean Motion in Greco-Latin Astronomy." *Archive for History of Exact Sciences* 50: 157–185.

Brunschwig, J.

(1978) Ed. *Les Stoïciens et leur logique.* Paris.

(1988) "La Théorie Stoïcienne du genre suprême et l'ontologie Platonicienne." In Barnes et al. (1988): 19–127.

Burnyeat, M. F.

(1979) "Conflicting Appearances." *Proceedings of the British Academy* 65: 69–111.

(1982) "The Origins of Non-Deductive Inference." In Barnes et al. (1982): 193–238.

(1983) Ed. *The Skeptical Tradition.* Berkeley.

Burstein, S. M.

(1978) *The Babyloniaca of Berossus.* (Sources and Monographs on the Ancient Near East 1:5). Malibu.

Burton, H. E.

(1945) Tr. "The Optics of Euclid." *Journal of the American Optical Society* 35: 357–372.

Cherniss, H.

(1951) "Notes on Plutarch's *De Facie in Orbe Lunae.*" *Classical Philology* 46: 137–158.

(1957) Tr. *Plutarch's Moralia: Vol. XII.* (Loeb Classical Library). London and Cambridge, Mass. (Includes Plutarch, *De Facie in Orbe Lunae.*)

(1976) Tr. *Plutarch's Moralia: Vol. XIII:* 2. (Loeb Classical Library). London and Cambridge, Mass.

Chilton, C. W.

(1971) *Diogenes of Oenoanda: The Fragments.* Oxford.

Collinder, P.

(1964) "Dicaearchus and the 'Lysimachean' Measurement of the Earth." *Sudhoffs Archiv* 48: 63–78.

De Lacy, P. H., and E. A. De Lacy.

(1978) Eds. *Philodemus on Methods of Inference.* (Scuola di Epicuro 1). Naples.

Diller, A.

(1949) "The Ancient Measurements of the Earth." *Isis* 40: 6–9.

Donini, P. L. and G. F. Gianotti.

(1982) "La luce della luna in Apuleio, *De deo Socratis* 1.117–119 Oud." *Rivista di filologia e di istruzione classica* 110: 292–296.

Edelstein, L.

(1936) "The Philosophical System of Posidonius." *American Journal of Philology* 57: 286–325.

Erler, M., and R. Bees.

(2000) Eds., *Epikureismus in der späten Republik und der Kaiserzeit.* Stuttgart.

Furley, D. J.

(1989a) *Cosmic Problems.* Cambridge.

(1989b) "The Dynamics of the Earth: Anaximander, Plato, and the Centrifocal Theory." In Furley (1989a): 14–26.

(1993) "Some Points about Stoic Dynamics." *Proceedings of the Boston Area Colloquium in Ancient Philosophy* 9: 57–75.

(1996) "The Earth in Epicureanism and Contemporary Astronomy." In Giannantoni and Gigante (1996) 1: 119–125.

(1999) Ed. *The Routledge History of Philosophy. Volume 2: Aristotle to Augustine.* London.

Giannantoni, G., and M. Gigante.

(1996) Eds. *Epicureismo Greco e Romano: Atti del Congresso Internazionale Napoli, 19–26 Maggio 1993.* 3 vols. Naples.

Goldstein, B. R.

(1967) *The Arabic Version of Ptolemy's "Planetary Hypotheses."* (Transactions of the American Philosophical Society n.s. 57:4). Philadelphia.

González, P. P. F.

(2000) "Ératosthène de Cyrène." *Dictionnaire des philosophes antiques* 3: 188–236.

Gottschalk, H. B.
(1980) *Heracleides of Pontus.* Oxford.

Goulet, R.
(1980) *Cléomède: Théorie Élémentaire. Texte presenté, traduit et commenté.* Paris.
(1994) "Cléomède." *Dictionnaire des philosophes antiques* 2: 436–439.

Gratwick, A.S.
(1995) "Alexandria, Syene, Meroe: Symmetry in Eratosthenes' Measurement of the World." In Ayres (1995): 177–202.

Hahm, D.
(1977) *The Origins of Stoic Cosmology.* Columbus, Ohio.
(1978) "Early Hellenistic Theories of Vision and the Perception of Colour." In Machamer and Turnbull (1978): 60–95.

Hankinson, R.J.
(1994) "Galen and the Logic of Relations." In Schrenk (1994): 57–75.

Heath, T.
(1913) *Aristarchus of Samos: The Ancient Copernicus.* Oxford.

Hill, D.E.
(1973) "The Thessalian Trick." *Rheinisches Museum für Philologie* 116: 221–238.

Huby, P., and G. Neal.
(1989) Eds. *The Criterion of Truth: Essays Written in Honour of George Kerferd.* Liverpool.

Ierodiakonou, K.
(1993) "The Stoic Division of Philosophy." *Phronesis* 38: 57–74.

Inwood, B.
(1991) "Chrysippus on Extension and the Void." *Revue internationale de philosophie* 45: 245–266.

Jacoby, F.
(1958) *Die Fragmente der Griechischen Historiker,* II.C.1. Leiden.

Jones, A.
(1997) "Studies in the Astronomy of the Roman Period: I. The Standard Lunar Scheme; II. Tables for Solar Longitude." *Centaurus* 39: 1–36 and 211–229.
(1999) *Astronomical Papyri from Oxyrynchus.* 2 vols. in 1. (Memoirs of the American Philosophical Society 223). Philadelphia.

Kidd, D.
(1997) Ed. *Aratus: Phaenomena.* (Cambridge Classical Texts and Commentaries 34). Cambridge.

Kidd, I. G.
(1978a) "Philosophy and Science in Posidonius." *Antike und Abendland* 24: 7–15.
(1978b) "Posidonius and Logic." In Brunschwig (1978): 273–283.
(1989) "*Orthos Logos* as a Criterion of Truth in the Stoa." In Huby and Neal (1989): 137–150.
(1997) "What is a Posidonian Fragment?" In Most (1997): 225–236.

Kieffer, J. S.
(1964) Tr. *Galen's "Institutio Logica": English Translation, Introduction, and Commentary.* Baltimore.

Kuhrt, A.
(1987) "Berossus' *Babyloniaka* and Seleucid Rule in Babylonia." In Kuhrt and Sherwin-White (1987): 32–56.

Kuhrt, A., and S. Sherwin-White.
(1987) Eds. *Hellenism in the East: The Interaction of Greek and Non-Greek Civilizations from Syria to Central Asia after Alexander.* Berkeley and Los Angeles.

Lamberton, R., and J. J. Keaney.
(1992) Eds. *Homer's Ancient Readers: The Hermeneutics of Greek Epic's Earliest Exegetes.* Princeton.

Lloyd, G. E. R.
(1978) "Saving the Appearances." *Classical Quarterly* n.s. 28: 202–222; repr. with introd. in Lloyd (1991): 248–277.
(1982) "Observational Error in Later Greek Science." In Barnes et al. (1982): 128–164; repr. with introd. in Lloyd (1991): 299–332.
(1987) *The Revolutions of Wisdom: Studies in the Claims and Practice of Ancient Greek Science.* Berkeley.
(1991) *Methods and Problems in Greek Science.* Cambridge.

Lloyd, G. E. R., et al.
(1978) Eds. *Aristotle on the Mind and the Senses.* (Proceedings of the Seventh Symposium Aristotelicum). Cambridge.

Long, A. A.
(1988) "Socrates in Hellenistic Philosophy." *Classical Quarterly* n.s. 38: 150–171; repr. in Long (1996): 1–34.

(1992) "Stoic Readings of Homer." In Lamberton and Keaney (1992): 41–66; repr. in Long (1996): 58–84.

(1996) *Stoic Studies.* Cambridge.

Maass, E.

(1898) Ed. *Commentariorum in Aratum Reliquiae.* (Includes Achilles, *Isagoga*, and texts referred to as *Aratea.*) Berlin.

Machamer, P., and R. Turnbull.

(1978) Eds. *Studies in Perception.* Columbus, Ohio.

Mansfeld, J.

(1994) *Prolegomena: Questions to be Settled before the Study of an Author, or Text.* (Philosophia Antiqua 61). Leiden.

Mansfeld, J., and D. T. Runia.

(1997) *Aetiana: The Method and Intellectual Context of a Doxographer, Volume One: The Sources.* Leiden.

Marcovich, M.

(1986) Review of Theiler. *Gnomon* 58: 110–120.

Michaelides, S.

(1978) *The Music of Ancient Greece: An Encyclopedia.* London.

Moffatt, A.

(1984) Ed. *Maistōr: Classical, Byzantine and Renaissance Studies for Robert Browning.* (Byzantina Australiensia 5). Canberra.

Most, G. W.

(1997) Ed. *Collecting Fragments: Fragmente sammeln.* (Aporemata 1). Göttingen.

Mugler, C.

(1959) "Sur l'origine et le sens de l'expression καθαιρεῖν τὴν σελήνην." *Revue des études anciennes* 61: 48–56.

Neugebauer, O.

(1941) "Cleomedes and the Meridian of Lysimachia." *American Journal of Philology* 62: 344–347.

(1955) "Apollonius' Planetary Theory." *Communications on Pure and Applied Mathematics* 8: 641–648; repr. in Neugebauer (1983): 311–318.

(1959) "The Equivalence of Eccentric and Epicyclic Motion according to Apollonius." *Scripta Mathematica* 24: 5–21; repr. in Neugebauer (1983): 335–351.

(1975) *History of Ancient Mathematical Astronomy.* 3 vols. Berlin.

(1983) *Astronomy and History: Selected Essays.* New York and Berlin.

Newton, R. R.

(1980) "The Sources of Eratosthenes' Measurement of the Earth." *Quarterly Journal of the Royal Astronomical Society* 21: 379–387.

Pease, A. S.

(1955) Ed. *Marci Tulli Ciceronis De Natura Deorum.* Cambridge, Mass.

Pedersen, O.

(1974) *A Survey of the Almagest.* Odense.

Petruccioli, S.

(2001)Ed. *Storia della scienza: I. La scienza greco-romana.* Rome.

Plug, C., and H. E. Ross.

(1994) "The Natural Moon Illusion: A Multifactor Angular Account." *Perception* 23: 321–333.

Préaux, C.

(1973) *La lune dans la pensée grecque.* (Académie royale de Belgique: Memoires de la classe des lettres. Ser. 2, 61:4). Brussels.

Quine, W. V.

(1975) "On Empirically Equivalent Systems of the World." *Erkenntnis* 9: 313–328.

(1987) *Quiddities.* Cambridge, Mass.

Rawlins, D.

(1982a) "Eratosthenes' Geodesy Unraveled: Was There a High-Accuracy Hellenistic Astronomy?" *Isis* 73: 259–265.

(1982b) "The Eratosthenes-Strabo Nile Map: Is It the Earliest Surviving Instance of Spherical Cartography? Did It Supply the 5000 Stades Arc for Eratosthenes' Experiment?" *Archive for History of Exact Sciences* 26: 211–219.

Reinhardt, K.

(1921) *Poseidonios.* Munich.

(1926) *Kosmos und Sympathie.* Munich.

(1953) "Posidonius von Apamea." *Real-Encyclopädie* 22:1: 558–826.

Romeo, C.

(1979) "Demetrio Lacone sulla grandezza del sole (P. Herc. 1013)." *Cronache Ercolanesi* 9: 11–35.

Roseman, C. H.

(1994) *Pytheas of Massalia "On the Ocean."* Chicago.

Rosen, E.

(1981) "Nicholaus Copernicus and Giorgio Valla." *Physis* 23: 449–457.

Ross, H. E.
(2000) "Cleomedes (1st century A.D.) on the Celestial Illusion, Atmospheric Enlargement, and Size-Distance Invariance." *Perception* 29: 863–871.

Ross, H. E., and C. Plug.
(2002) *The Mystery of the Moon Illusion*. Oxford.

Sandbach, F. H.
(1985) *Aristotle and the Stoics*. (Cambridge Philological Society Suppl. Vol. 10). Cambridge.

Schnabel, P.
(1923) *Berossos und die babylonisch-hellenistische Literatur*. Leipzig.

Schofield, M.
(1978) "Aristotle on the Imagination." In Lloyd et al. (1978): 99–140.

Schöne, R.
(1897) Ed. *Damianos Schrift über Optik*. Berlin.

Schrenk, L.
(1994) Ed. *Aristotle in Late Antiquity*. (Studies in Philosophy and the History of Philosophy 27). Washington, D.C.

Schumacher, W.
(1975) *Untersuchungen zur Datierung des Astronomen Kleomedes*. Cologne.

Scott, D. J.
(1988) "Innatism and the Stoa." *Proceedings of the Cambridge Philological Society* Ser. 3, 33: 123–153.

Sedley, D. N.
(1976) "Epicurus and the Mathematicians of Cyzicus." *Cronache Ercolanesi* 6: 23–54.

Sharples, R. W.
(1990) "The School of Alexander?" In Sorabji (1990): 83–111.
(1994) Tr. *Alexander of Aphrodisias Quaestiones 2.16–3.15*. London and Ithaca.

Smith, A. M.
(1996) Tr. *Ptolemy's Theory of Visual Perception: An English Translation of the 'Optics' with Introduction and Commentary*. (Transactions of the American Philosophical Society n.s. 86:2). Philadelphia.

Smith, M. F.
(1993) Ed. and tr. *Diogenes of Oinoanda: The Epicurean Inscription*. Naples.

Sorabji, R.

(1988) *Matter, Space, and Motion: Theories in Antiquity and Their Sequel.* London.

(1990) Ed. *Aristotle Transformed: The Ancient Commentators and Their Influence.* London.

Taisbak, C. M.

(1973–74) "Posidonius Vindicated At All Costs? Modern Scholarship Versus the Stoic Earth-Measurer." *Centaurus* 18: 253–269.

Tarrant, H.

(1993) *Thrasyllan Platonism.* Ithaca and London.

Thorp, J.

(1990) "Aristotle's *Horror Vacui.*" *Canadian Journal of Philosophy* 20: 149–166.

Todd, R. B.

(1973) "The Stoic Common Notions." *Symbolae Osloenses* 48: 47–75.

(1976) *Alexander of Aphrodisias on Stoic Physics.* (Philosophia Antiqua 28). Leiden.

(1982) "Cleomedes and the Stoic Concept of the Void." *Apeiron* 16: 129–136.

(1984) "Alexander of Aphrodisias and the Case for the Infinite Universe." *Eranos* 82: 185–193.

(1985) "The Title of Cleomedes' Treatise." *Philologus* 129: 250–261.

(1988) Review of Sandbach (1985). *Ancient Philosophy* 8: 304–309.

(1989) "The Stoics and Their Cosmology in the First and Second Centuries A.D." *Aufstieg und Niedergang der römischen Welt* 2.36.3: 1365–1378.

(1992) "Cleomedes." *Catalogus Translationum et Commentariorum* 7: 1–11.

(1995) "Peripatetic Epistemology Before Alexander of Aphrodisias: The Case of Alexander of Damascus." *Eranos* 93: 122–128.

(2001) "Cleomedes and the Problems of Stoic Astrophysics." *Hermes* 129: 75–78.

(2004) "Physics and Astronomy in Post-Posidonian Stoicism: The Case of Cleomedes." *Aufstieg und Niedergang der römischen Welt* 2.37.5 (in press).

Toomer, G.
(1974–75) "Hipparchus on the Distances of the Sun and Moon." *Archive for History of Exact Sciences* 14: 126–142.
(1984) Tr. *Ptolemy's Almagest.* New York.

Toulmin, S.
(1967) "The Astrophysics of Berossos the Chaldean." *Isis* 63: 65–76.

Verbrugghe, G. P., and J. M. Wickersham.
(1999) *Berossus and Manetho: Native Traditions in Ancient Mesopotamia and Egypt.* Ann Arbor.

Wasserstein, A.
(1978) "Epicurean Science." *Hermes* 106: 484–494.

Whittaker, J.
(1987) "Platonic Philosophy in the Early Centuries of the Empire." *Aufstieg und Niedergang der römischen Welt* 2.36.1: 81–123.

Wolff, M.
(1988) "Hipparchus and the Stoic Theory of Motion." In Barnes et al. (1988): 471–545.

PASSAGES FROM CLEOMEDES IN COLLECTIONS OF TEXTS

BOOK ONE

I.1.3–7: *SVF* 2.529; **3–16**: F276 Th(eiler); **7–17**: *SVF* 2.534; **20–38**: *SVF* 2.537 (summary); **43–48**: *SVF* 2.537; **43–54**: F277 Th.; **64–67**: *SVF* 2.541; **68–74**: *SVF* 2.546; **68–80**: F278 Th.; **81–82, 89–103**: F278 Th.; **96–103**: *SVF* 2.540; **113–149**: *SVF* 2.538 (summary); **150–152**: *SVF* 2.557 (selections); **153–166**: *SVF* 2.557; **159–192**: F279 Th.

I.2.1–19: F280 Th.; **36–40**: F281 Th.

I.4.30–43: F282 Th.; **90–131**: F210EK; **90–131**: F283 Th.; **132–146:** F284 Th.; **197–213**: F285a Th.

I.5.114–145: F286 Th.; **128–134**: *SVF* 2.455

I.7.1–49: F287 Th.; **1–50**: F202EK; **121–I.8.18:** F288 Th.

I.8.79–99: F289 Th.; *SVF* 2.572; **158–162:** F19EK; F274 Th.

BOOK TWO

II.1.2–524: F290a Th.; **51–56**: F114EK; **269–286**: F115EK

II.3.61–II.4.107: F291 Th.

II.4.95–107: F123EK

II.5.1–7: F292 Th.

II.6.168–191: F293 Th.

II.7.1–4: F294 Th.; **11–14**: F275 Th.; T57EK

GENERAL INDEX

INDEX LOCORUM

Editions are not identified for major authors; cross-references to editions are to those cited in Part I of the bibliography. *CAG* = *Commentaria in Aristotelem Graeca; Suppl. Ar.* = *Supplementum Aristotelicum.*

Compositor: Integrated Composition Systems
Text and Display: Janson
Printer and Binder: Sheridan Books, Inc.

ng ethnic identity according to situational circumstance. In the art trade, hants manipulate their ethnic identity according to what they perceive as ng economic advantage. Although, as noted earlier, many of the traders in lateau market place in Abidjan are Wolofs from Senegal they will often urists that they are from Côte d'Ivoire and usually, more precisely, that are Baule. Wolof traders claim to be from Côte d'Ivoire in order to satisfy ourist's quest for authenticity. Traders say that tourists prefer to buy enirs from Ivoirian merchants – the authenticity of the experience being tened if they buy from "local" or "native" sellers (cf. MacCannell 1976: 60). The reason many Wolofs claim specifically to be members of the e ethnic group is twofold. First, since much of the art that tourists buy is d in the Baule style (e.g., wooden Baule face masks and statues), a r's claim to Baule ethnicity makes him a more legitimate spokesman for bjects he sells. Second, since it is a well-known fact that the President of Republic of Côte d'Ivoire, Félix Houphouët-Boigny, is of Baule heritage, laiming Baule identity the trader is also drawing a symbolic connection een himself, the President, and the state. When making a sale, traders will etimes make direct reference to the President. If they are trying to sell a e mask, for example, they might say, "This is a mask from the President's ge." Or, when selling a Dan doll, I once overheard a trader telling a p of tourists, "These are [like] the dancers that came out to entertain the ident. I saw them on television" (10/11/87). The practice of "ethnic arication" is so widespread in the market place, that even a popular Côte oire guide book finds reason to warn tourists about the manipulation of nal and ethnic identities: "[M]any of the Senegalese try to pass themselves s Ivoirians in order to lend authority to their pittance of knowledge about ian art" (Rémy 1976: 89, my translation from the French).[17]

raders who do not try to hide their Senegalese identity – i.e., those who it openly to being Wolof – emphasize instead in the banter of bargaining "pan-African" identity. Remarks of this nature include: "We are all cans in the marketplace"; "We are all brothers"; "All Africans are the e." When I interviewed Mulinde Robert, a Nigerian merchant who now in the United States, I asked him why he became involved in the African rade. He answered, succinctly, "Well first of all, I am African" (9/12/89).

The fabrics of identity

ause the authenticity of an African art object is often measured not only he quality of the object itself but also by the characteristics (credibility, tation, knowledge, appearance) of the person selling the object, African art ers who have extensive contact with Western buyers sometimes choose their e carefully in order to heighten or underscore their status as "authentic"

Chinese in Southeast Asia, the Asians in East Africa, the Syrians and Lebanese in North and West Africa, and the Jews in Medieval Europe provide some of the best documented examples (Cohen 1971; Foster 1974; Curtin 1984). African ethnic groups who have migrated to West Africa from other parts of the continent also play a crucial role in all aspects of trade and marketing. Among the best-known West African ethnic groups specializing in cross-cultural commerce are the Wolofs of Senegal, the Hausa of northern Nigeria and southern Niger, and the Dioula or Malinké of the western Sudan.[12] These groups have migrated throughout the globe, conducting their business – as ethnic minorities – in all parts of Africa, Europe, North America, and (more recently) Southeast Asia and Japan. All of these groups, not surprisingly, control major shares of the contemporary African art market.[13]

One of the reasons ethnic minorities play such a prominent role in transnational commerce is that their alien status dismisses them from many of the social obligations or constraints which could potentially hinder the successful operation of economic exchange. If, for example, a trader were fully part of a village society and subject to the controls and moral obligations of the community, he would be expected to be generous in the "traditional way" to those in need. It would be difficult for him to refuse credit, for example, and it would not always be possible to collect debts. In short, it would be hard to reap a profit from the very network of kin, neighbors, and friends on whom a trader's life was dependent and with whom he was socially and culturally intermeshed (Bonacich 1973: 585; Foster 1974: 441).

Although ethnic minorities are involved, for different reasons, in a huge variety of commercial enterprises, a trader's status as an outsider is almost a necessary precondition for participating in the African art trade. If members of the local population attempted openly to buy and sell art objects on their own, they would inevitably face the scorn of the community. In many cases, they would probably be punished severely if they were found selling sacred objects.

As a means of diverting local outrage, village elders who are forced, by economic need, to sell sacred goods to traders often report to their community that the pieces were stolen by the traders. Traders say that as a result of this practice some itinerant merchants have even been killed by local populations. In one exceptional case with which I am familiar, a village community reported the theft of a mask to the local police. The matter was brought to authorities in Abidjan who were able to trace the mask to a Wolof trader who had already left for France with the object. The trader's associates in Abidjan were able to contact him in Paris before the mask had been sold, and the object was shipped back to Abidjan and returned to the village. The Wolof trader was never reimbursed for the cost of the mask, but he avoided a serious legal confrontation (which he would almost surely have lost). This is the only widely reported instance of its kind. In this case, extensive media coverage and the Ivoirian government's recent campaign for nationalism through the preservation of local artistic traditions (see below) were largely responsible for the return of the sacred icon.

Outsiders in inter-ethnic commerce have the advantage of being able to leave an area quickly if they are caught in illicit or sacrilegious activities. Furthermore, they have the advantage of not being enmeshed in the moral or religious fabric of the local community with whom they trade. As Georg Simmel once put it, the itinerant trader, whose status in a community is defined as that of outsider or stranger, "is not tied down in his action by habit, piety, and precedent" (1950: 405; cf. Levine 1979). Abdurrahman Madu said:

> In the past, most Ivoirians who came to the market place were afraid to touch the masks. They thought that they would die if they came in contact with them. To these people the masks were fetishes. For those of us who sell masks, however, we can touch them all we want. We don't give a damn. Even if they tell me that there's a certain kind of mask that prevents you from sleeping at night, I could take it home with me and use it as a pillow. Because I don't believe in these things. You have to believe in something in order for it to be effective. If you don't believe in it, then nothing will ever happen to you. (6/25/91)

Since a disproportionate number of art traders – whether they be Wolof, Hausa, or Mande – are Muslim in faith, their commercial activities are not hindered by conflicting ethical interests or religious beliefs. Immersed solely in the economics of the trade, Muslim traders are completely detached from the spiritual aspect of the objects they sell. As Alhadji Kabiru stated when discussing a Baule monkey figure (*asri kofi*) that was displayed in his stall: "To me this thing represents money. To the Baule it's a god. A Baule could die if he touched this. Yet to me it means absolutely nothing. It's simply a piece of wood – no different from this countertop [in my stall]. It means nothing to me . . . To me this is just money, it's not a fetish" (6/19/91).

Muslim traders have no interest in collecting the African art they sell. Most traders decorate their home with Western products – images of urban prosperity clipped from newspapers and magazines, wall calendars with glossy pin-ups, colorful posters, or banners supporting African and European sports teams. As Malam Yaaro stressed in one of our conversations, "If you see a Muslim selling [African] art, you can be sure there is a non-Muslim somewhere who is buying it. I would never spend money on art just to display it on a table . . . I wouldn't want my children to see it or to know anything about it" (6/22/91).

Indeed, many of the art traders are critical of the populations from whom they buy art – viewing them variously as pagan, idolatrous, animistic, and superstitious.[14] As Yaaro went on to say:

> These people speak to their [art] objects but they never
> object somewhere for a year or so, and then return, you'll
> hasn't eaten, it hasn't drunk, it hasn't spoken. But if the ob
> would speak back when spoken to! Only the people who
> things in this way. We Muslims have never done that. (6/2

Both the Qur'an and the hadith[15] take a decidedly an
view toward representational art – and, in particula
the creation of idols and their use in religious practic
René Bravmann notes, "is uncompromising on the s
art, and its judgments are leveled not only at all t
that is, painters and sculptors – but also at all type
which representational forms are possible" (1974: 1
condemns the fashioning of representational art an
forms, the activities of art traders should not only b
with their own religious beliefs, but indeed the deco
materials could actually be looked upon favorably
traders.[16]

Situational ethnic identity

In his classic monograph *Political Systems of H*
Edmund Leach argued that ethnicity is based on subje
cal ascription that have no necessary relationship to
of cultural discontinuities (cf. Bentley 1987). His argu
as a direct challenge to socio-geographic models of "
Melville Herskovits (1930) on the mapping of ethni
which complexes of cultural traits that are identified
to map cultural or ethnic divisions within a prescri
Subsequent to Leach's work on ethnicity and ethnic gr
area approach has largely fallen into disfavor amo
Fredrik Barth concluded in his work among the Swat
'culture areas' . . . becomes inapplicable. Different et
types will have overlapping distributions and disconf
1088).

The insight that ethnicity is claimed rather than p
primordial way) has been widely adopted in social sci
identity (Wallerstein 1960; Barth, ed. 1969; Skinner 1
Van Binsbergen 1981; Maybury-Lewis 1984; Comaro
Such studies view ethnicity variously as a conscious e
economic interest, as a fiction constructed by lead
followers, or as the by-product of a dynamic process
The departure from fixed ethnic attributes allows

African traders.[18] Thus, many of the traders have taken to wearing long flowing robes made of cotton damask and finely embroidered silk which have come to symbolize the traditional style of Muslims in Africa (cf. Mazrui 1970).[19] The fact that Islam is (theoretically) incompatible with the representational art which the traders sell, does not seem to undermine the effectiveness of their clothes in communicating to tourists their role as legitimate *Africans.*

In subtle ways, traders also use personal adornment to differentiate themselves from other Africans in the market place. Many of the more successful traders, for instance, especially in the Plateau market place, have attached small Akan brass pendants to the key rings that hold their car keys and/or the keys to the padlocks on their storage trunks. The traders explain that by continually handling the pendants which hang from their key rings, they give the brass a shiny patina which imitates indigenous handling and age. After several months, the traders say, they can saw off the loop on the pendant and sell the object as though it were a fine old Akan goldweight. During the period of field research, however, I never witnessed a trader removing a pendant from his key ring, let alone sawing off a pendant's loop. In fact, many traders could point to the various pendants on their ring and recount when they purchased them and for how long they had been there (usually several years). Rather than a mechanism for artificial patination, I would postulate, the pendants ought more correctly to be regarded as symbolic markers which differentiate the stallholders and prosperous market-place traders from the suppliers and other Africans who frequent the market place. Many of the traders displayed their key rings rather ostentatiously by allowing the pendants to hang from their trouser or robe pocket.

Traders returning from Europe or America symbolize their high status as international dealers by wearing expensive European-tailored clothes, fine leather shoes or trendy imported "high-back" sneakers, expensive watches, designer sunglasses, and heavy gold chains. When the traders walk through the market places, looking to buy supplies for their next trip abroad, their status as successful international merchants is clearly signaled by their fancy Western attire. It is these same dealers, ironically, who sometimes switch to wearing traditional Muslim robes when they are in Europe or America selling to Western buyers.

The construction of ethnic artifacts

At the demand end of the African art market, ethnicity functions as a form of commodity – which can be packaged, marketed, and sold to foreign buyers. Many of the art objects traded in the market places are classified according to their ethnic style: *Dan* masks, *Senufo* figures, *Baule* combs, etc. In most cases, it is ethnicity alone which is the most intrinsic element in the definition and classification of art objects. When tourists buy art in the market place, one of the first questions they ask the trader is, "What tribe is this from?" Some traders are able to recognize, with great accuracy, the ethnic attribution of an art object. Others, however, have not the slightest clue where an object may be from. Whether or not a trader knows in which ethnic style an object is carved, he will *always* provide the buyer with an ethnic attribution – i.e., failure to do so might jeopardize the sale.

For those who are unable to recognize ethnic attributions, the tendency is to identify *everything* as Baule: "It's Baule, just like the President" (11/20/87). Objects which are asymmetrical or unusual in form are often identified as Lobi (since most Lobi art is of distorted or unequal proportions). I once asked a stall apprentice to identify a very poorly executed Dan mask (it was clearly an attempt to copy the form of a Dan mask, but was probably carved by a novice workshop-artist from another part of West Africa). The trader examined the mask and said that it was "Dan-Lobi." Clearly, he recognized the artist's attempt at capturing a Dan aesthetic, but, at the same time, he attributed the mask's lopsidedness to the work of a Lobi. When I asked him how the Dan and the Lobi could come together, since they were separated by so much geographic distance,[20] the trader said, without skipping a beat, "No, this is from before-before [*avant-avant*] when the two groups were really one" (2/10/88).

I would argue that the importance of ethnicity in the classification of art objects in the market is largely a result of the dissemination of Western scholarship and its particular vision of African art. Until fairly recently, most publications on African art were organized around the theme of "style regions" which divided the arts of Africa into discrete and easily identifiable ethnic zones. In her critique of this system of organization, Sidney Kasfir summarizes what is sometimes called the "one tribe, one style" paradigm by noting that according to this view "every tribe is a universe unto itself and . . . furthermore, the art of one tribe is quite meaningless to the members of another tribe" (1984: 171). Historian Jan Vansina offers a similar critique: "The identification by 'tribe' rested on the ingrained European belief that each then-catalogued ethnic group differed from all others in its customs and especially, apparently, in its visual arts, while all members of the 'tribe', on the other hand, wrought art objects in the same style" (1984: 29; see also Ravenhill 1988). The idea that there exists a direct relationship between ethnic identity and aesthetic style grows out of a Durkheimian assumption that "primitive" art is the creation of a collective mind instead of an individual artist.

In many publications on African art, writers have tended to use a core group of objects (from Western museums and private collections) which have come to symbolize, through repeated use in different publications, the quintessential aesthetic forms of individual ethnic groups.[21] Anything which deviates from these "accepted" archetypes are judged in the art market as either inauthentic or unsaleable (see Chapter 5).

Hence, within the narrow parameters of minimal aesthetic deviation, an object is made to stand for the art of an entire ethnic group or culture. The result of this type of classification, Clifford has observed in a slightly different context, is to "create the illusion of adequate representation of a world by first cutting objects out of specific contexts (whether cultural, historical, or inter-subjective) and making them 'stand for' abstract wholes – a 'Bambara' mask, for example, becoming an ethnographic metonym for Bambara culture" (1988: 220; see also Stewart 1984: 162–65). Thus when a buyer purchases an object from a particular ethnic group, s/he is carrying away a slice of an entire cultural system.

The stereotypical view of ethnic attributions, which underlies the classification of art in both the tourist and collectible markets, constitutes a process which Dean MacCannell (1984) has termed "reconstructed ethnicity" and one which Nelson Graburn (1984) has called "secondary ethnicity." In this type of ethnic identity formation, the principal goal is for "the maintenance and preservation of ethnic forms for the entertainment of ethnically different others" (MacCannell 1984: 385). Ethnic attributes become commodities which can be bought and sold on the global market – ethnicity being reduced to a kind of typology of bare essences. In a famous essay in *Mythologies*, Roland Barthes linked this type of ontological reductionism to the demands of the tourist industry itself – a special brand of cultural knowledge which is produced and reproduced in tour books and brochures. "For the *Blue Guide*," writes Barthes, "men exist only as 'types'. In Spain, for instance, the Basque is an adventurous sailor, the Levantine a light-hearted gardener, the Catalan a clever tradesman and the Cantabrian a sentimental highlander. We find again here this disease of thinking in essences, which is at the bottom of every bourgeois mythology of man" (1982: 75). Ethnic groups, like their arts, can be colorfully mapped out on a poster from the Ministry of Tourism.

Masks, ethnicity, and the state

The Ivoirian government has had very little involvement in the African art market. In 1982, the Ministry of Tourism opened a branch, called the Direction de la Promotion de l'Artisanat d'Art (DPAA), which specializes in the promotion of Ivoirian art. The office published an eighty-page booklet, *Artisanat d'art de Côte d'Ivoire* which contains a wide selection of indigenous arts and crafts – everything from carved wooden masks to woven rattan floor lamps. The office was intended to facilitate access for foreign buyers to Ivoirian souvenirs and arts. As the booklet explains, "The National Office of Arts and Crafts exports its products to the whole world and may be consulted at any time about the authenticity of Ivoirian Handicrafts" (Anon. n.d.a: 9–10). The DPAA was expected to increase the profits of Ivoirian artists by cutting directly into the earnings of professional art traders. This too is made clear in the introduction to the booklet: "We inform you that the National Office of Arts and Craftsmen [sic] is a promotion office which initiates and coordinates the different commercial and technical actions between craftsmen and customers. Thus, due to Government subsidies the Office profits are low and reduced as compared to the profit margins of private dealers" (Anon. n.d.a: 10).

On the whole, the DPAA has been completely ineffectual in achieving its goals. Foreign buyers have not chosen to place their orders through the government agency, but have continued to rely on their contacts in the market places. One of the unintended consequences of the publication of the DPAA booklet was that traders in the market place have used the booklet as a means of authenticating the objects they sell. Rather than cut into their profits, therefore, the DPAA has actually helped traders in their work: "See," said a Wolof merchant to a French tourist while pointing to the government publication, "it's just like the one they published in this book" (3/3/88).

The Ivoirian government has made a second unsuccessful foray into the African art world through its planning and organization of large-scale masked festivals. These public "masquerades" were intended to fulfil the government's dual projects of (1) promoting international tourism in light of the country's most severe economic recession,[22] and (2) fostering national unity in the face of growing ethnic factionalism and tension. Although, as I shall argue, the ideological frameworks underlying these two goals are in many ways diametrically opposed to one another, the use of masks and masked dancing is an attempt on the part of the Ivoirian state to bridge the differences between these two nation-stabilizing strategies and mute their potential contradictions.

Masks and masking in Côte d'Ivoire are found in different forms in a variety of coastal and inland communities. Many of the estimated sixty ethnic groups in the country have their own style of mask carving and their own repertoire of masked dancing and performances. Although some aspects of masking are shrouded under a veil of secrecy and used only in the context of secret society activities, many forms consist largely of public displays intended purely for general entertainment. While these secular forms of masking are often carried out at the local level, they are sometimes incorporated into public events organized by members of both regional and national government. A meeting of town mayors, a visit to a village by a district (*préfecture*) administrator, or a national tour by a high-ranking minister or diplomat are all events that would call for the performance of a masked festival. Although certain forms of secular masking probably found expression at the village level in pre-colonial times, I would argue that most public displays of masking became associated with political and bureaucratic events during colonial rule. Huge masked festivals, for example, were organized each summer by the French to celebrate Bastille Day, while smaller masked festivals were often held at the ground-breaking reception for the construction of administrative buildings, at official ceremonies for the naming of city streets, or at the unveiling of colonial monuments (see Gorer 1935: 322–28).

Together with their function in national politics, masks and masking in post-colonial Côte d'Ivoire have, in recent years at least, played a critical role in the promotion of international tourism and the marketing of African art. Within the last decade, the mask has been appropriated by the Ivoirian state as a symbol of national identity or character.[23] As Duon Sadia, the Ivoirian Minister of Tourism, noted in a 1987 interview: "Because Côte d'Ivoire does not possess pyramids or grand ancient monuments like Egypt or Mexico, and because it does not have an abundance of wildlife like some of the countries in East Africa, Côte d'Ivoire has chosen to promote itself through its only indigenous product, Ivoirian man himself – with his culture and his traditions, of which masks and masking are an integral part" (Bouabré 1987b: 8). In another interview, the Minister of Tourism further clarified the specific function of masks in the development of the modern Ivoirian polity by noting that:

> We now declare that the trademark [of Côte d'Ivoire] will be the mask, for it is representative of this country, rather pleasing to the eye, and enshrouded with an air of mystery. The mask could arouse the curiosity of foreign tourists and lead them to visit our country. We have [therefore] chosen the mask for we believe that it integrates several aspects of our culture and civilization. The mask encapsulates the traditional arts of Côte d'Ivoire, and represents the strength and history of our nation. (Philmon 1982: 13)

The promotion of tourism through the marketing of the image of the mask represents, in point of fact, a radical departure in the rhetoric of the Ivoirian state. Less than a decade before this recent campaign, for example, Félix Houphouët-Boigny, President of the Republic and founder of independent Côte d'Ivoire, declared to a congress of the National Democratic Party: "We are fed up with having Africa relegated, through the futile gaze of the observer, to a land of sunshine, rhythms, and innocuous folklore" (quoted in Boutillier, Fiéloux, and Ormières 1978: 5). For Houphouët-Boigny, in his first years of power after independence, both national integration and international economic success were to be found in the promotion of modern industrial technologies rather than in a return to traditionalism or the recreation of a "primitivist" aesthetic.[24]

Hence, in light of this philosophy, how can one explain the state's sudden shift toward traditional cultural resources and, in particular, its appropriation of the mask as a symbol of national, multiethnic pride? I would argue that this return to traditionalism is a direct result of the nation's financial collapse following the failure of its cash-crop export economy – beginning sometime in

1980 (Brooke 1988). That is to say, as long as Côte d'Ivoire enjoyed economic prosperity through its production and export of cacao and coffee, the state used its success in the international economy as a device for rallying nationalist sentiment. It needed nothing else. Following the dramatic collapse of the price of cacao and coffee in the world market, however, politicians scrambled to find not only a new source of foreign income but also a new gathering point for nationalist sentiment. The mask was thought by some to be capable of achieving both. On the one hand, it fueled the Western imagination through its mystery and exotic appeal. On the other hand, it reconciled growing ethnic divisions by elevating the symbolism of the mask – with its plethora of ethnic styles and interpretations – to a single, national icon.

The first attempt by the government of Côte d'Ivoire to promote tourism and national solidarity through the use of African art was the festival of masks in the town of Man which was held on April 14–15, 1979. The festival was organized by Bernard Dadié, the Minister of Cultural Affairs. On the whole, the festival was poorly attended, and it received very little coverage from the Ivoirian press (only three short articles in the semi-official daily newspaper *Fraternité Matin*).

The second masked festival was organized by the Minister of Tourism, Duon Sadia. It too was held in the town of Man from February 11–15, 1983. In the second festival at Man, there was a more overt effort on behalf of the government organizers to use African art as a symbol of Côte d'Ivoire and as a mechanism for attracting the financial benefits of tourism. The masked festival at Man, Duon Sadia noted at a press conference held at the luxurious Hotel Ivoire in Abidjan, "will be the equivalent of Carnival in Rio, with an added element of the profound soul and mystery of 'non-commercialized' Africa" (Anon. 1983: 19).[25] The 1983 festival of the masks at Man was again reported by the press to be an overall failure. Very few tourists went to the festival. And the mask-bearers, who felt they were being treated without sufficient respect, refused to appear on stage. A delegation, consisting of three national ministers and a district representative, had to plead in public with the masked dancers to come out and perform for the small gathered crowd (Djidji 1983: 11).

The Ivoirian state's appropriation of the mask reached its epitome in the summer of 1987, when the Ministries of Tourism and Culture jointly organized a national masked festival. Promoted under the name "Festimask," the festival was funded by the state at an estimated cost of $500,000. Unlike previous state-sponsored masked festivals which were organized by district administrators with the exclusive participation of local ethnic groups, the Festimask attempted to bring all the ethnic groups of Côte d'Ivoire into a single event which, not surprisingly, was held in the President's home town of Yamoussoukro, in the center of the country.[26] The location of the festival was moved from Man to Yamoussoukro for several purposes.[27] The official reason reported in the national newspaper for holding the festival in Yamoussoukro was because of its proximity to the economic capital and port city of Abidjan – thereby, the argument went, encouraging more expatriates and more tourists to attend the festival of masks. However, the unstated reason for the site of the event, I would argue, was to link the festival of masks and, more generally, the symbolism of masks and masking to the national government through its association with Houphouët-Boigny's natal village and place of retreat.

When the masked festival was moved to Yamoussoukro in 1987, it became not only a vehicle for promoting international tourism, it was also used as a means of stressing national unity. Since the end of the colonial period, many burgeoning African nations have had to push for national unity in the face of internal ethnic factionalism. Although cultural pluralism may be profitable within the realm of the international art market, it is often perceived as a major obstacle in the domain of centralized state politics. As Immanuel Wallerstein noted in 1960, "The dysfunctional aspects of ethnicity for national integration are obvious. The first is that ethnic groups are still particularistic in their orientation and diffuse in their obligations . . . The second problem, and one which worries African political leaders more, is separatism, which in various guises is a pervasive tendency in West Africa today" (1960: 137–38).

Until recently, post-colonial Côte d'Ivoire had a history of successful national integration. In a country made up of approximately sixty different ethnic groups, this record of success is an impressive triumph. One of the reasons which accounts for successful integration of ethnic groups in Côte d'Ivoire is the rapid growth and expansion of the Ivoirian economy – the so-called Ivoirian "miracle" which took place from 1960 to 1980.[28]

Because a majority of Ivoirian nationals were reaping the benefits of favorable transnational trade, it was to their (economic) advantage to remain united under a national economic cause (Dozon 1985: 53–54). Since the economy has weakened, however, in the past several years, it could be argued that ethnic factionalism has become an increasing concern to the representatives of the centralized Ivoirian state. Viewed in this context, then, the masked festival at Yamoussoukro was yet another way of promoting nationalist sentiment in the face of growing ethnic factionalism. The Festimask respected ethnic heterogeneity, i.e., each masked performance was associated with a different and unique ethnic style, while, at the same time, it brought disparate ethnic groups together into a single, united cause.

The Festimask stresses national unity in at least two ways. First, it aims to bring the ethnic distinctions embedded in styles of art into a single "folkloric" category. All masks, said the organizers of the festival, are to be thought of as members of the PDCI (Partie Démocratique de Côte d'Ivoire). And, all masks are to be considered Ivoirian patriots struggling for the good of the modern nation-state (Gnangnan 1987). Secondly, the festival of masks strives to bring the concerns of the older generation (the so-called *mentalitées traditionelles* of the rural population) into step with national concerns, such as the promotion of international tourism and the President's long-standing campaign for West African regional peace. In the context of Festimask, the mask is a tool of the modern nation-state that serves "rational" political goals while being presented to both nationals and foreigners as a kind of "traditionalizing instrument" (Moore and Myerhoff 1977: 8–9; see also Ranger 1983). At a press conference held to clarify the role of the mask in the nationalist party, the Ivoirian Minister of Tourism, Duon Sadia, said:

> When we say that the mask must become militant, we mean to signal that the mask must no longer transmit the knowledge of the ancestors in a mechanical way without any explanations. The mask must become a spokesman – communicating in the common language of our culture – for the message of peace. The performance [of Festimask] is not intended to caricature our traditional values, but rather it is aimed to preserve these traditions by adapting them to the exigencies of the modern world. (Bouabré 1987b: 8)

The Festimask was thus intended to collapse divisions in *both* space (i.e., ethnic geography) and time (i.e., generational differences).

According to Ernest Gellner, there are at least three pre-conditions for the flourishment of state nationalism: (1) that a population be culturally homogenous without internal ethnic sub-groupings; (2) that a population be literate and capable of authoring and propagating its own history; and (3) that a population be anonymous, fluid, mobile, and unmediated in its loyalty to the state (1983: 138; see also Handler 1988). International tourism in most of the developing world hinges on the exact opposite criteria from those which underlie the foundation of state nationalism. First, international tourism demands that a population be as culturally and ethnically diverse as possible. In Côte d'Ivoire, for example, the tourist art market is driven by the production of a large variety of supposedly autochthonous and stereotyped ethnic arts.[29] Second, international tourism seeks to discover a population that is *il*literate, and without a sense of historical knowledge or a proper understanding of its geographic place within the world system. And third, international tourism calls for the existence of small-scale populations in which there is no anonymity, in which whole societies recognize each and every one of their members, and in which long-distance communication is not possible among putatively isolated groups. In essence, therefore, the demands of state nationalism and the demands of international tourism are situated at opposite poles in the realm of possibilities concerning the individual's relationship to society.

The organization of Festimask was an attempt by the Ivoirian government to satisfy *simultaneously* both the monolithic requirements of effective state nationalism and the polymorphic demands of successful international tourism.

By elevating the mask to a national icon, the state was attempting (1) to subvert ethnic differences; (2) to emphasize an indigenous form of national literacy and ethnohistorical consciousness; and (3) to create a national category of aesthetic identity through the hidden and anonymous face of the mask. At the same time, however, the state was also trying to encourage international tourism by stressing both the visual diversity in ethnic material productions and the exoticism of the masked dance itself.

Although the aims of the Festimask were both complex and diverse, its results were unambiguous. Tourists, art traders, and nationals all judged the event as a complete failure. Tourists stayed away from the Festimask because, I was told by one, they anticipated a large, staged, "tourist" event. The art traders who had attended the festival in order to ply their wares to tourists were very disappointed by the poor turn out. And nationals were disgusted with the Festimask because they felt they had been treated without respect – like pawns in a commercial venture. As one of the elders who attended the Festimask put it to a representative from the Ministry of Information:

> My son, we went to Yamoussoukro, and we were happy for we had been invited to the village of our President . . . But you should know that nobody took care of us; nobody even provided us with food, and that just isn't normal. Not only were we not greeted by the organizers of the festival, as is the custom, but when we [finally did get some food] it was their leftovers that we were sent to eat. (Anon. 1987b)

I would argue, in sum, that the masked festival failed in the eyes of both Ivoirian nationals and foreign tourists for the same reason. In both instances, the Festimask was viewed as an inauthentic event because it had been, as it were, too "modern" in its tactics and too insensitive to the demands of "custom." The appropriation of the hidden face by the hidden hand resulted in a particular form of the commoditization of ethnicity, in which neither the producers nor the consumers were willing to strike a bargain.

5 The quest for authenticity and the invention of African art

I understand you can't buy real African art anymore.

An American tourist in Plateau market place, Abidjan (1987)

The concept of "authenticity" is among the most problematic and most difficult issues in the study of African art. Yet, despite its central relevance, and the frequent use in the literature of such terms as "real," "genuine," and "authentic," the subject of authenticity has received surprisingly little attention by scholars in either the fields of anthropology or art history.[1] The definition of authenticity in African art draws upon connections among such disparate issues as cultural or ethnic "purity," historical timing or periodicity, and artistic or commercial intentionality. In this chapter, I will review the definitions of authenticity in the literature on African art which have been set out by Western academics, dealers, and collectors. I will then consider the way in which African art merchants have interpreted the Western notion of authenticity, and how, in particular, they have tailored the concept to fit their own understanding of African art. In addition, I will discuss the fascination with age in the Western appreciation of African art, and the artificial reproduction of antiques which results from this fascination. I will examine both the classification of aesthetics and the construction of the category "art," and their relevance to the contemporary African art trade. I will analyze the recirculation of "trade beads" as an example of how the familiar gets transformed into the exotic. And, I will end with some brief thoughts on, what I call, the "crisis of mis-representation" in modern transnational exchange.

The authenticity of African art

The definition of authenticity in African art that is most current among academics and dealers alike combines elements concerning an object's condition and history of use, intended audience, aesthetic merit, rarity, and estimated age. Thus, an authentic piece of African art, dealer Henri Kamer tells us, "is by definition a sculpture executed by an artist of a primitive tribe and destined for the use of this tribe in a ritual or functional way" (1974: 19). In defining authentic African art, it is commonly asserted that there should be no intention of economic gain on the part of the artist. "The sculptor who creates these fetishes and masks," Kamer goes on to say, "does so without any thought of profit, in the same spirit that an inhabitant of the Cyclades executed an idol in marble 5,000 years ago" (1974: 19). Or, using a slightly more veiled language, African art connoisseur Raoul Lehuard has said that, "In order for a sculpture to be authentic, it must not only be derived from a formal truth, but its language must also be derived from a sacred truth" (1976: 73). The age of an object generally conditions its commercial viability. "In judging authenticity," writes American collector/dealer Herbert Baker, "except for a few recently discovered tribes or areas, pre World War II seems to have validity inasmuch as there was little or no commercial market" (1973: 7).

Drawing on similar themes, but using a rather more "technical" form of discourse, African art scholar Malcolm McLeod defines an "authentic" African art object as "any piece made from traditional materials by a native craftsman for acquisition and use by members of local society (though not necessarily members of his own group) that is made and used with no thought that it ultimately may be disposed of for gain to Europeans or other aliens" (1976: 31). And, in a similar vein, African art appraiser Carl Provost is of the opinion that:

> An authentic piece must be produced for group use by an artist belonging to it or to a related group; the basic materials (with the exception of materials from outside sources incorporated into the object for specific purposes of decoration, protection, or magic) utilized in the creation of the piece must be indigenous to the region, and the object must function within the group in accordance with its traditions. (1980: 141)

Regardless of whether an object dates from this century or the last, it is always judged inauthentic by Western evaluators if it has not been used in a "traditional" manner. "The most obviously authentic works on which all would agree," writes art historian Frank Willett, "are those made by an African for use by his own people and so used" (1971: 216). From an anthropological perspective, William Bascom reaches the same conclusion when he notes that: "A piece may have been carved many years ago but if it was never used, perhaps because the customer who commissioned it died unexpectedly, it is not 'authentic'" (1976: 316).

African art traders understand very well the parameters of "authentic" African art as defined by Western collectors and appraisers. The language they use when speaking about objects is influenced heavily by their comprehension of Western taste, and its associated perception of authenticity. Almost nothing in the market is ever sold as "new." If pushed by a disbelieving customer, a trader might eventually concede that something he is trying to sell is only "a little old" or that "It's not ancient, but it wasn't made today." He would never state outright, however, that an object is brand new. If age is in doubt, then the condition and usage of an object would be stressed: "The mask has danced," or "It's been around," or "It's been used until it's been tired out [*fatigué*]". If all else fails, then the uniqueness or rarity of an object is played up in the course of a sale: "I don't see why you didn't take this [object]. It's very important. You mustn't leave it" or "There isn't another one like this anywhere in the market place. If you find one like this, I'll give you this one as a gift."[2]

Although African art traders have incorporated into the vocabulary of their discourse the Western-oriented themes of age and antiquity, condition and usage, and uniqueness and rarity, I never heard a trader actually use the word "authentic" (in French, *authentique*). Most of the traders divide the objects they sell into two broad categories: old (*ancien*) and copy (*copie*). If asked to analyze the terms further, most traders with whom I spoke said that all the objects in the market should in fact be referred to as copies. A more accurate way of classifying the types of art objects they sell is to draw a distinction between an older copy and a more recent one. Dramane Kabba explained:

> At the beginning, there was only one of everything that you now see in the market place. When the Europeans came, they took these things with them and put them in the museums and in the books. After that time, everything became a copy [of what was in the books]. A copy can be a hundred years old or it can only be a few years old, but everything is a copy of those first objects which are now in the museums and in the books. (3/2/88)

The trader's statement reveals an understanding of authenticity that is, at its core, significantly different from the concept of authenticity which is held widely in the West. First, according to Kabba's remarks, the original (i.e., the "authentic") objects of African art are all in Europe – they have been removed from circulation and they are, therefore, unattainable.[3] This aspect of the trader's concept of authenticity reverses the spatial/temporal chronogeography of the Western concept. While the Western version of authenticity focuses on pre-colonial Africa as the locus of "genuine" African art (i.e., a vision that is oriented toward distance in time), the trader's version of authenticity focuses instead on contemporary Europe (i.e., a vision that is oriented toward distance in space). For the Westerner, authentic African art only existed in the past, *before* European contact. For the trader, authentic African art only exists in the present, *after* European contact, when the objects were taken out of Africa and declared by the Western authorities to be authentic. The trader's formulation of authenticity thus depends upon contact with Europe to discover the inherent value of an African art object.[4]

A second reversal of the Western concept of authenticity occurs at yet another level of the trader's statement. Unlike the Western collector or dealer who focuses on the object itself as the source of authenticity, the trader views authenticity as something which emanates directly from the pages of a book. This point turns on its head Walter Benjamin's celebrated notion of the "aura" of an authentic work of art. "The presence of the original," wrote Benjamin:

> is the prerequisite to the concept of authenticity . . . Confronted with its manual reproduction, which was usually branded as a forgery, the original preserved all its authenticity; not so *vis à vis* technical reproduction . . . [T]hat which withers in the age of mechanical reproduction is the aura of the work of art . . . By making many reproductions it substitutes a plurality of copies for a unique existence. (1969: 220–21)

For Kabba, in contrast to Benjamin, it is the book itself which holds the "aura" of truth about an object. It is the book which provides an original with which to measure or gauge the quality and accuracy of all subsequent "copies."[5]

The authority of the text, and the distance which is felt between the trader and that which is represented in the book, is plainly illustrated by the miscommunication in the following encounter. The owner of an African art gallery in Abidjan published a catalogue of a collection of masks and statues that were to be sold from her gallery. An American collector of my acquaintance purchased a mask from her gallery, and displayed it in his Abidjan home. Abdurrahman Madu was at the collector's house one day trying to sell him some art. The collector pointed out to him the mask that hung on the wall, and showed him the photograph of the same mask that was reproduced in the book. Madu compared the photograph with the object and told the collector that he had really done well: "The mask looks exactly like the one in the book." When the collector explained that the mask was the *actual* piece that was in the book, the trader replied, "Yes, I see what you mean, the resemblance is almost perfect. It's an excellent copy" (2/18/88).

The death of culture and the birth of fakes

A fascination for antiquity and things from the past runs deep in Western culture. An *objet d'art*, like a fine bottle of wine, is said to improve with the passage of time. Visible signs of age are demonstrated by the outward effects of decay – scratches, tears, chips, cracks, bruises, corrosion, and patina. These marks of wear, writes historian David Lowenthal, "are prized as personal links with the past, like the arms of an old rocking-chair whose lacquer and staining had succumbed to long and constant rubbing" (1985: 152; cf. Shils 1981: 63–77). As the supply of antiques dwindles, however, some artists and forgers are tempted by economic motivation to simulate the effects of natural age. Roman fondness for original Greek art, for example, is said to have led to the production of "weathered" copies. In the Renaissance, the young Michelangelo was persuaded to age artificially the surface of a marble sculpture of Cupid by burying it for a time in the dirt. "When it was done," Vasari writes, "Baldassare de Milanese caused it to be shown to Pierfrancesco, who said 'If you buried it, I feel sure that it would pass for an antique at Rome if made to appear old, and you would get much more than by selling it here'" (1927 ed.: 4: 113). And, in a similar fashion, Giovanni Baglione reported in 1642 that after darkening a painting with smoke, Terenzio da Urbino added variegated layers of varnish and dressed a decrepit frame with shabby gilt "so that his work eventually looked as though it were really old and of some value" (quoted in Arnau 1961: 43).

One aspect of the Western image of Africa which resonates throughout the African art collecting world is the notion that authentic Africans, and by extension authentic objects of African art, no longer exist. Like the societies themselves, contemporary art objects produced in Africa are considered inauthentic approximations of traditional forms, sullied, as it were, by the degenerative impact of Western influence. Real African art, the argument goes, consists of old objects which were manufactured in the pre-contact or pre-colonial era for indigenous use. "The problem begins," art historian William Rubin explains, "when and if a question can be raised – because of the alteration of tribal life under the pressure of modern technology or Western social, political, and religious forms – as to the continuing integrity of the tradition itself" (1984: 76). Commenting on this passion for appropriating "genuine" cultural products, James Clifford has suggested correctly that "authenticity in culture or art exists just prior to the present, but not so distant or eroded as to make collection or salvage impossible" (1987: 122; see also Handler 1986).

Both the museum and the art market thrive on this particular vision of an African world, whose rapid spin into decay is set immediately in motion by the contagion of Western contact. In the context of a Western museum, one notices a reverence for the past which is expressed both in the language of explanatory text and in the symbolism of display. Even when referring to living populations, for example, museum label copy is often written in past tense (Clifford 1985: 171). Like objects suspended mysteriously in space on mounts of translucent Plexi, cultures are suspended enigmatically in time within the vacuum of preteritive language. Objects that show signs of Western influence are judged impure and unworthy of either serious scholarly research or art collection with an eye toward investment. Ladislas Segy, a once prominent dealer in African art, illustrates the collector's negative attitude toward Western influence when he writes: "[W]hen actual European penetration undermined the ideological background for carving[,] the artwork [of Africa] degenerated" (1958: 23). And, in a similar vein, art critics Paul Guillaume and Thomas Munro noted as early as 1926 that, "the coming of the white man has meant the passing of the negro artist; behind remains only an occasional uninspired craftsman dully imitating the art of his ancestors, chipping wood or ivory into a stiff, characterless image for the foreign trade" (1926: 13).

The concept of the "death of culture" was used by early anthropology as a means of legitimating its burgeoning status as a science of preservation, where artifacts stood as visible symbols salvaged from the ravages of a decaying modern world. The same concept, however, was also necessary to twentieth-century European artists who could only "discover" African art in a context freed from the encumbrance of a living African artistic counterpart with an alternative voice, and a different evaluative language. For artists like Picasso and Braque, the producers of African art were silent not only because they lived in lands that were oceans away, nor simply because they communicated in unfamiliar tongues, but more profoundly because their world view was stifled, quite conveniently, by the West's summary dismissal of whatever constituted "authentic" African culture. Finally, the "death of culture" concept reappears once again in the context of the contemporary art market. Here, it functions to inflate price by creating an artificially limited supply. Commenting on this process in both the "modern" and "primitive" art markets, Joseph Alsop has written:

> On the most superficial level, it is a truism of the art market that the works of a dead but still-admired master are more valued than the works of a master still alive and currently productive. Death limits the supply; the law of supply and demand begins to operate; and so $2,000,000 comes to be paid for a Jackson Pollock, significantly described as one of the last of Pollock's major works to be available. In the same fashion, collectors', museums' and the market's interest in American Indian artifacts has been growing rapidly for a good many decades. Yet it seems most unlikely that this would be the case if the West were still dotted with trading posts where [one could cheaply obtain] admirable examples of quill work and beadwork, featherwork, blankets, pottery, wood carvings, and the like. While that was the situation, no one was interested in Indian artifacts except for a tiny number of ethnologically minded persons.[6] (1982: 21–22)

Emphasis on age not only creates a limited supply which is beneficial to those who have a stake in controlling the African art market, but it also denies artistic capacity to those who could benefit by producing for the contemporary trade. Modern work, including so-called tourist or "airport" art, is considered a demon in the cult of antiquity. In a widely circulated book on African art, Elsy Leuzinger commented: "If tourists want masks and sculptures, then they shall have them, as many as they wish! But what is produced is of most questionable value: works without any cultural roots or artistic content" (1960: 209). It is only in recent years, in fact, that tourist art has been the subject of any serious anthropological inquiry (e.g., Ben-Amos 1971; Graburn, ed. 1976; Jules-Rosette 1984). And, with few exceptions, it is still considered an inappropriate subject for art history (see Adams 1989; Kasfir 1992).

If the "death of culture" increases art market profits by limiting the supply of authentic art, it also creates in its own wake the birth of its worst enemy – an industry of fakes, forgeries, and frauds (Allison, *et al.* 1976; Maurer 1981; Schoffel 1989). Devised to deceive the consumer, the fake object of African art is far more troublesome to the collector than the petty production of tourist crafts. African art historian William Fagg once referred to a fake as, "a work of the devil and a sin against art" (quoted in Sieber 1976: 22). And, in another instance, fakery was even placed on an equal footing with a parasitic tropical disease: "Mr. Kahan [a New York City art dealer] goes to the African continent two or three times a year, where he must fend off both new strains of dysentery and a profusion of art fakes (two out of every three pieces)" (Anon. 1982: 15).

Those who fake – both European and non-European art merchants and artists – capitalize on the "death of culture" by creating objects that imitate signs of age, ritual wear, and indigenous use. Armed with a highly sophisticated technology of *re*-production, the faker can simulate a thick accretion of soil and soot which would result from object storage in the rafters of a village cooking hut, the "sweat marks" on the inside of a mask which testify to its extensive use and wear (i.e. suggesting that the mask, as they say, has actually been "danced"), layers of feather and blood encrustation which build up after years of sacrificial libations, and a gleaming surface patina which results from decades of extensive object handling. Even an object's collection history can be forged by the faker's art. Inagaki, a Japanese mount maker who died in Paris shortly after the Second World War, was known for the exquisitely crafted mounts which he made during the early decades of this century for dealers and collectors of African art. Art forgers have imitated Inagaki stands in order to underscore an object's aura of authenticity, thereby falsely dating to the prewar period the sculpture which the mount supports (S. Vogel, ed. 1988: 4).[7]

Taxonomy and hegemony

In his review of the MoMA exhibition, *"Primitivism" in 20th-Century Art*, Clifford concluded that whatever else it may have represented the event stood as a document of a "*taxonomic* moment" in the history of art; a moment when African art and modern art were brought together on an "equal" footing; a moment when the status of so-called "primitive" objects was profoundly elevated and redefined (1985: 170). However, like any moment in a system of classification, the order and division among types is not permanent. Like the ephemeral power of a "big man" in the New Guinea Highlands, African art's moment of glory can be wiped easily away and replaced by another art form which is raised by its fall.

Taxonomies are always constructed by interested individuals to tell a particular story. They are never objective, neutral, or without intentionality. "[I]n nation-states," Shelly Errington observes, "objects associated with the past require an interpretation to place them within the historical narrative that tells what they are and what their relation is to the present" (1989: 53). A contemporary example is the display of Aztec, Maya, and Olmec objects in the National Museum of Mexico, where artifacts are pressed into service as evidence of the state's glorious past (*ibid.*: 53; cf. Anderson 1983: 178–85; Wallis 1991).

African art has been the subject of numerous, often radically different, systems of classification. Cultural critic and philosopher V. Y. Mudimbe has noted that what "is called savage or primitive art covers a wide range of objects introduced by the contact between Africans and Europeans during the intensified slave trade into the classifying frame of the eighteenth century" (1988: 10). Though appearing fixed, and thereby natural, the frames of classification are constantly being negotiated and reconstructed (see Blier 1988–89). Dividing the world roughly in half, nineteenth-century evolutionists viewed the arts of one portion of humanity as a yardstick with which to measure the arts of the other. Their classification was intended to tell a story of human progress and the triumph of Western civilization. Twentieth-century artists reclassified African art in order to validate, and even heighten, their own modernist enterprise. Theirs is a story of discovery, affinity, and the generous reconciliation of two disparate artistic traditions. Today, African art dealers in Europe and America rank the arts using economic criteria. They confer value on certain objects and withhold it from others. Their system of classification establishes the boundaries of the market, and prevents those on the fringe from participating in the trade.

Whatever the intent of a particular taxonomy of African art, the point is that objects are stripped of their original meanings and manipulated to fit the agenda of the moment. Power differentials are always at play. As Clifford says: "The relations of power whereby one portion of humanity can select, value, and collect the pure products of others need to be criticized and transformed" (1985: 176). The shift of African objects from artifact into art has little to do with an objective change in the quality of the objects themselves but is a direct result of what several authors have characterized as a peculiar form of Western generosity which presents itself in a guise of cultural relativism. Sally Price writes: "The 'equality' accorded to non-Westerners (and their art), the implication goes, is not a natural reflection of human equivalence, but rather the result of Western benevolence" (1989: 25). And later, she goes on to add, the "Western observer's discriminating eye is often treated as if it were the only means by which an ethnographic object could be elevated to the status of a work of art" (1989: 68). In a similar tone, Clifford notes: "Turning up in the flea markets and museums of late nineteenth-century Europe, these objects are destined to be aesthetically redeemed, given new value in the object system of a generous modernism" (1985: 172).

The classification of African art has had a direct impact both on the market for old objects and on the current production of art for international trade. Taxonomies fix African art in stereotypical forms so that the market does not reward the production of artistic novelty. "[I]ndividual creativity beyond bounds set by Westerners," explains Jeremy MacClancy, "is generally unwanted. An object difficult to categorize can be difficult to sell" (1988: 172). Raoul Lehuard, a noted African art collector in France, confirmed MacClancy's observation when he declared: "An object is a fake when its forms are extraneous to the statuary repertoire of a given ethnic group" (1976: 74).[8] The art of Anoh Acou, a Baule woodcarver working in the West African beach resort of Grand Bassam, offers a case in point. Rather than imitate other tourist arts, which either copy "traditional" masks and statues or depict images of wildlife that stand as Western symbols of idyllic Africa, Anoh Acou carves wooden suits, formal shoes, baseball caps, crushed aluminum cans, "boom box" radios, and telephones. The market, however, does not reward his artistic talent and inventiveness. Images of progress and change do not have a place in the West's vision of Africa; they cannot be situated easily in the taxonomy of non-Western art (Jules-Rosette 1990: 29–30).

Constructing the category art

African art entered the West with a certain degree of trepidation. Unsure of its proper place in the hierarchy of the world of goods, Westerners first assigned African art to the category of curiosity, where objects were worthy neither of scientific investigation nor aesthetic appreciation. Natural and artificial objects gathered from the conquest and exploration of foreign lands were arranged together in the "cabinets of curiosities" belonging to Europe's elite. Throughout this early period, ethnographic specimens were collected haphazardly and displayed in a cluttered mass. "Because there was no *Systema Naturae* . . . by which to arrange them and precise terminology with which to discuss them" writes Adrienne Kaeppler, "few [collectors] took them seriously" (1978: 37). The objects were valued principally as trophies – icons of conquest attesting to unbridled Western power in the Age of Discovery (see Steiner 1986a).

In the second half of the nineteenth century, when the field of anthropology began to evolve from amateur avocation to scientific enterprise, the status of African art was elevated from artificial curiosity to ethnographic artifact. Random and unordered collections were given new meaning as they were subsumed within a then burgeoning museological branch of anthropological science. Pitt-Rivers, for example, who was among the first to promote the systematic collection, storage, and display of ethnographic artifacts, wrote in 1875 that the "products of human industry, are capable of classification into genera, species and varieties, in the same manner as the products of the vegetal and animal kingdom" (1875: 307). Artificial curiosities were thus admitted to the kingdom of science – having been assigned their own version of a *Systema Naturae*.

The conceptual reclassification of material culture was followed quickly by a physical relocation of the objects themselves. In 1884, for example, the artifacts collected by Pitt-Rivers were placed in a museum which had been built especially for them at Oxford (Gerbrands 1990: 17). African art was thus moved from private display to public domain – from the wonder cabinets of the leisured class to the popular halls of ethnographic museums and world expositions. In their new context, African art objects were perceived as visual windows through which one could catch a fleeting glimpse of "primitive man." Though making a pivotal transition from personal prize to public specimen, the objects of anthropology were still intertwined with those of what were then considered to be associated scientific disciplines – e.g., paleontology, entomology, gemology, and geology. Commenting, for example, on the status of American Indian art at the American Museum of Natural History, Edmund Carpenter has written: "On public display was an incredible wealth of Northwest Coast art. Yet every piece was classified and labeled as scientific specimen. Tribal carvings were housed with seashells and minerals as objects of natural history" (1975: 11). While developing their plans for the opening of the Musée d'ethnographie du Trocadéro, Paul Rivet and Georges-Henri Rivière emphasized in 1929 that "primitive" artifacts were to be understood as "objects of knowledge" that were meant to inform instead of please (Jamin 1985: 60–63). As such, it was argued, the artifacts were privileged in their capacity for revealing dispassionate ethnographic truths. According to Marcel Griaule, in a manual he prepared for his team of researchers on the 1931–33 Dakar–Djibouti expedition, ethnographic objects were to be collected and stored systematically since they provide "an archive more precise and more revealing than the written word itself" (quoted in Jamin 1982: 90).

African art made a second fundamental category shift sometime in the early years of the twentieth century when a group of Parisian artists first discovered the art that was embedded in the category primitive. "We owe to the voyagers, colonials, and ethnologists," writes Rubin, "the arrival of these [African] objects in the West. But we owe primarily to the convictions of the pioneer modern artists their promotion from the rank of curiosities and artifacts to that of major art, indeed, to the status of art at all" (1984: 7). For these twentieth-century artists, objects of African art did not derive their meaning from an ethnographic context but were of interest only when *dépaysé* – stripped of all contextual references and indigenous symbolic meanings (Jamin 1982: 89).

This second shift in the conceptualization of African art was, however, less encompassing than the first. That is to say, while the early anthropologists were democratic in promoting *all* artificial curiosities to scientific specimens, artists and (later) art historians were more discriminating in their selection of which objects would be permitted to pass from artifact into art. Items that were thought to speak across cultural barriers were judged as works of art, whereas those considered impervious to cross-cultural interpretation remained locked in the artifact class. Currently, one of the most articulate proponents of the art/ artifact distinction is Susan Vogel, founder and director of The Center for African Art in New York City. She takes the view that a

> pedestrian bundle of sticks communicates only within its own culture where understanding may depend, for example, on the knowledge that it contains branches from specific medicinal plants with particular powers . . . [A carved] figure, on the other hand, has an expressive dimension and, if it is good, I venture that it will communicate something of its import even when we do not know what its intended content is. (1981b: 76)

Vogel is well aware that her proposition becomes problematic the moment we are called to judge for ourselves whether or not the figure is "good." If we do not understand the intended content of the work, how are we to know on which scale to weigh its symbolic load?

Although there are some noteworthy exceptions, participation in the art/ artifact distinction tends to be divided along academic disciplinary lines. Art historians, on the one hand, generally legitimate their scholarly interests (and negotiate their research funds) by insisting that the formal aesthetic character of African art is entirely capable of being analyzed and taught through the lens of Western art historical discourse. Most anthropologists, on the other hand, justify their *raison d'être* by arguing that objects can be understood only within an appropriate ethnographic context. Those favoring a universal category of art, disparage anthropologists for their attention to ethnographic detail which, they say, suffocates the object's aesthetic below a mass of academic dogmatism. Furthermore, they suggest that emphasis on cultural context diminishes an object's capacity to *stand alone* as a genuine aesthetic creation. "The motives for this insistence [on context]," writes James Faris, "while perhaps honorable (and of paramount importance to the discipline [of anthropology]), could be argued to deny objects of the Other any potential on their own – any freedom from the security of context – almost as if in their emancipation they might be revealed as, well, primitive" (1988: 778–79).

Capitalizing on the category art

The debate over definitions of art and artifact has implications which stretch beyond the geo-politics of academic disciplinary frontiers. Indeed, the dichotomy is of central concern to participants in the African art market – where the distinction between an artifact and a work of art can mean a significant difference in an object's price tag. "The continuum from ethnographic artifact to *objet d'art* is clearly associated in people's minds," writes Price, "with a scale of increasing monetary value and a shift from function (broadly defined) to aesthetics as an evaluatory basis" (1989: 84). Western dealers of African art drench the objects they sell in the radiance of the category art – displaying objects in pristine, context-free conditions and often criticizing the pedestrian craft of anthropology.

Drawing on a metaphor implied in the philosophical writings of Arthur Danto (1981), Errington has proposed that the category art be envisioned as a container (such as a bag) into which objects are stuffed until there exists such tremendous diversity and quantity that the container threatens to burst. No participant in the art world would want to destroy the bag (since, after all, the category art is meaningful only if it can be juxtaposed to whatever else is *not* art), yet because the category takes on greater significance and value the more limited it is in scope, all the participants in the art world want *their* class of objects to be the last one dropped into the bag – after which it would be sealed shut. Errington writes:

> Curators of Renaissance paintings may shut the Art bag long before African masks can be dropped into it; curators of African masks and Benin bronzes, for their part, might insist that the items they care for should be inside the bag, but that nineteenth-century American Indian baskets and beaded shoes should be outside; dealers in "quality" "non-Western art" may insist that Navaho rugs and old Pueblo pottery should be dropped into the bag but draw the line at mere "curios" like post-World War II Zuni "fetishes," Alaskan stone carvings, and Southwestern silver jewelry made explicitly for the tourist market. (Errington forthcoming)

Research on the contemporary African art market reveals an ever-broadening expansion of the category African "art." As the supply dwindles on the African continent of what dealers and collectors classify as "authentic" African art – ideally masks and figural statues created for indigenous use before the advent of colonialism – the definition of collectible art grows in Europe and America to encompass what previously had not been appraised as legitimate art commodities.[9] Thus, more and more objects are dropped into the imagined bag. Once considered products of recent "culture contact" (and therefore as inauthentic works of art), so-called "colonial" figures – mostly polychrome statues representing either Africans in Western attire or Westerners in colonial garb – are now making the transition to the category art (see Chapter 6).

One former "artifact" which recently crossed the proverbial threshold into the world of "art" is the African metal currency. At an East Side Manhattan gallery, a recent exhibition, "Symbols of Value: Abstractions in African Metalwork," featured various forms of indigenous African iron wealth. Although previously dealing in medieval and ancient art, the gallery owner ventured into the world of African art by presenting metal currency "not as ethnic materials, but as abstract sculpture" (Reif 1988: 37) with "its strong silhouettes, powerful shapes, and subtle textures" (Ward 1988: [1]). The abstract qualities of the objects were underscored in the gallery by intense boutique lighting and dramatic framing of single pieces each on its own pedestal. The material's transition to the category art was further signaled by the *new* economic values assigned to the currencies: from $400 for an X-shaped item to $12,000 for one in accordion-fold form. The headline of a *New York Times* review of the exhibition, "It May Look Like a Hoe, But It's Really Money," would perhaps have been more enlightening to African readers had it read "It May Look Like Money, But It's Really Art."

African stepped house ladders, used in village contexts to access the upper opening of a granary or the roof of a home, have been making their way onto the international market in the last few years. A handful of galleries in Europe and America began featuring these forked ladders in their catalogues and advertisements sometime in 1987. In contrast to the typical dirt-encrusted wooden surface of house ladders found in villages, the ladders in the galleries were cleaned, highly polished, and tastefully mounted on unobtrusive black metal bases. They were indeed well on their way to making it into the category art. Spurred by this new demand, African traders in Côte d'Ivoire began collecting vast quantities of house ladders, especially where they are most commonly found in villages throughout the region of Korhogo (Plate 18). Although the "authenticity" of ladders is not yet universally acknowledged, a number of Abidjan-based traders report having made several important wholesale transactions with Western gallery owners.

Recently, one Hausa storehouse owner of my acquaintance had acquired approximately 200 common wooden pestles – objects used by women in villages to pound rice, yams, manioc, etc. (Plate 19). Although the pestles had not yet been "discovered" by Western art collectors or dealers, Alhadji Moussa was hoping to corner the demand if it should ever arise. Some of the other traders in his storehouse that day were teasing Moussa. "Whites are crazy [*fou*] but this is really too much. These things are just ugly Alhadji!" (6/27/91). The jury is still out on the admittance of pestles to the category art.

A good example of an object's successful transition to the category art is perhaps best illustrated when one follows the commoditization of the Baule slingshot – a process which has occurred during the last few years in the Côte d'Ivoire art market. In November 1987, an illustrated coffee-table size book entitled *Potomo Waka* appeared in the window displays and shelves of many Abidjan bookstores. The book contains over one hundred glossy color photographs of sculpted slingshots from the private collection of one of the book's coauthors, Giovanni Franco Scanzi.[10] Each slingshot which is reproduced in the book is depicted on its own full color page, and each is framed in

18 Wooden house ladders in a trader's storehouse. Treichville quarter, Abidjan, June 1991.

the vibrant hues of a different backdrop. An Italian entrepreneur specializing in the sale of overstocked European promotional items (ashtrays, pens, and cigarette lighters blazoned with company logos), Scanzi has been conducting business in Côte d'Ivoire for over twenty years. He only began collecting slingshots about six years ago, however, when he was encouraged to do so by his friend Antoine Ferrari, a longtime Abidjan-based African art dealer/collector.

Because slingshots had rarely appealed to Western collectors, not many were available on the market when Scanzi began forming his collection. Market-place traders confirm that before his interest in slingshots, few of the itinerant suppliers who collect art in rural areas regularly brought back such items for sale in Abidjan.[11] Over a period of approximately three years, Scanzi collected more than a thousand slingshots purchased from about twenty different suppliers. Any trader who arrived in Abidjan knew that he could either sell his slingshots directly to the Italian collector or to a middleman who would take them to his home.

In the preface to *Potomo Waka*, the authors carefully construct a case for the "authenticity" and "artistic" value of the objects in their book. The first point they make, is meant to underscore the aesthetic quality of slingshots. They do so, largely by disparaging any interest in the ethnographic context from which the objects were removed:

> The Baule catapult is a typical example of an object about which ethnography has almost nothing to reveal . . . If the ethnographer can throw some extra light on the understanding of works of art from the Third World, this is and should remain a minor detail. Does Western art really need explanations from ethnographers to be understood? Of what importance is it to know that Rembrandt came from a patrilineal type of family or know the method of cultivating soil in "quatrocento" Italy![12] (Delcourt and Scanzi 1987: 10–11).

Here, the intended message of the text is clear: slingshots belong to the universal category art and, as such, can be sold at prices which befit their aesthetic character. The second point, which the authors emphasize in the preface, is the "authenticity" of the objects in their book. In so doing, they take great pains to emphasize that wooden slingshots are not products of the colonial era, but rather pre-date the advent of any European contact:

> Some people refuse to accept Baule catapults [i.e., slingshots] as authentic African works of art because they incorporate rubber strips of European manufacture. Even though it is not possible to be certain that Africans used local rubber for the catapults they made in West Africa at the beginning of the 20th century, before the advent of the motor car and its rubber inner tube, it is more or less certain – according to an investigation carried out by Mr. Scanzi – that the Dogon tribe used large catapults powered by animal gut in pre-colonial times. The theory that catapults have only made their appearance since the use of rubber from European sources is [therefore] difficult to sustain. (1987: 9)

19 Wooden pestles in a trader's storehouse. Treichville quarter, Abidjan, June 1991.

The reason the authors are so concerned with providing "scientific" proof that slingshots originated in the pre-colonial era is, of course, that they are trying to create a market for the objects within the accepted Western definition of "authenticity" in African art – i.e., which demands that the style has been conceived in an environment untainted by European influence. Not only does Scanzi carefully attempt to construct a pre-colonial past for the African slingshots which he now sells, but when he started his collection he refused to purchase slingshots that had been painted.[13] As a result, slingshots with paint were sanded down and restained with potassium permanganate[14] before being presented for sale at his home (Plate 20). At present, Scanzi's entire collection is up for sale at art galleries in Europe and West Africa.

With the appearance of *Potomo Waka* in Abidjan stores and hotel gift shops, tourists and local buyers have been eager to purchase slingshots as part of their African art collections. In the market places, the price of an average slingshot jumped from a range of about 1,000–6,000 CFA ($5–20) in early 1987 to 3,000–45,000 CFA ($15–150) in late 1988 (Plate 21). Because the value of slingshots increased so dramatically, traders responded quickly and in several ways. First, traders in Abidjan commissioned slingshots from local carvers (Plate 22). Most were purchased unfinished from the artists and either stained

20 Slingshots stained with potassium permanganate drying at the Plateau market place. Abidjan, March 1988.

or painted by the traders themselves. Second, because they generally follow the shape of a branch and – at least some of the simpler varieties – are relatively easy to carve, traders with entrepreneurship and some basic artistic skills began to sculpt their own slingshots as they sat during the day in the market place waiting for clients to arrive. Third, traders learned to convert broken wooden artifacts into slingshots. A small Baule statue with broken legs, for example, was transformed into a slingshot by substituting a forked pinnacle for the absent legs (Plate 23). The creator of the object explained to me that, "These days there is a better chance of selling a slingshot than a broken statue" (4/12/88).[15]

When I showed *Potomo Waka* to some of the traders in the Plateau market place, they made two sorts of remarks. On the one hand, many were able to identify specific slingshots which either they or one of their intermediaries had sold to the Italian collector. On the other hand, a number of traders remarked that they had sold Scanzi other slingshots – much higher quality and much finer – none of which, they said, were reproduced in the book. "I don't understand why he would not put those [slingshots] in the book. The whole idea of writing a book like this," said Sheriff Ousman, "is to show the best pieces in the collection so that other people will know what you have" (12/8/87). When I showed *Potomo Waka* to an African art bookseller in New York, I asked her

21 A pile of slingshots displayed among other objects at the stall of a Plateau market-place trader. Abidjan, April 1988.

what she thought of the remarks made by the traders about the author not having put in the best pieces from his collection. She told me that their observations were not only plausible but indeed predictable. That is to say, the author probably selected inferior pieces on purpose. There are two economically motivated reasons for which one would publish a book with inferior examples from a large collection of objects which were eventually going to be sold. First, she explained, by having the objects reproduced in a catalogue, the seller adds value to some of the lesser quality pieces in his collection. In other words, he can capitalize on the objects in the book simply because they are the objects in the book. Second, after he has sold most of the objects which are reproduced in the book, he can slowly begin to bring out some of the better quality pieces from his larger collection. These pieces are now valuable precisely because they are not in the book. The seller can point to the objects in the catalogue and tell his client: "See this piece here, it's even better than anything in my book." For the same reasons, the bookseller added, "Most African art dealers love books that illustrate inferior pieces. They can turn to any object in the book and say that what they have for sale in their gallery is of much finer quality" (12/22/88).

A second example of the manipulation of the category art can be found in the African art framing (*encadrement*) business which has been carving out its

22 Carver finishing the detail work on a wooden slingshot. Abidjan, March 1988.

23 Broken Baule statue "transformed" into a slingshot by replacing the broken legs with a forked pinnacle. Abidjan, July 1988.

niche in the Côte d'Ivoire art market during the past several years. Within a relatively short span of time, at least four framing galleries have opened in Abidjan (Plate 24). The work they sell consists essentially of three-dimensional African art objects which are encased in wooden box frames (*coffrets*). The gallery owners (all of whom, it is interesting to note, are European) buy African objects from the market-place traders. Most of what they buy would be classified as *copies* or *nyama-nyama.* The objects are mounted on fabric covered boards, and enclosed in box frames, usually with a decorated trim (Plate 25).

One of the original, and most prominent, framers of African art in Abidjan is Mary-Laure Grange Césaréo. Approximately a hundred examples of her work were featured in an exhibition at the Abidjan Hilton in February 1988. She is the owner of the Galerie Equateur in the Marcory quarter of Abidjan. The gallery, which opened in 1982, originally sold a variety of African art objects – largely contemporary replicas aimed at the popular export trade. In 1987, however, the focus of the gallery was shifted toward box frames.[16] Césaréo says that she is "neither an expert in African art, nor a collector of African art. I have not put together an exhibit of art objects, but rather have specialized in the valorization [*mise en valeur*] of works of art" (*Guido* 1988a: 66). Although she does not market her merchandise as "antique," it is interesting to note that she ages artificially the surface of the wooden frames – thereby following a Western convention for the framing of antique paintings. "A number of the frames were aged artificially [by my staff] with kaolin and other materials, using techniques with which I myself am not familiar" (*Guido* 1988a: 66; my translation from the French).

The framing of African art illustrates, almost *ad absurdum*, the "process of aesthetization" (Baudrillard 1972) which occurs when African art is transplanted to the West. The process of framing goes one step beyond the art historical tradition of exhibiting isolated African objects in the putatively "neutral" context of a gallery or art museum – allowing the well-lit object to speak for itself without the support of interpretive ethnographic text. The frame around the object functions to invent a whole new context, or literally a constructed *cadre*, for the object in which the piece is insulated both from its original milieu and from the potential aesthetic ambiguities which might result from allowing it to speak for itself.[17] As one of the African art framers suggested in an interview: "We try to create something of value out of these African objects. By putting an object in a frame, we provide it with an *environment* that is better suited [for display] in an apartment" (7/6/91). In other words, the *encadrement* substitutes the interpretive framework of an ethnographic discourse with a physical frame which encodes its appropriated content by unequivocally announcing to its viewers that "this is art."

24 Employee at an African art framing gallery constructing a box frame for a display of Akan goldweights. Abidjan, July 1991.

The domestication of African art

Collections of African art tend to efface the social history of their own production. As if muted by the prophylactic glass of a display case, objects in galleries are silent as to how they got to where they are. With varying amounts of detail, accompanying explanatory text informs the viewer of an object's meaning and use in its original cultural context. Yet, it says nothing about how or why the object was collected, from whom or for how much it was purchased, or even by whom it was made. Writing about an exhibit of Sepik River art at New York's IBM Gallery, Clifford remarked:

> The history of collecting was not included in the presentation. While the names of individual collectors were sometimes provided, the circumstances, priorities, funding, institutional and political contexts for the objects' physical move from New Guinea to Switzerland to New York were deemed irrelevant to their presentation as "tribal art". In their salvaged, aesthetic-ethnographic status they exist outside such mundane historical dimensions. (1987: 123–24)

In her research on the history of the African Hall at the American Museum of Natural History, Donna Haraway has pointed out a similar process in which Africans, who were involved in supplying museum expeditions with the wild animals which were shipped overseas, were conveniently overlooked in the historical record – their guidance, contribution, and sacrifice having been written out of authorship. "Behind every mounted animal, bronze sculpture, or photograph," she writes, "lies a profusion of objects and social interactions among people and other animals, which in the end can be recomposed to tell a biography embracing major themes for 20th-century United States. But the recomposition produces a story that is reticent, even mute, about Africa" (1984/85: 21; also see Simpson 1975; Coquery-Vidrovitch and Lovejoy 1985).

Although African art is silent on the subject of its collection, appropriation, and ultimate passage to the West, objects are never tacit about their subsequent excursions through Western hands. "An African figure that was once owned by Henri Matisse or Charles Ratton or Nelson Rockefeller is unrelated," writes Price, "to a sculpture by the same artist that was not" (1989: 102). An object's list of previous owners constitutes what is called its pedigree. The longer the pedigree, and the more illustrious the caretakers in the line of descent, the more prestigious (and more valuable) the object. Commenting on a Cameroonian Bangwa sculpture (known in the West as the "Bangwa Queen") which was recently sold at auction in New York, the head of Sotheby's Tribal Art Department, Bernard de Grunne, noted that "[p]art of the value of this carving is its pedigree" (Reif 1990: 32). The figure was collected at the turn of the century in Cameroon by a German explorer (not surprisingly described as "the first white man to reach the African Bangwa kingdoms"). It was deposited in

25 A framed Dan mask (right) and brass bracelet and goldweights (left) on display at an African art framing gallery. Abidjan, July 1988.

1898 at the Museum für Völkerkunde in Berlin, where it remained until it was acquired in 1926 by a German collector who then sold it a few years later to Charles Ratton, a Paris dealer. The Princess Gourielli (Helena Rubinstein) bought the queen figure from Ratton in the 1930s, at which time it was exhibited at the Museum of Modern Art's first African art show in 1935 (Northern 1986: 20). Around the same period, the figure was photographed by Man Ray, who posed a nude model at its feet. The sculpture is now in the possession of an unnamed collector who bid recently at auction $3.4 million, the highest recorded price for an object of African art.

Drawn from the semantic domain of bred animals – where authenticity is also judged by purity of descent – the term pedigree neatly captures what might be called the domestication of African art. In its wild, undiscovered state African art is raw, meaningless, and without value to the Western collector. Tamed, through appropriation and the controlled reproduction of ownership, African art becomes assimilated to the broader category art. Another term which is related to the judgment of authenticity and also drawn from the domain of animal life is vetting. Literally meaning to submit an animal to medical examination, the word vetting is used by art dealers to refer to the scrutinization of objects for the purpose of detecting post-production alteration or fakes. Recently, in the antique trade, the term was found to be indecorous. "The majority opinion is that, while vetting is no doubt appropriate for horses and cattle – the noun, after all, is derived from that beastly word 'veterinarian' – it has no place in the civilized society to which antique dealers cater and sometimes belong" (Swan 1989: 96). In the African art trade, where the canons of civilized society perhaps do not always apply, the term and practice have not been met with the same opposition.

Trade beads and the recreation of exotica

Colored glass beads of European manufacture were introduced in West Africa by seafaring traders as early as the fifteenth century. In the coastal enclaves of European trade, beads functioned as a medium of exchange through which Europeans could acquire African palm oil, ivory, gold, and slaves. Along with their economic role, imported beads also functioned in the realm of international diplomacy – currying favor of chiefs along vital links in political and commercial networks. The systematic production in Europe of glass beads for foreign consumption began in Venice during the thirteenth century. Production and export increased steadily throughout the Age of Discovery. As a result of growing demand, competition to Venetian manufacturers eventually came from bead makers in Holland, as well as Bohemia and Moravia (provinces of modern-day Czechoslovakia). With the invention of new technologies and the rapid expansion of global markets, European bead production reached its peak in the middle of the nineteenth century (Francis 1979; Dubin 1987: 107–14).

European trade beads were incorporated into the social, political, and economic symbolism of many West African cultures. They have served as emblems of social rank, indicating the power and wealth of the individual wearer. On ceremonial occasions they were worn by elders and other dignitaries to symbolize status and prestige. Young women who had completed initiation could wear strands of imported glass beads or beaded aprons during dances performed in celebration of their coming of age. Along with other forms of personal adornment, such as clothing, tattooing, and scarification, beads functioned as signs of group membership (Steiner 1990). They were integrated into all aspects of the aesthetic of dress, and were appendaged to the visual arts as decorations on masks and statues.

Some beads, especially the large Venetian chevrons, were more valued than others and were reserved for members of the ruling class. Reporting on the use of foreign beads in the kingdom of Swazi, Hilda Kuper notes that when "trade goods – particularly factory manufactured fabrics and china beads – were incorporated [into the aesthetics of local politics], the king and in some districts the chiefs established connections with individual traders, some of whom were instructed not to sell the same designs to the families of others" (1973: 355).

Since the introduction of monetary currencies during the colonial period, the use of European trade beads as a medium of exchange has essentially ended – save perhaps as a form of ritualized bridewealth payment in certain communities. Especially among urban women, trade beads have lost much of their visual appeal as a source of adornment. In present-day Côte d'Ivoire, for example, where Western fashion is promoted by the sale of French magazines and weekly broadcasts of American television serials, most African women have little interest in adorning themselves with beads dating to the era of the slave trade. Instead, young women wear modern European-style gold, silver, metal, and plastic jewelry, as well as locally manufactured ornaments of gold-plaited filigree. African women who still buy European trade beads in the local markets are either medicinal or ritual specialists who continue to value the beads for their spiritual potency (Cole 1975).

The market price of trade beads is established by African merchants through economic calculation which takes account of both scarcity in local supply and demand by foreign buyers. One exception – which throws further light on continued African demand in specialized sectors – is a particular type of tabular bead, with a blue and white "face" design on a black background, which is disproportionately expensive in comparison to similar beads. The unusually high price of this bead is a result of intense local demand by women of the Akan ethnic group who prize the bead for its symbolic and curative powers (Picard 1986). Since they know that only a select group of local patrons would

26 American gallery owner buying trade beads from a Hausa merchant in the Treichville market place. Abidjan, December 1987.

27 Young Hausa trader with a display of trade beads at his market-place stall. Treichville quarter, Abidjan, July 1988.

be willing to pay the inflated price, traders in the various market places usually keep these type of beads separate and hidden below their stalls.

With the exception noted above, *African* demand for trade beads has, in the past several decades, been replaced largely by *European* demand. Hence, although the consumption of European beads has diminished among Africans, trade beads continue to circulate along specialized routes of transnational commerce (Plate 26).[18] The same beads which for centuries were prized by Africans are now sought by those inhabiting the very shores from which the beads originated. While the beads themselves have not really changed – some perhaps are smoother or more rounded from use, others may be slightly chipped – the meaning of the beads has been drastically altered.

Today, in the market places of Abidjan (especially in Treichville) beads of European origin are displayed in vast quantities for sale to foreign tourists, dealers, and collectors (Plate 27). Traders depend on the sale of beads to tourists and other visitors. One of the techniques the traders use to successfully market the beads in this arena is to establish that other foreigners (just like the present buyer) have also purchased and worn this type of "exotic" jewelry. In the center of one stall in the Cocody market place in Abidjan, for instance, a trader had framed a cover from *Elle* magazine (April 20, 1987) which shows a glamorous European model covered in African trade beads and beaded bracelets. The message to the buyer is clear: if it is fashionable enough for her, then, surely, it must be fashionable enough for *you*.[19]

Most traders also have a regular clientele of dealers and private collectors. A number of bead enthusiasts who reside as expatriates in Abidjan have developed close links to some of the major bead suppliers. In return for a slightly higher price or an occasional loan, bead collectors urge their suppliers to guarantee them first selection from a newly arrived lot of beads. Like the Swazi sumptuary decrees which Kuper described as regulating the sale of beads to Africans by European suppliers, Europeans are now regulating, through economic incentives, the distribution among themselves of these same beads by, what have now become, African suppliers.

In both their original and present markets, the beads have been valued for their "exotic" qualities. To the Africans who acquired imported beads in exchange for the riches of their land, the beads were desired because they came from far away, they were unavailable locally, and they were produced in colors and patterns which were not feasible in the repertoire of local technology. They were loaded with symbolic prestige because they represented contact and rapport with a powerful outsider. To the travelers who now buy old trade beads in the markets of West Africa, their appeal, at least in part, stems from the fact that their long presence in Africa has *again* made them exotic. Once more, they have been packed with a symbolic charge. This time, however, their symbolism communicates encounter with a romanticized vision of traditional, pristine Africa. Hence, at one time trade beads were popular because they were foreign and European, now they are again fashionable because they are considered "ethnic" – and, therefore, by definition, *still foreign*.

The assimilation or, what might be called, the "naturalization" of foreign imports is not unique to the history of beads. It is a process that has occurred in various contexts and at different times. In his tours of Great Britain during the reign of William and Mary, Daniel Defoe observed that the wearing of fine East Indian calicoes and the furnishing of homes with delicate China ware had become so integral to English taste that they came to represent "Englishness" more than they did "otherness" (Bunn 1980: 306). A similar process has also occurred in Africa where women traders give indigenous names to the imported European factory cloths which they sell in local markets. The names establish group-specific meanings and metaphors which transform a foreign commodity into an item of local "production" (Touré 1985).

Epilogue: fake masks and faux modernity

In the Plateau market place, I once witnessed the following exchange between an African art trader and a young European tourist. The tourist wanted to buy a Dan face mask which he had selected from the trader's wooden trunk in the back of the market place. He had little money, he said, and was trying to barter for the mask by exchanging his Seiko wrist watch. In his dialogue with the trader, he often expressed his concern about whether or not the mask was "real." Several times during the bargaining, for example, the buyer asked the seller, "Is it really old?" and "Has it been worn?" While the tourist questioned the trader about the authenticity of the mask, the trader, in turn, questioned the tourist about the authenticity of his watch. "Is this the real kind of Seiko," he asked, "or is it a copy?" As the tourist examined the mask – turning it over and over again looking for the worn and weathered effects of time – the trader scrutinized the watch, passing it to other traders to get their opinion on its authenticity.

Although, on one level, the dialogue between tourist and trader may seem a bit absurd, it points to a deeper problem in modern transnational commerce: an anxiety over authenticity and a crisis of *mis*representation. While the shelves in one section of the Plateau market place are lined with replicas of so-called "traditional" artistic forms, the shelves in another part of the market place – just on the other side of the street – are stocked with imperfect imitations of modernity: counterfeit Levi jeans, fake Christian Dior belts, and pirated recordings of Michael Jackson and Madonna. Just as the Western buyer looks to Africa for authentic symbols of a "primitive" lifestyle, the African buyer looks to the West for authentic symbols of a modern lifestyle. In both of their searches for the "genuine" in each other's culture, the African trader and the Western tourist often find only mere approximations of "the real thing" – tropes of authenticity which stand for the riches of an imagined reality.

"We in Africa prefer things that were made by Europeans than things made by ourselves," Abdurrahman Madu explained. "When a European comes to Africa he buys African things – boubous, textiles, masks, and so on. He likes these things because they are African, they are not from where he comes. The same holds true for us. We want what the Europeans have . . . If I had more money I would collect European art, and the European would continue to collect our art. That's just how it is" (6/25/91). Viewed in the context of a global dialectic, the international consumer's search for authenticity underscores the idea that far-away collectors reinvent their objects of desire. While Western notions about the authenticity of African art are constructed by privileging aesthetic forms imagined to have existed in the past – worlds that never were but might have been – African beliefs about Western authenticity are projected into the future – worlds that aren't yet but someday could be.

6 Cultural brokerage and the mediation of knowledge

. . . the men who become dealers are not, as a class, quite fools.

Desmond Coke, *Confessions of an Incurable Collector* (1928)

As performers we are merchants of morality. Our day is given over to intimate contact with the goods we display and our minds are filled with intimate understandings of them; but it may well be that the more attention we give to these goods, then the more distance we feel from them and from those who are believing enough to buy them.

Erving Goffman, *The Presentation of Self in Everyday Life* (1959)

In a recent issue of a glossy coffee-table French art magazine there appears a full-page advertisement for Jean-François Gobbi's Galerie d'Art. A black-and-white photograph shows Monsieur Gobbi, in his Paris gallery, dressed in a dark suit, sporting an oversized cigar, and surrounded by his *objets d'art*. The text, below the photograph, reads as follows:

The world's largest museums and collectors await your valuable paintings. The problem, however, is that you don't know where they are – these museums and collectors who are ready to pay good money for your works of art. Jean-François Gobbi, *he*, knows where they are: they are all his clients. Today, the demand for masterpieces is so great that museums and collectors don't even discuss price . . . Now if you think you will some day come in contact with a big museum or an important collector willing to pay top dollar for your masterpiece, don't call Jean-François Gobbi. Otherwise dial 266-50-80. (*Galerie des arts*, 1985, my translation from the French)

In the art market, as in any other large-scale business enterprise, the success of the middleman demands the separation of buyers from sellers. Social, legal, and bureaucratic barriers are erected to maintain the distance between the primary suppliers of art and their ultimate consumers.[1] "The producers and consumers of the art," Bennetta Jules-Rosette has written, "live in quite different cultural worlds that achieve a rapprochement only through the immediacy of the artistic exchange" (1984: 8). It is the art trader who provides the linkages at crucial points in a series of cross-cultural exchanges. The art trader, in this sense, fits Eric Wolf's classic definition of the middleman, i.e., as a person "who stand[s] guard over the critical junctures or synapses of relationships" (1956: 1075). The middleman is a mediator, whose principal role can be seen as one of "limiting the access of local persons to the larger society" (S. Silverman 1965: 188). The role of the middleman comes into existence so as to bridge a gap in communication.[2] The "bridge," however, must be well guarded for, as F. G. Bailey puts it, "perfect communication will mean that the middleman is out of a job" (1969: 167–68).[3] Art dealers, in general, earn their livelihood as middlemen – moving objects across the institutional obstacles which, in some cases, they themselves have constructed in order to restrict direct exchange. The art market in Côte d'Ivoire is characterized by precisely this sort of network of relations, in which a host of African and non-African middlemen forge temporary links in a transnational chain of supply and demand.

Supplies of art objects from villages are tapped by professional African traders who travel through rural communities in search of whatever they believe can be resold (see Chapter 2). Almost all the suppliers, and even most of the wayfaring traders, have little or no idea where these objects are destined, why they are sought after, and for what price they will ultimately be bought.[4] After being collected in villages, objects are moved from town to town until they are eventually sold to a trader who has direct contact with Western clients. These traders are based in the larger towns and capital of Côte d'Ivoire. Because their primary network of relations extend outward into the world of Western buyers (rather than inward into the world of local village suppliers), they are dependent on village-level traders for their supply of goods. They themselves never buy art directly in villages.

Through their relations with Western buyers, urban traders have partial understanding of the world into which African art objects are being moved. Their experience enables them to discern certain criteria underlying Western definitions of authenticity. They know, through trial and error, which items are easiest to sell, and they can predict which objects will fetch the highest market price. Using this refractured knowledge of Western taste, traders manipulate objects in order to meet perceived demand. In this chapter, I will illustrate three ways in which African art objects are manipulated by African traders: (1) presentation of objects; (2) description of objects; and (3) alteration of objects.

Presentation of objects

The presentation of an art object for sale is not something which is carried out in a haphazard way. Careful preparations are made before an object is shown to a prospective buyer. The context in which an object is placed, and the circumstances surrounding its putative "discovery," weigh heavily in the buyer's assessment of quality, value, and authenticity. The presentation is thus a key element in the success of a sale. If an object is uncovered in its original or "natural" setting, it is believed to be closer to the context of its creation or use and, therefore, it is less likely to be judged as a fake.[5] The very process and act of discovery confirms the collector's sense of good taste. In a travelogue narration of a voyage to Côte d'Ivoire in the 1970s, one African art collector described his visit to the Hausa quarter in Korhogo in the following way: "While my travel companions were all taking naps, I *ventured* into the *obscurity* of a hut, which was *lost* somewhere in the heart of a labyrinth of identical alleys, where I saw a magnificent Senufo hunter's tunic covered with talismans: fetish horns, leather pouches, feline teeth, and small bones; but no figural sculptures worthy of my attention" (Lehuard 1977: 28, my translation from the French, emphasis added). As the narration moves on to western Côte d'Ivoire, the same collector comes across a Muslim trader who says he has masks for sale but that they are at his home which is a one hour walk away from town. "That is no obstacle," the collector tells the trader, and the two set off on foot. "In the [trader's] home," writes the collector, "where the *light of day scarcely shines*, the masks and statues are aligned in front of me. But nothing is really old. A Bété statue is frankly new. I bargain for two Dan masks (they cost me 400 FF) and I regain the path" (Lehuard 1977: 32, emphasis added). Finally, in another article describing his travels through Burkina Faso, the author provides his readers with the following advice: "For those who collect antiques, but cannot afford the cost, or do not possess the training to '*hunt*' for objects in the bush, it is always possible to *excavate* through the stock of the urban traders who are set up alongside the Hotel Ran in Ouagadougou. Although there is a preponderance of modern sculptures, it is sometimes possible to ferret out a rare gem" (1979: 19, emphasis added).[6] Throughout these travel narratives, the author's repeated allusions to the themes of discovery and concealment cater to a peculiar vision of the African continent – the West's unending search for the elusive beat in the heart of darkness.

The manipulation of context through the calculated emplacement of objects is a widespread practice among art dealers around the world. One of the classic tricks of the French antique market, for example, was to plant reproduction furniture in old homes. In the autobiography of André Mailfert, one of the most successful makers of reproduction French antique furniture, who worked in Orléans between 1908 and 1930, the author unveils the following scheme. A dresser was made to fit the specifications of a dealer who knew he could sell such a piece to a client who was soon scheduled to arrive from abroad. Once completed, the dresser was taken to the home of an elderly woman who lived on the outskirts of Paris. A wall in the front hall was thoroughly cleaned, the dresser was placed in front of it, and with the help of an air compressor, dust was sprayed on the wall so as to leave a noticeable mark around the edge of where the dresser now stood. The dresser was scratched in appropriate places, the drawers were filled with odds and ends taken from the woman's closet, the accumulation of decades of floor wax was quickly applied to the brass rim of the dresser legs, and a yellowed envelope, which dated from thirty years earlier, was casually tucked under the marble top to support a corner which had been made to warp. The customer was told that an old woman needed to sell an antique dresser in order to pay her taxes. He was brought to the woman's home, whereupon, after scrupulous inspection, he bought the dresser for 12,000 FF. For the use of her home the woman received 1,000 francs, the cabinet maker was paid for the price of the reproduction dresser, and the dealer kept the rest (Mailfert 1968: 54–57).[7]

In Africa, as in Europe and America, one of the key factors in the presentation of an art object is to create an illusion of discovery. Because of the barriers, which I noted earlier, that separate art consumers from art suppliers, the buyer almost never has the opportunity to uncover an object by chance or recognize its potential value *before* it enters the market, or *before* it becomes a commodity. However, part of the collector's quest, I would argue, is to discover what had previously gone unremarked.[8] As one European art critic put it, "the charm [of an art object] is bound up with the accident of its discovery" (Rheims 1961: 212). From the perspective of the Western collector, African art traders are perceived as mere suppliers of raw materials. It is the gifted connoisseur, not the African middleman, who first "sees" the aesthetic quality of a piece and thereby "transforms" a neglected artifact into an object of art (Price 1989: 7–22).[9]

African art dealers are well aware of the discovery element in Western taste. Depending on the circumstance, traders will sometimes pretend not to know much about the goods they sell in order to let the buyer believe that s/he is getting something about which the seller does not recognize the true value.[10] The principal art market place in Abidjan, located in the Plateau quarter, is constructed in such a way as to facilitate the illusion of discovery. The front of the market place consists of a series of adjoining stalls, with layered shelves, upon which are displayed quantities of identical items (Plate 28). These items are marketed largely as contemporary souvenirs made for the tourist trade. In the back of the market place, which is accessed by only three narrow paths, there are a number of large wooden trunks in which some of the traders store their goods (Plate 29). Many of the objects kept in the back of the market place are similar to those displayed in the front. The difference, however, is that they are *presented* to the buyer as unique items.[11]

Drawing on the dramaturgical idiom in the sociology of Erving Goffman, Dean MacCannell, in his book *The Tourist: A New Theory of the Leisure Class* has noted the interplay between front and back regions in the construction of authenticity at tourist sites. "Just having a back region," writes MacCannell,

28 Hausa trader at his stall in the front of the Plateau market place. Abidjan, November 1988.

"generates the belief that there is something more than meets the eye; even where no secrets are actually kept, back regions are still the places where it is popularly believed the secrets are" (1976: 93). "[E]ntry into this space," he concludes further on in the text, "allows adults to recapture virginal sensations of discovery" (1976: 99). By permitting the buyer to penetrate the back region of the market place, the African art trader thus underscores the value and authenticity of the object he is showing.

Like antique dealers in France, African traders also plant objects in appropriate settings. Hamadou Diawara, an elderly Malian art trader, recounted an instance in which he tried to sell a mask to an American buyer. The buyer carefully examined the mask, but refused to purchase it, telling Diawara that he doubted it was "real." Several months later, the buyer returned to Abidjan. Diawara planted the very same mask (which had not yet been sold) in a village located not far from the capital. The buyer was brought to the village and taken with great circumstance into the house where the mask had been placed. The object was examined, as best it could, in the dim light of the house. This time, the customer bought the piece and, in Diawara's words, was "very happy with what he had found" (11/7/87).

29 Hausa trader with wooden trunk in the back section of the Plateau market place. Abidjan, May 1988.

Description of objects.

Verbal cues affect judgment of authenticity and success of sale in ways similar to object emplacement. In other words, what we are told about a work of art conditions what we see (cf. Berger 1972). In Europe, for example, the title of a painting is sometimes changed to fit the current taste of collectors. In the 1950s, when macabre themes in art were more difficult to sell than light-hearted and amusing works, the title of Van Gogh's painting described in earlier catalogues as *The Cemetery* was changed in an auction catalogue to *Church under Snow* (Rheims 1961: 209). Another example of verbal embellishment comes to us from fifteenth-century Europe, where obsolete foreign coins were transformed (with a thin gold or silver overlay) into religious icons. "If these 'improved' coins failed to tempt a purchaser," reports Frank Arnau, "he was informed that they were the pieces of silver which Judas received for his betrayal of Our Lord. Countryfolk proved to be particularly interested in this type of relic" (1961: 27).

In Africa, traders communicate through verbal means two types of information about the objects they sell. To Western collectors and dealers, traders communicate information relating to the object's market history (how it was acquired, where it comes from, etc.). To tourists, traders provide information regarding the object's cultural meaning and traditional use. Both types of information are constructed to satisfy perceived Western taste, and are intended to increase the likelihood of sale.[12] In the language of mediation studies, one might say that the trader's description of an object is phrased at the level of metacommunication or "communication about communication." The sender's messages, to borrow Gregory Bateson's phrase, are "tailored to fit" according to his ideas about the receiver, and they include instructions on how the receiver should interpret their content (1951: 210).

African art objects in the West are sometimes sold with a documented "pedigree" which consists of a list of previous owners. A quick glimpse at auction prices would indicate that a mask once owned by Vlaminck or Rockefeller is worth far more than a similar mask owned by an unknown collector.[13] Before an object leaves the African continent, however, its true pedigree is carefully hidden from prospective buyers. African traders are aware that the Western collector is concerned with neither the identity of the artist nor a history of local ownership and exchange. In fact, an object is worth far more if it is perceived by the buyer to have been created by a long-departed artist, and to have come directly out of a village community.

As a result of the economic structure of the art market in Côte d'Ivoire, however, most purchasers of African art are not the first Westerners to have seen a particular object. Since an art object does not have a fixed monetary worth, traders sometimes test an object's value by bargaining with different potential clients. In some instances, "important" pieces are even tested in the markets of Europe and America before they are returned to Africa, where ultimately they are sold to local collectors. In an attempt to hide the true channels of trade through which art objects are moved, traders will always tell the client that s/he is the *first* Westerner to have an opportunity to buy a given piece.[14] Even though (as I noted above) most traders who have direct contact with Westerners are not the ones collecting art at the village level, they will either indicate that they themselves have just arrived from purchasing an object in a village, or that a villager has just brought an object directly to them.[15]

Malam Yaaro, a Hausa dealer specializing in the art of Ghana, recently shipped to New York City twenty-six Asante stools (Plate 30). Upon arrival, an African colleague asked to select six stools from the lot in order to show them to one of his clients. After the six stools were picked out, Yaaro contacted a New York gallery owner who purchased all twenty of the remaining stools. A few days later, the trader who had taken the six stools returned with the merchandise, informing Yaaro that his client did not want to buy the stools. Yaaro immediately contacted the gallery-owner who had bought the original lot, and told him that a smaller shipment of six very fine stools had just been

30 Hausa trader unloading a shipment of Asante stools at Kennedy airport. New York, June 1989.

sent to New York by an old Dioula woman. The true itinerary of the objects was kept hidden from the prospective buyer for several reasons. First, the seller did not want to anger the gallery owner by revealing that six of what might have been the finest stools had been removed from the original lot. Second, the seller did not want the buyer to know that the six stools had already been rejected (for whatever reason) by another Western buyer. And third, the seller wanted to create the illusion that the goods were freshly brought from Africa and had not yet even been seen by other Western eyes. In fact, by saying that an old woman transported the objects, the seller was further communicating to the buyer that not even another African trader had yet seen the stools.

While collectors are interested in learning the path of objects from village to market, tourists in Africa are concerned with learning the traditional meanings and functions of African art. Traders provide tourists not only with objects to take back home, but also with the knowledge of what the objects mean in their traditional contexts (cf. Spooner 1986: 198).[16] Indeed, it could be argued that for many tourists their experience in the market place would not be complete unless the traders told them something "interesting" about the objects they wanted to buy.[17] Jonathan Culler's remarks about tourist *sights* could easily be applied to an analysis of tourist *arts*. He writes:

> To be fully satisfying the sight needs to be certified as authentic. It must have markers of authenticity attached to it. Without those markers, it could not be experienced as authentic . . . The paradox, the dilemma of authenticity, is that to be experienced as authentic it must be marked as authentic, but when it is marked as authentic it is mediated, a sign of itself, and hence not authentic in the sense of unspoiled. (1981: 137)

Traders are caught in the webs of this "paradox of authenticity." On the one hand, by offering authenticity markers, traders provide an invaluable service to tourists. Yet, on the other hand, their role as middlemen and economic intermediaries denies the tourist direct access to a wellspring of "genuine" cultural encounters with the actual art producers and users themselves.

Most of the descriptions which traders use to embellish the objects they sell are derived from market-place lore – stories which are passed from one trader to another, or which are simply overheard in the banter of market-place discourse. Most of the explanations revolve around specific "key symbols" or stereotypes of African art, which include (1) the religious/ritual nature of the art; (2) the association of the art with kingship and royalty; and (3) the antiquity and traditional foundation of the art.[18] Some of the objects sold in the market are described as religious icons or sacred symbols: "This is the most sacred object" of a given ethnic group; or that the object is "carved from the most sacred piece of wood." Some objects in the market are also described as being associated with royalty or leadership: "This is the chair of the king" or "this is the mask of the big chief."[19] On this particular point, one Western observer has remarked wryly, "What pieces lacked in authenticity is compensated for by tall tales of their origins and use. Were these stories true, one would have to believe that Africa has as many chiefs as France has counts and marquis" (Imperato 1976: 73).

Some objects in the market are also described variously as "old," "an antique," "not from today," or from "before before." I once witnessed a Wolof trader selling a contemporary carving of an African hunter to a skeptical tourist who kept insisting that the object was not really old. In response to the buyer's incredulity, the trader was forced to revise his definition of age. Rather than concede to the buyer that the piece was, in fact, new, he simply shifted the locus of "antiquity" from one semiotic domain to the next. The result was that he never appeared to be *directly* contradicting himself. The trader began the bargaining by telling the client that the sculpture itself was old: "This is a very old piece" (*C'est une pièce très ancienne*). When pushed by the tourist to revise his claim, the trader explained that what he meant to say was that the workmanship was old (*ancien travail*) – in other words, he said, it was a replica of the type of work that was done in the past. Finally, when pushed even further to admit that it did not even represent a "traditional" style, the trader said, in exasperation, that at least the carving represented an old man (*c'est un vieux*).

Some of the market lore which is circulated among traders has nothing to do with "traditional" usage – they are stories which are invented purely to entertain the buyer. Among the items to which such stories are attached are small wooden masks which reproduce on a miniature scale the style of larger ones. Though most of these masks are attributed to the Dan of western Côte d'Ivoire, there also exist on the market miniature masks carved in the styles of several other ethnic groups. In their original context, miniature masks were integrated into a system of belief in which they functioned as spiritual guides and personal protectors (Fischer and Himmelheber 1984: 107; Steiner 1986b). In the market, however, they are known by the name "passport" masks. Asan Diop, a Wolof trader in one of the art markets of Abidjan, explains the function of "passport" masks to a group of European tourists: "Before the whiteman brought paper and pen to Africa, these small masks were the only form of identification that we Africans could carry with us. Each person owned a carving of himself and each tribe had its own kind of masks. This is the only way people could cross the frontier between tribal groups." The tourists were all amused by the trader's story, and one asked rhetorically "but where did they put the rubber stamp?" Collapsing the conventional categories of tradition and bureaucracy, the trader's explanation of the mask is phrased in terms which tourists can easily assimilate. It mocks the true meaning of the masks, reaffirms the tourist's sense of technical and cultural superiority, and provides an entertaining tale with which to return home.

Alteration of objects

Together with presentation and description, traders also satisfy Western demand through the material alteration of objects.[20] These activities include the removal of parts, the restoration of fractures and erosions (Plate 31), the addition of decorative elements (Plate 32), and the artificial transformation of surface material and patina (Plate 33). The simplest kind of material alteration consists of removing an object from its base. Objects from local private collections and galleries are sometimes put back on the market in the hands of African traders. Gallery owners, who are having trouble selling certain pieces in their Abidjan showrooms, may choose to liquidate a portion of their stock by consigning objects to an African trader. The moment the trader receives the art, he removes and discards any base on which the dealer may have mounted the object. I questioned a trader as to why he should remove the base from a piece which was so unbalanced that it could not otherwise stand. He told me that his clients would never buy something from him which had been mounted. The presence of the mount indicated that the object had already been "discovered" by another collector. Its removal reaffirmed the image that traders collect art directly from village sources.[21]

To satisfy the demand among certain Western clients for strong evidence of age and ritual use, traders replicate the shiny, worn patina which results from years of object handling. They reproduce surface accumulation of smoke, soot, soil, and dust. And, they imitate the encrustation of blood, feathers, and kola nuts which results from repeated sacrificial offerings. Because it is inexpensive and easy, traders tend to spew chewed kola nuts on a variety of objects. The process is undertaken on pieces that might normally receive such a treatment in its original context, such as masks and statues (Plate 34), but the principle is also extended to objects which would never receive kola sacrifice – such as wooden hair combs (Plate 35).

When collectors of African art began judging the quality of masks by the amount of wear on the inside surface of the mask (i.e., the so-called "sweat marks" or "grease marks" which form on the surface of the wood where the dancer's forehead, nose, and cheeks would rub), traders responded by rubbing oil (sometimes the residues which ooze from the "sea bean")[22] into the inside of a mask. More recently, many tourists have been alerted (by tour books and tour guides) to look at the inside of a mask for the tell-tale signs of use which testify that it is indeed a "real" African dance mask. Unlike the collectors, however, most tourists are not experienced enough to know that the marks are often evidenced by subtle discolorations and delicate variations in surface patination which might only be visible under certain light conditions or when viewed at particular angles. To ensure that the tourists do not overlook the presence of these marks, traders (or

31 A carver repairing the arm on an Asante female figure which was damaged during shipment from the Kumase workshop where it was produced. Plateau market place, Abidjan, January 1988.

32 Storehouse assistant embellishing Dan masks with padded cowrie-covered headdresses. Treichville quarter, Abidjan, November 1987.

33 Small Akan brass boxes stained with potassium permanganate to dull the surface finish. Plateau market place, Abidjan, July 1988.

commissioned artists) use sandpaper to boldly inscribe the stamps of authenticity.[23]

During the early eighteenth century in Paris, French collectors of Oriental art were offended by the nakedness of certain Chinese *celadon* porcelain figures. In the hope of fetching higher prices, Parisian dealers responded by dressing up the figures in French clothes (Rheims 1961: 167). In a parallel (but inverted) case, African art traders involved directly with European buyers have remarked that their clients are more likely to buy naked Baule statues than those whose waist has been covered by a carved wooden loin cloth or *kodjo*. Collectors prefer to buy Baule statues that have no loin covering (Plate 36), or those with loin protectors made of actual cotton fabric which is affixed to the surface of the wooden sculpture (Plate 37). These type of Baule figures are thought by collectors to pre-date those that have loin cloths carved into the sculptural form itself (cf. S. Vogel 1981a).[24] Traders have responded to this preference among collectors by systematically removing, with the use of a chisel or knife, the wooden loin cloths which cover the Baule figures (Plate 38). The unstained wood which is left as a result of this process is restained with an appropriate dye, and an old piece of fabric is tied around the figure's waist to cover the damage. I have witnessed some collectors, who are now aware of this practice, peeking under a statue's loin cloth,

34 Dan wooden face masks covered with kola nut compound to imitate the surface texture of a mask that had received "traditional" sacrifices. Man, June 1988.

35 Asante combs splattered with a kola nut residue. Plateau market place, Abidjan, March 1988.

before buying it, to see whether its sculptural nakedness was genuine or spurious.

Traders also commission artists and craftsmen to convert, modify, beautify, improve, and repair a whole variety of objects. Utilitarian items with no obvious aesthetic motif are embellished with figural or geometric patterns. The handle of a well-worn wooden spoon, for example, can be decorated at one end with a carved representation of a human face. A comb with no adornment is embellished with a geometric motif on its handle. A beater bar (or heddle comb) with no decoration has a mask carved on its large wooden base. Old wooden bowls are "transformed" into circular face masks by gouging eye holes, carving a nose and mouth, and affixing cowrie-laden leather thongs across the chin. In so doing, an undecorated (and potentially unsaleable) bowl becomes a sculptural form which – because of the wear on the bowl – looks like it is old and used.

Objects which have decayed because of exposure to insects or harsh elements are reworked to satisfy demand.[25] A Lobi figure with a heavily eroded head and body has the features of its face recarved so that it is once again recognizable as a human being with a facial expression. The rest of the body is left untouched – showing the age and wear on the figure.[26] Combs with broken teeth are repaired. Chairs with broken legs or backs are

36 Baule figure with beaded waistband and necklace. Height 29 cm. Photograph by Hillel Burger.

37 Baule figure with cotton loin-cloth affixed around its waist. Height 32 cm. Photograph by Charles Lemzaouda.

fixed. Statues with missing arms, feet, or legs are restored to their original form.

Finally, a poignant example of object alteration can be found in one of the recent trends in the Ivoirian art market, namely in the sale of so-called *colon* or "colonial" statues (Plate 39). Wooden carvings of *colon* figures (representing either Europeans or Africans in Western attire) are found in societies throughout West Africa (Lips 1937; Jahn 1983). Though bearing elements of European design (clothing, posture, and various accoutrements) these statues were not originally conceived for the market, but rather for indigenous use. Among the Baule, according to research conducted by Philip Ravenhill, statues in fashionable dress were used in the same manner as other wooden statues to represent a person's "spirit lover" in the other world. "A Baule statue in modern garb," he writes, "is neither a replica of a European nor the expression of a wish for a European other-world lover, but rather a desire that the 'Baule' other-world lover exhibit those signs of success or status that characterize a White-oriented or -dominated world" (1980b: 10).

During the colonial period, modern polychrome statues, such as Baule spirit mates clothed in European dress, were not generally sold in the African art market. A Wolof trader, who has been selling African art in central Côte d'Ivoire for over forty years, recounts the following:

> My father began as an art dealer in Senegal in 1940. In 1945 we moved to Côte d'Ivoire and set ourselves up in the town of Bouaké. At the time, *colon* statues had no value whatsoever in the art market. In the region of Bouaké, where there were many such carvings, we called them "painted wood" and would give them as gifts to customers who purchased large quantities of other merchandise . . . But some clients even refused to take them for free.[27] (Quoted in Werewere-Liking 1987: 15)

During the late 1950s, toward the end of French colonial rule in Côte d'Ivoire, foreign administrators, civil servants, soldiers, and other colonial expatriates began commissioning portraits of themselves – as souvenirs to take back home. This gave rise to a whole new genre of "tourist" art which grew out of an indigenous tradition of representing Africans in Western attitudes or attire (cf. Gaudio and Van Roekeghem 1984).

The colonial style of carving has reached new heights of popularity during the past several years. Following a series of well-publicized auctions held recently in London and Paris (Chauvin 1987; Melikian 1987; Roy 1987) – where *colon* figures sold for significantly more money than they ever had before – the value of *colon* statues in Côte d'Ivoire (mainly those carved in Baule and Guro styles) has been inflated dramatically and the production of replicas has swelled (Plate 40).[28] In addition, the publication of an illustrated book, *Statues colons*, by an Abidjan-based writer and gallery owner, as well as two articles on the subject of *colon* figures in the weekly Abidjan entertainment

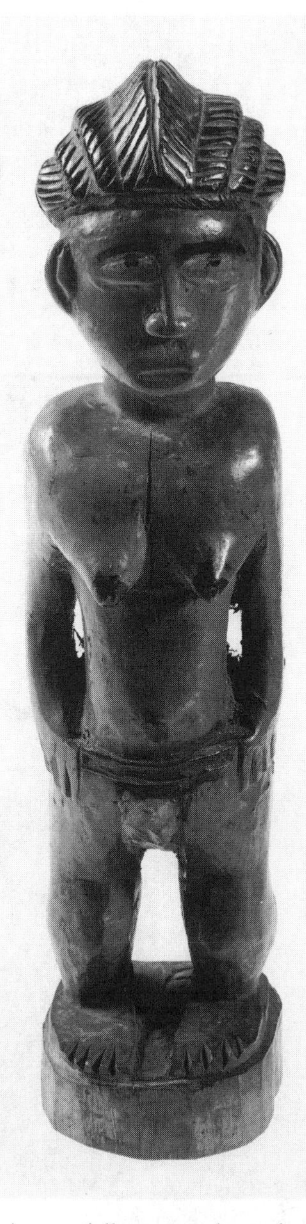

38 Baule figure showing partially removed wooden "loin-cloth." Height 17 cm. Private collection. Photograph by Hillel Burger.

39 Baule "colonial" figure wearing Western-style cap, shirt, shorts, sandals, and wristwatch. Height 24 cm. Private collection. Photograph by C. B. Steiner.

40 Guinean workshop artist with "colonial" figures. Port de Carena, Abidjan, June 1988.

guide (*Guido* 1987: 42–46; 1988b: 74–75), has further increased the demand for such carvings by European expatriates and tourists alike.

When traders commission *colon* statues from workshops they sometimes specify the style they want. One trader in the Plateau market place, for example, sold an equestrian colonial-style figure for a considerable margin of profit. A few days later, he sent word to a workshop in Bouaké that he wanted to order six *colon* figures on horseback with riders wearing military helmets. When they do not specify a particular style, artists offer a whole range of different statues – figures carrying bibles and rifles, soccer balls and tennis rackets. When carvings such as these are purchased from the artists, they are always painted in lustrous enamel colors (Plate 41). Traders have found, however, that brightly painted objects do not sell as well as faded, older-looking ones – i.e., buyers like to believe that the statues were made during the colonial era, and that the paint has eroded naturally through time. Thus, when a trader purchases a *colon* figure from a workshop, he will remove (or pay someone else to remove) a layer of paint with sandpaper (Plate 42). The object will then be stained with potassium permanganate. This treatment of the object produces a darkened surface, with flaked paint, which can often be marketed as antique.[29] The process of artificial aging underscores the separation of art suppliers from art buyers. Most of the artists producing *colon* statues for the

41 Workshop apprentice painting "colonial" figures. Bouaké, December 1988.

42 Dioula trader sanding down the paint from a lot of newly arrived "colonial" statues. Plateau market place, Abidjan, June 1988.

market have no idea that their works are being transformed by traders, and furthermore, I believe, they would not really understand why such a transformation would add value to the piece.[30]

The commoditization of *colon* figures has altered the discourse of their meaning. As the intended audience shifts from the African village community to the Western export trade, the message that this style of carving conveys is changed radically.[31] Originally conceived as icons signifying the incorporation of new status represented by imported materials, these seemingly similar statues now symbolize the impact of colonialism in Africa. Unlike most African art which is brought into the West and made to stand for the exotic nature of a romanticized Other, *colon* figures stand for the Other's relationship to the West. In the context of the transnational African art trade, the *colon* figure is perhaps the ultimate "postmodern" creation – as defined recently by literary critic Terry Eagleton:

> There is, perhaps, a degree of consensus that the typical postmodernist artefact is playful, self-ironizing and even schizoid . . . Its stance towards cultural tradition is one of irreverent pastiche, and its contrived depthlessness undermines all metaphysical solemnities, sometimes by a brutal aesthetics of squalor and shock. (Quoted in Harvey 1989: 7–8)

Rather than achieve their value or authenticity by an emphatic denial of foreign contact, *colon* figures are interpreted by their buyers as a celebration of modern Western expansionism. Like the "vintage imperialism" or "safari-look" created by fashion designers at Banana Republic and Ralph Lauren (Brown 1990), the art traders in Côte d'Ivoire are constructing a marketable fantasy of the colonial experience. In its new context, the *colon* statue is *still* a symbol of social status not, however, because it represents the appropriation of the West by Africa, but rather because its very ownership by a Westerner signifies the reappropriation of Africa and is thus prized as an image which pays homage to the conquest of the continent.

Trade and the mediation of knowledge

The concept of "cultural brokerage" found its way into anthropological discourse sometime in the late 1950s. The term has principally been applied to the realm of politics, where culture brokers are described as middlemen bridging a socio-cultural distance in patron–client political relations. Building on the work of Fredrik Barth, Robert Paine (1971) describes the cultural broker as a middleman capable of attracting followers who believe him able to influence the person who controls the political authority (the patron or, from the Spanish, *patrón*). Paine draws a useful contrast between a *broker* and a *go-between*. The go-between, he says, is someone who faithfully relates instructions or messages between two separate parties. The broker is someone who manipulates, mediates, or "processes" the information which is being transmitted between the two groups (1971: 6). The go-between is thus understood as a direct liaison or relay in a complex system of network communications. The broker is one who interprets, modifies, or comments on the knowledge which is being communicated. "[W]hile purveying values that are not his own," writes Paine, "[the cultural broker] is also purposively making changes of emphasis and/or content" (1971: 21). Hence, rather than simply facilitate the relationship between two different groups separated by social, economic, or political distance, the broker actually constitutes, molds, and redefines the very nature of that relationship. In the literature on cultural brokerage, the brokers who mediate between two different cultural groups are described as being "bi-cultural" in their knowledge of the two cultures, in their ability to communicate with both cultures, in their living style and physical appearance, and in their capacity to value certain elements from each culture (Briggs 1971: 61–2; Nash 1977: 40–1). The broker is perceived as someone who has "vested interests that are separate from yet dependent upon those groups to which he is intermediate" (Paine 1971: 99). Although he must understand the phenomenology of the worlds which he bridges, the broker "must now allow himself to be caught in the web of his clients' symbols, even when he undertakes to interpret these symbols to third parties" (Paine 1976: 80).

African art traders fit very nicely the model of the "cultural broker." Like the political mediator described by Paine and others, the African art trader is a cultural mediator between two groups brought in contact through common economic pursuits. Like the political mediator, the art trader is (in the language of Victor Turner) betwixt and between the groups he "brings together."[32] As links in a long chain of distribution, art traders control neither the supply nor the demand which they mediate. They can neither create a stock of objects necessary to satisfy the market, nor can they create a market for the objects they have in stock. The two principal components of a market system (supply and demand) are controlled by forces external to the trader's world.[33] Supply is dependent on the availability of objects from village sources and on the production potential of contemporary artists. Demand is largely set by Western publications, museum exhibitions, auction records, and the tourist industry. Like the *bricolage* of Lévi-Strauss's famous myth-maker, the art trader constructs a product from raw materials and conceptual tools which are limited and pre-determined by elements outside his immediate control. The best the trader can do is manipulate the perception of the objects he has at hand in order to meet what he believes are the tastes and demands of the Western buyer. Although I respect and espouse E. P. Thompson's (1966) assertion (*contra* Althusser) that historical subjects are as much determining as determined in shaping the(ir) world through lived experience, I also believe that the intentionality of historical life is not without limitations – i.e., culture offers not only a range of possibilities, but must also be understood as something which presents a limit on the possible (cf. Ulin 1984: 148–53). Thus, although traders fashion and market images of Africa and African art, these images are constrained by the buyer's *a priori* assumptions about what is being bought – the images are constructed to *satisfy* demand rather than to *create* demand.[34]

Separated by oceans of geographic distance and worlds of cultural differences, African traders and Western collectors are brought together in a fleeting moment of economic exchange. From their brief encounter, the Western buyer departs with an object of art which s/he will integrate into a world of meaning comprehensible only through Western eyes. The African trader walks away with yet another impression of Western taste which will become part of his repertoire of understanding how Westerners perceive African art.

Conclusion: African art and the discourses of value

> The African in general has no desire to produce more than he needs for subsistence level. That is the experience, I think, of all of us who have lived here long enough, and I think we realize that putting silver in his pocket is not the African's first aim and object . . . [Indeed] one of the biggest problems is the high leisure preference of the African.
>
> Legislative Assembly of Southern Rhodesia, 1960
> (quoted in Montague Yudelman, *Africans on the Land*, 1964)

> The art [of Africa] has degenerated [in the last few years] and veritable studios have opened to meet the demand of tourists. All that was good in African sculpture has disappeared in these examples conceived solely for sale . . . Contemporary works are far from having the same value as the older ones which were made at leisure, without any consideration for monetary profit.
>
> Denise Paulme, *African Sculpture* (1962)

It was long held in the West that "economic man" was not to be found among the indigenous populations of sub-Saharan Africa. During the colonial period, scores of documents and testimonials were written by administrators and other Western observers which reported on the putative lack of interest among Africans for matters of money, market, and work. In general, it was believed among the colonizers that Africans were unable to appreciate either the value of free enterprise or the economic potential of wage labor. In 1911, for example, the French colonial governor of Côte d'Ivoire complained to his compatriots back in the metropole, that "the indigenous populations are excessively lazy and will never devote themselves to regular work" (Angoulvant 1911: 167). In short, it was thought, by the foreign powers ruling overseas, that *homo antiquus*, not *oeconomicus*, was the chief inhabitant of the African colonies (cf. Miracle and Fetter 1970; Perinbam 1977).

During the same period of time – when colonial capitalist ventures were being stifled by an alleged unresponsiveness of Africans to economic incentives – the collectors and connoisseurs of African art in Europe and America were already beginning to complain that the traditional spirit of African art was being lured away by the seductive spirit of capitalism. They lamented the loss of artistic traditions, and declared that the quality and aesthetic merits of art objects in Africa were suffering an abrupt decline as a result of the hedonistic calculus of the African artist and his no better accomplice the African trader – both of whom were reproached for their excessive eagerness to reap the financial benefits of the Western market economy.

On the one hand, then, if they were unwilling to participate fully in the wage labor of a colonial mode of production, Africans were criticized by Westerners and viewed as traditional, non-market oriented, and blind to the wondrous possibilities of a mercantile capitalist economy. Yet, on the other hand, if they were engaged in the production or distribution of art for economic gain, Africans were *equally* criticized by Westerners, and viewed as commercial, debased, or even worse, "greedy." While those in the West who bought, sold, and collected African art cherished the "pre-capitalist" principles of so-called traditional African society, those involved in the production and export of cash crops, and other colonial commodities, found these same principles to be wholly incompatible with their own expansionist goals.

Although these two Western attitudes toward African productivity are indicative of a whole range of contradictions in the colonial (and indeed post-colonial) relationship between Africa and the West, for the purposes of this discussion I wish to single out one point in particular, namely that these differences stem, at least in part, from a collective misconception of the nature of value in the Western art world – that is, a total disavowal of the economic structures that underlie both the production and consumption of art or, what may more generally be called, cultural capital. "[W]e all go to great pains," Mark Sagoff has noted, "to distinguish the value of art as art from the value of art as an investment" (1981: 320). In other words, the aesthetic value of art is always seen as separate from, or rather above, its economic value.

In his book *Rubbish Theory*, a seminal study on the creation and destruction of value in the world of material goods, Michael Thompson draws attention to those who stand guard at the carefully monitored boundaries between these two discursive domains – what he calls an "art aesthetic," on the one hand, and a "commercial aesthetic," on the other. "Those persons who are particularly concerned with the manning of the controls on the transfers between categories," writes Thompson, "operate almost entirely in terms of aesthetic values, refusing to countenance the vulgarities of economics, and directing all their energies towards the paramount duty of maintaining purity by identifying rubbish and preventing it from getting anywhere it does not belong" (1979: 115).

According to Thompson's theory of value as applied to the art market, both collectors and producers of art have a vested interest in denouncing the gallery system and its seemingly vulgar economic attribution of value. For the art collectors, it is necessary to obscure the relationship between the art aesthetic and the commercial aesthetic in order to certify their capacity to recognize an object's "pure" artistic value untainted, as it were, by the inelegance of economic speculation. Although the two systems of value are obviously linked via the art object itself, the collectors must treat them as *separate* categories in order to legitimate their claim. As for the producers of art, Thompson goes on to argue, they too must deny the connection between the art aesthetic and the commercial aesthetic. If an artist's work is rejected by the established gallery system, then the artist can capitalize on the art-commercial distinction by creating anti-establishment objects – e.g., conceptual, auto-destructive, or unpossessable objects that achieve value through their repudiation of a commercial aesthetic. However, if the artist's work is accepted into the gallery system, then the artist just "adopts the gallery owner's line that the two aesthetics . . . are simply different aspects of some whole and that the best art works, judged by the art aesthetic, will inevitably command the highest prices" (Thompson 1979: 122).

No one has analyzed this conceptual negation of the economic value of art with greater insight and understanding than sociologist Pierre Bourdieu. In a general statement on the subject, Bourdieu says, "The challenge which economies based on disavowal [*dénégation*] of the 'economic' present to all forms of economism lies precisely in the fact that they function, and can function, in practice . . . only by virtue of a constant, collective repression of narrowly 'economic' interest and of the real nature of the practices revealed by 'economic' analysis" (1980: 261). Writing specifically about the European art market, Bourdieu goes on to say that the "art business, a trade in things that have no price, belongs to the class of practices in which the logic of the pre-capitalist economy lives on" (1980: 261). Adopting a Durkheimian language, he further suggests that the "world of art stands in opposition to the world of daily life, like the sacred to the profane" (Bourdieu and Derbel 1990: 165; see also Bourdieu 1968: 610–11).[1]

Drawing both on Thompson's deconstruction of the axiology of aesthetics, and on Bourdieu's analysis of the "misrecognition" (*méconnaissance*) of cultural capital, I will demonstrate below how the Western (e)valuation of African art builds itself in direct opposition to both use value and exchange value. In the first instance, I will argue that Western discourse on African art denies the instrumentality – material, metaphysical, or spiritual – of an African object once it has reached the Western context. In its new setting, the art has no "function" other than to attest to its "non-functional" role. In the second instance, I will argue, that the discourse denigrates economic benefit as a basis of, or motivation for, African art connoisseurship and collecting. In other words, cultural capital can only be maximized if economic capital is disavowed. The discussion will then conclude by locating the place of the

African art trade between these two systems of value or, what might be called, these two enchanted worlds – the spirit world from which the objects are artfully removed and the art world into which the objects are spirited.[2]

The disavowal of use value

African art objects when displayed in the West are elevated to the category of art by denying them their former utility or use value[3] – baskets and calabashes are displayed on pedestals not balanced on the head, face masks are suspended without motion from mounts on the wall not danced in the open space of a village square, and statues are intended for disinterested contemplation not religious veneration.[4] Indeed, it could be argued that an object's aesthetic merit in the art world is heightened, or perhaps even made possible, by the very fact that its value transcends any practical function thereby shedding its former utility.

Writing in the late nineteenth century, Georg Simmel touched upon this idea briefly in *The Philosophy of Money*. Though phrased in evolutionary terms, his observations nonetheless translate remarkably well to the process of valuation that occurs in the African art market today – a process which operates through transformations in space rather than in time. Simmel wrote:

> The beautiful would be for us what once proved useful for the species, and its contemplation would give us pleasure without our having any practical interest in the object as individuals . . . [W]hereas formerly the object was valuable as a means for our practical and eudaemonistic ends, it has now become an object of contemplation from which we derive pleasure by confronting it with reserve and remoteness, without touching it. (1978, ed.: 73–74)

Raising a similar point in order, however, to make a quite different argument, Bourdieu, more recently, has remarked that:

> [art] objects are not there to fulfil a technical or even aesthetic function, but quite simply to symbolize that function and solemnize it by their age, to which their patina bears witness. Being defined as the instruments of a ritual, they are never questioned as to their function or convenience. They are part of the "taken for granted" necessity to which their users must adapt themselves. (1984: 313)

Finally, following a parallel train of thought, for the purpose of making yet a different argument, literary critic James Bunn has suggested that, "Because objects may be given different turns as determined by their contexts, they are either useful or wasteful; in waiting they are potential . . . Removed from the context of use, tools become objects of art that display themselves as failed metaphors, as 'utensils' warped from one category of means to another of forestalled ends" (1980: 313).[5] Whether one constructs this general argument with the aim of demonstrating the necessity of "distance" in the creation of value (Simmel), revealing the class structure that underlies the consumption and appreciation of art (Bourdieu), or elucidating the transformative sequence of tropes through which "use" and "beauty" unfold (Bunn), the essential point remains constant: i.e., that the disavowal of use value is a prerequisite for admitting what has the potential of being utilitarian into the realm of aesthetics.

One may conclude from this discussion that in the West, at least, use and beauty are not threads in a single fabric of meaning but are viewed, more commonly, as separate elements of value competing toward radically different ends.[6] In Oliver Goldsmith's novel *The Citizen of the World*, a fictional memoir written in the last century, the central character, a Mandarin philosopher traveling in England, recounts the following visit to the home of a British collector of *chinoiserie*. The passage illustrates clearly not only the contempt for utility in the Western world of "art," but interestingly it also places the Western negation of use value within the framework of a self/other or civilized/primitive dichotomy.

> "I have got twenty things from China," [said the lady], "that are of no use in the world. Look at those jars, they are of the right pea-green: these are the furniture." Dear madam, said I, these though they may appear fine in your eyes, are but paltry to a Chinese; but, as they are useful utensils, it is proper they should have a place in every apartment. "Useful! Sir," replied the lady, "sure you mistake, they are of no use in the world." What! are they not filled with an infusion of tea as in China? replied I. "Quite empty and useless upon my honour, Sir." Then they are the most cumbrous and clumsy furniture in the world, as nothing is truly elegant but what unites use with beauty. "I protest," says the lady, "I shall begin to suspect thee of being an actual barbarian." (1891: 51, cited in Bunn 1980: 313)

Like Chinese jars housed in a British parlor, African art objects in the West are divorced from their proper function and original meaning. When taken out of context, African art is no longer valued "instrumentally" or "as a means to some end," but rather it is valued "for its own sake" or "as an end in itself" (cf. B. Smith 1988: 126).[7] When collectors or tourists ask a trader in the market place what function an object has (or had) in its indigenous setting, they are not asking the question in order to replicate its use in their own environment, but rather they are seeking to uncover the function of the object in its prior "life,"[8] thereby allowing the obsolescence of its past to testify, and indeed to celebrate, a loss of utility and functional value.

Because most of the traders in the art market are Muslim, and do not therefore believe in the sacredness or ritual efficacy of the objects they sell (see Chapter 4), they are generally not distressed by the desacrilization of African art in the West. That is to say, they accept the notion of aesthetic contextualization as a natural result of the commodifying process. In one instance, however, which I recorded in the Plateau market place, a group of Muslim traders became noticeably disturbed by a purchase which had been made by a French art dealer. One of the market-place traders had sold the dealer an Islamic "protective" belt (*guru*) – a narrow thong from which are suspended about a half dozen small leather pouches containing Qur'anic scriptures and verses (*laya*).

After witnessing the purchase of the belt, one of the Muslim traders, who had gathered around the dealer, asked him why he had bought this sash of Qur'anic charms. The dealer simply said that he had bought it because he "liked the way it looked." Then Ali Bagari, another Muslim trader who had joined the congregated crowd, added, in an acrimonious tone, "But you didn't have the belt commissioned for yourself, so you don't know what the charms are used for. They might be for hunting. They might be for travel. What good will it do you, if you don't know what the charms are for?" (6/7/88). A moment of tension was broken by group laughter – Bagari shook his head, silently voicing his bewilderment and disapproval. Because the belt – unlike most other things that are sold in the art market – represented to the Muslim traders something which had a meaningful religious use value, and therefore more than either a simple economic exchange value or an aesthetic value, the boundary that normally separates these different systems of value was suddenly brought into question – the process of commoditization having been momentarily jarred.

The disavowal of exchange value

Bourdieu's remarks about the negation of the economy in the collection of works of art, are plainly confirmed by an analysis of the African art market in the West – not only in the attitude of its participants toward the production of art for sale (i.e., tourist or export art), but also in the way they judge those who sell or collect art purely for financial gain and investment. Writing in 1979, for example, art dealer/collector Irwin Hersey pointed to what he perceived as a disturbing trend in the African art market which was moving toward the collection of art for unabashed monetary gain. "For the first time," he writes, "the field [of African art collecting] is being *invaded* by investors who are more interested in appreciation of the value of the objects than in appreciation of its beauty" (Hersey 1979: 1; emphasis added). Evoking similar concerns, another African art dealer remarked:

> I'm getting a little tired of people coming in and saying, "What's the best investment piece you've got here?" I've always enjoyed selling beautiful things to people who appreciate them. Now, however, it's a whole other ballgame. These people couldn't care less about whether an object is a great work of art or not. All they're interested in is how much money they can make on it. It's really terribly discouraging. (Quoted in Hersey 1979: 1–2)

Both of these statements underscore without question Bourdieu's contention that "too obvious a success in the market or what is worse too obvious a desire for such success leads to cultural delegitimization because of the overall struggle between cultural and economic capital" (Garnham and Williams 1980: 220–21).

Following a general pattern of inflation in the international art market, ownership of African art today has become linked closely to economic investment. Once considered a thrifty substitute for the ownership of modern art,[9] African art (when purchased from a "reputable" auction house or gallery) now constitutes a major financial venture with high monetary stakes. Yet, on the whole, as Bourdieu predicts, collectors emphatically deny that they collect for economic gain. Like the functional context (use value) of African art which is said to interfere with the collector's pure aesthetic response to the *objet d'art*, the economic dimension of art (its exchange value) is also said to compete with a true sense of artistic enjoyment.[10] "Allow yourself to experience the object itself," gallery owner Ladislas Segy once advised his customers, "If you know what an object costs, you experience it differently" (quoted in Anon. 1988: [2]).

The commoditization of African art objects in the twentieth century would, I imagine, have fascinated and perhaps even astonished Karl Marx. Objects that were once the subjects of "fetishism" in a world putatively dominated by an organic unity between persons and their products, now become the subject of a new form of fetishism – commodity fetishism – which results from the calculated alienation of production from consumption and the overestimation of transcendent worth in the pseudo-sacral space of the international art market.[11] The mystification of value that Marx analyzed so brilliantly in the context of late-modern industrial capitalism, has in its post-modern manifestation become nearly a parody of itself.

Consider, for example, the words of one American collector as he explains his personal desire for anonymously produced African objects:

> It doesn't bother me . . . looking at a piece [of African art] that I don't know the name of the artist. I mean I like the object itself enough, that I don't need that kind of information. Part of my desire to collect magical, strange objects would be destroyed if I knew it was carved by Mr. X in such and such a village and [that] he spent so much time carving it. If I knew the whole process so exactly, it would take away some of the magic and mystery. (8/10/91)[12]

The collector's remarks underscore with crystalline clarity the general assertion that the trade in African art conceals or, as it were, mystifies the relationship between human labor (both production and exchange) and its products. The *work* of African art thus becomes socially repressed by a complicity of consumers who destroy in their imagery of the African art object all traces of production, and, in the end, celebrate the decontextualized results of dehumanized labor – the mysterious sparkle of the commodity cult.

Between two enchanted worlds

It has long been said that African art is produced in an enchanted world – a world in which a community of believers endow their ritual objects with spiritual values which are seemingly incommensurate with their physical properties. Oceans away, in a geographically separate and culturally distant corner of the globe, African art is being consumed today in a totally different, yet, I would argue, equally enchanted world – a world in which a community of believers endow objects of art with economic values that are seemingly incommensurate with their cost of production. And, at the same time, through a mystifying sleight of hand, these same devotees deny completely that their cultural capital is consecrated by, and indeed deeply embedded in, the wider economy. African art objects are moved, therefore, from a world that defies the logic of ontology to one which denies the logic of capital.

Somewhere between these two enchanted worlds is wedged, rather awkwardly, a lively commerce in African art objects – a commerce which moves in the hands of professional African traders objects of art and material culture from their context of creation to their point of consumption. Confronting directly the objects they sell, with no motives of sacralization or pretensions of economic disavowal, African art traders unveil a naked truth in the logic of art valuation – a truth which sometimes exposes itself rather unpleasantly to those who dwell in either of the enchanted realms.

Through a complex process of cultural and economic exchange, the African art trade reveals, while it also constitutes, the linkages and connections that make the world a system. Rather than view the process of commoditization as an impersonal force which arrives full-sail like a ship to shore, the study of the African art trade challenges us to view the penetration of capitalism in Africa as a series of personal linkages, forged one at a time by different individuals each with their own motives, ambitions, and set of goals. At all levels of the trade, individuals are linked to one another by their vested interest in the commoditization and circulation of an object in the international economy. Yet, ironically, the very object which brings them together does not hold the same value or meaning for all participants in the trade.

Notes

INTRODUCTION: THE ANTHROPOLOGY OF AFRICAN ART IN A TRANSNATIONAL MARKET

1 Eric Wolf has also noted the similarity between early diffusionist anthropology and recent studies of world systems and global political economy. "[The diffusionists] were not much concerned with people," he writes, "but they did have a sense of global interconnections. They did not believe in the concept of 'primitive isolates'" (1982: 13).
2 In order to respect the decree of 14 October 1985 by President Félix Houphouët-Boigny, the country name "Côte d'Ivoire" will not be translated into English. The original remarks made by the President on this subject were transmitted by the Foreign Broadcast Information Service. My thanks to Amelia Broderick of the United States Information Service in Abidjan for making available this document.
3 Dioula is a Mande language spoken in different forms by people throughout the western Sudan. Although the term "Dioula" refers to a specific ethnic group in northern Côte d'Ivoire, the Dioula language is more commonly a reference to a Dioula-related lingua franca – a language that is used as a medium of communication among people who have no other language in common (see Lewis 1971; Launay 1982).
4 The presence of a tape recorder caused such a flurry of activity whenever it was brought out – people either objecting to its presence or wanting to buy it – that I decided against the use of a recording device. Notes were either written at the time of interview and observation, or transcribed from memory at the end of the day. Some of the direct interview quotations that appear in the book, were collected during the filming of *In and Out of Africa* – a documentary video project on the African art trade that I collaborated on in 1991.
5 It is ironic to note that while admiring African art for its supposedly "communitarian," "antiauthoritarian," and "egalitarian" qualities, the modern primitivist artists and intellectuals had inadvertently brought these very objects into a competitive and hierarchical capitalist market economy.
6 The technical development of the long-distance steamship, which took place at about the same time, also helped of course to provide improved access for Europeans to the region (Baillet 1957).
7 For details on the life of Paul Guillaume, see Revel (1966) and Bouret (1970).
8 A second example of African objects being moved through industrial-commercial channels comes from Bohemia (modern day Czechoslovakia), where a German

manufacturer of glass beads, Albert Sachs, was acquiring African art as early as the 1980s through his company representatives in West Africa (Paudrat 1984: 147).

9 From a conversation with Pierre Verité, Paris, August 23, 1986.

10 For some recent views on the history of the study of African art, see Adams 1989; Ben-Amos 1989; Blier 1990; Drewal 1990; Gerbrands 1990.

11 The justification for the study of African art within the narrow parameters of a single village community suspended artificially in time and space is similar to the motivation for so-called "salvage ethnography" and "salvage archaeology" which were characteristic approaches to the anthropological study of Native Americans at the turn of the century in the United States.

12 Journalist Nicholas Lemann encountered this problem when he conducted an investigative report for an article in *The Atlantic* on African art traders in New York City. "Because I found that it was impossible to gain access to the runners as a reporter," he says, "I began buying art from them earlier this year" (1987: 28).

13 As in many art/antique markets, the "discovery" element plays a crucial role both in the assessment of value by consumers and in their perception of authenticity. If a collector believes that s/he is the first to whom a trader has shown a particular object, then there is a greater chance that s/he will be interested in buying the item. The more often an object is shown among a particular circuit of buyers, the less it is worth in the eyes of the local collectors. In this sense, although art objects are not perishable in the same way as fruits or vegetables, art traders share some of the risks of a perishable commodity market.

14 I take my lead here from Pierre Bourdieu who once remarked that the "world of art [which is] a sacred island systematically and ostentatiously opposed to the profane, everyday world of production, [offers] a sanctuary for gratuitous, disinterested activity in a universe given over to money and self-interest" (1977: 197). Like theology in a past epoch, he concludes, the art world provides "an imaginary anthropology obtained by denial of all the negations really brought about by the economy" (*ibid.*: 197). This densely stated point raises a theme which Bourdieu takes as the subject of his later book *Distinction: A Social Critique of the Judgment of Taste* (1984).

15 In *Primitive Art in Civilized Places* (1989), Sally Price offers a phenomenology of the Western collector's idyllic image of "primitive" cultures frozen in time. Many collectors of African art share in the vision of this empowered gaze which strives to suspend artificially the history of Africa and its art. In *The Art of Collecting African Art*, Susan Vogel describes one collector's views which support emphatically Price's observations: "This first experience with African art reveals a touch of the romantic nature that [this collector of African art] feels goes hand in hand with his collecting. Further evidence of this can be heard in the reasons he gives for feeling that going to Africa is not necessary to his collecting. He views his objects, he says, as being conceived in the pre-colonial aesthetic that he admires. He adds that 'if Addidas [sic] sneakers and Sony Walkmen were absent from the Ivory Coast, I might reconsider my position, but, at present, my romantic vision of pre-colonial Ivory Coast is too fragile to tamper with'" (S. Vogel, ed. 1988: 58).

16 In a supposedly objective report about the African art trade in the United States, one journalist could not stop himself from putting a patronizing, negative spin on a trader's explanation about his understanding of the aesthetics of African art.

"Smoking cigarettes continuously and dropping ashes haphazardly to the floor of his one-room dwelling in New York," writes the reporter, "Zango spoke most emphatically between telephone reservations for plane flights to Chicago and Los Angeles that contrary to what many white scholars and anthropologists have written, he is one African who does indeed possess an aesthetic appreciation of art separate from art as an adjunct of ritual. 'I know art long time,' he emphasized" (Griggs 1974: 21).

17 Although the study of the African art trade has received little systematic attention in either anthropological or art historical research, a few exceptions include a number of short descriptive essays which have been written about art markets in various parts of Africa, see Elkan (1958), Crowley (1970, 1974, 1979), Blumenthal (1974), Himmelheber (1975), Robinson (1975), and Ravenhill (1980a). For an excellent bibliographic overview of different aspects of the African art market, see Stanley (1987).

1 COMMODITY OUTLETS AND THE CLASSIFICATION OF GOODS

1 The term "market place" is used here to describe the physical site where market transactions occur. It refers to an official, enclosed area where seller access is restricted and stall fees are collected. The term "market" is used to denote the social institution of the market itself – i.e., any domain of economic interactions where prices exist which are responsive to the supply and demand of the items exchanged (cf. Plattner 1985: viii). For another distinction between "market" as physical site and economic process, see Agnew 1986: 17–27.

2 Sundays are considered to be very slow market days; only one quarter to a half of the traders show up at the market place on that day.

3 Compare Cohen (1965: 14, fn. 3) on the use of similar protective charms by Hausa cattle merchants in Nigeria. The *malam* (pl. *malamai*) is an educated person trained in the teachings of Islam.

4 Abidjan was formerly the political capital of Côte d'Ivoire. However, as part of a major effort toward decentralization, the official capital was moved in 1983 to Yamoussoukro, the natal village of President Félix Houphouët-Boigny. Foreign embassies and political consulates, however, have been permitted to stay in Abidjan – which remains the focal point of the Ivoirian economy (Mundt 1987: 16).

5 In other major urban centers in Côte d'Ivoire (i.e., Bouaké, Korhogo, and Man), one finds a similar diversity of sale outlets but on a much smaller scale.

6 Because Bouaké is located in the center of Côte d'Ivoire, it has one of the largest market places in the region. People from all over the country travel to Bouaké to buy their supplies.

7 Construction in the city of Abidjan began during the years 1899–1903, when an epidemic of yellow fever threatened to wipe out the European population in the coastal city of Grand Bassam, which was the official colonial capital of Côte d'Ivoire until 1934. Abidjan grew from a village of 1,400 inhabitants in 1912, to 17,000 in 1934. The 1985 census recorded a population of over 2 million in Abidjan. In its original layout, sketched in 1925 by a corps of French military and civilian engineers brought to Côte d'Ivoire to design the new capital, the city of Abidjan was divided into three distinct neighborhoods separated by the different

waterways of the Ebrié lagoon. The Plateau Quarter, which was located on the mainland and was the most elevated section of the city – thereby providing the most temperate and healthful climate for expatriates – was reserved for European housing, a military camp, the future site of the Abidjan railroad station, and the colonial administrative offices. A second section of the city, near the banks of the lagoon, was to be used as the port and the chief commercial district of Abidjan. And a third area to the east of the Plateau on the island of Petit Bassam was designated as the *cité indigène*, where African workers and their families were to be housed. Although Côte d'Ivoire gained its independence from France in 1960, the division of Abidjan has retained much of its original segregation. The Plateau has become the business district and the center of political administration, and the *cité indigène*, which is now in a section of Abidjan called Treichville Quarter, has remained a residential neighborhood for working-class Africans. Among the several new districts which have grown around the urban core of Abidjan is the Cocody Quarter which was planned in 1960 to accommodate European, Lebanese, and high-income African residents (Le Pape 1985; Mundt 1987: 15–16).

8 Every day, the women travel to the market place with baskets full of fresh produce. After setting out rows of careful displays (fruits and vegetables meticulously stacked in small pyramidal piles), they store their empty baskets on the tin roofs of the wooden stalls. Looking down from the second floor of the market building one sees only a vast sea of upside-down woven baskets.

9 Of the two largest indoor market places which cater to an African clientele, Treichville and Adjamé, only the market place in Treichville is visited regularly by tourists. The market place in Adjamé has gained a reputation for being a dangerous place – where criminals and pickpockets prey on foreign visitors. Although Treichville is by no means free of urban crime, it is considered to be more safe than Adjamé, and thus it attracts more tourists and expatriate collectors alike.

10 As an American Embassy brochure put it, "Even without a purchase, the interplay of business at the African market and the warmth and friendliness of the African people is worth seeing first-hand" (Anon. n.d.b: 13).

11 This feature of the Cocody market place qualifies the business space as a purpose-designed marketing arena rather than the more general-purpose premises which are characteristic of Treichville and Plateau.

12 The public garden or park was built to separate the Plateau into two discrete neighborhoods: the government area in the northern part of the Plateau, and the commercial district to the south (Mundt 1987: 16). About half the garden is occupied by the art market place. The rest of the park is used as public leisure space – for sitting, sleeping, and walking; it serves as an area for Muslim prayer, and as a gathering spot for street performers. The park contains two Lebanese-operated snackbars and a public urinal.

13 The construction and expansion of the Hotel Ivoire on the other side of the Ebrié lagoon, which was begun in 1957, drew too much business away from the once prestigious Hotel du Parc. The building remains standing in Plateau, but its doors are sealed with plywood and it is uninhabited. Business in the Plateau market place has suffered a major decline since the closing of the Hotel du Parc.

14 It is with sadness that I report the closing of this market place in the beginning of 1990. Not long after I left Côte d'Ivoire (1988), the city government began

systematically to close down and fence in all large, public areas within the city of Abidjan. Gardens and monuments in the center of traffic circles (*rond points*), open-air lots with food concessions and public tables and benches, and the Plateau's *jardin public* which housed the art market were all sites targeted for closure and eventually barricaded with concrete walls and wire fencing. Because the period beginning in early 1990 was a time of political unrest – students protesting openly on the campus of the national university as well as on the city streets – I believe the Ivoirian administration was trying to eliminate all large public areas where students could potentially congregate and protest. All this was happening, after all, in the shadow of the widely reported Chinese student uprising in Tien-am men Square.

To compensate the traders who were displaced by the market's closure, the Abidjan City Hall allocated 52 million CFA ($170,000) toward the construction of a new market place, called "Centre Artisanale de la Ville d'Abidjan," in an industrial suburb known as Zone Trois. The new market place, which was completed in spring 1991, houses twenty concrete structures which are each divided into four stalls (*magasins*). Each trader was asked to pay a deposit fee of 45,000 CFA ($150) followed by a monthly rental of 15,000 CFA ($50). Although most of the traders were able to pay the deposit, none have been able to keep up with the monthly fee. The traders say that because the new market place is located so far outside of the central part of town, the few remaining tourists that were not scared off by the recent student uprisings are not even able to find their way to the new market place. One trader who maintained a prosperous business in the Plateau earned only 5,000 CFA ($17) from the time he moved into the new market until fall 1992. Were it not for his recent, moderately successful ventures in the New York art market, this particular trader would obviously have been forced out of the art trade.

15 According to Falilou Diallo, a Wolof trader in the Plateau market place, the reason the Wolof were able to dominate the market for so many years is because of the preferential treatment they received from the French. "Since Dakar was the capital of A.O.F. [Afrique Occidental Française] during the colonial period, the Senegalese could move wherever they wanted. When my father began selling in this market place in 1950, the Malians and Ivoirians used to run around and do chores for the Senegalese. When they realized that there was money to be made in art, then they started selling it on their own" (10/26/87). Another Wolof trader told me that Ivoirians were not able to sell art because they were too involved in the religions which sanctify the arts. It took the Senegalese – outsiders who did not believe in the sacredness of the arts – to show the Ivoirians that "what they had was actually worth money" (12/1/87).

16 The government policy of "Ivoirianization" was developed in the 1970s largely in reaction to French, Syrian, and Lebanese control of the export-import economy. The policy was aimed specifically to replace foreign control of capital and industry with control by Ivoirian citizens. Profits made in agriculture were used to create jobs for skilled Ivoirians, mainly by buying modern industries and technologies abroad (Den Tuinder 1978: 7–8; Monson and Pursell 1979). Although the art market was not specifically targeted as an area of state controlled Ivoirianization, the creation of the syndicate and the pressure placed on Wolof traders must be understood within the broader framework of strategic national economic policy. Barbara Lewis signals a similar pattern of Ivoirianization in the fruit and vegetable market in Treichville.

"[T]he non-Ivoirians who dominated market activity during the colonial period," she writes, "are still an important presence in the marketplace, though political pressures have reduced their dominance" (1976: 138).

17 Each major market town in Côte d'Ivoire (Bouaké, Man, and Korhogo) has its own sub-unit of the syndicate, with local representatives and elected officials. The role of syndicates or unions has a long history in Côte d'Ivoire. The Syndicat Agricole Africain, which was established in 1944 by Félix Houphouët-Boigny and was later replaced by the national political party (PDCI), was largely responsible for the independence of Côte d'Ivoire and a break with the European-planter-dominated economy which existed during French rule. Other prominent trade unions which were also instrumental in national independence include, the Syndicat des Agents du Reseau Abidjan-Niger, the Syndicat des Cheminots Africains, the Syndicat des Commerçants et Transporteurs Africains, the Syndicat des Fonctionnaires Indigènes de la Côte d'Ivoire, and the Syndicat du Personnel Enseignant Africains (Mundt 1987: 124–25).

18 An elderly Hausa trader who sells art in the Plateau market place was in the process of buying old colonial currency from an itinerant supplier, when a group of youths wandering through the market place spotted the exchange and (thinking the old bills were valuable) demanded that the traders give them some of the money. When they refused to comply, the youths stopped two policemen who were walking by, and told them that the traders were dealing in counterfeit currency (an issue which had been widely publicized in the recent press). The police arrested the two traders and incarcerated them in the Plateau police station jail. The next day, members of the syndicate stepped in to explain that the two men were only trading in antique colonial currencies. The case was pleaded before the precinct chief, who eventually dropped the charges and set free the two Hausa merchants.

19 In theory, the annual membership fees are invested to finance syndicate costs. However, during the period of field research, the syndicate had no funds. The past president of the syndicate was asked to temporarily step down, while he was being investigated for embezzlement. When a trader died, the acting-president of the syndicate went around the market place asking members to donate whatever they could toward the funeral expenses of their deceased colleague.

20 The separation of artists from Western consumers is also underscored by the artists themselves. In her book *Statues colons*, Werewere-Liking quotes one workshop artist from Bouaké saying, "We could carve better: create interesting forms with a high quality finish . . . But the traders want things which are easier to make and [therefore] less expensive . . . Unfortunately, we have no contact with the people who could better reward our work by paying us what they are paying the traders" (1987: 19, my translation from the French).

21 The traders were divided on their opinion as to whether or not the perimeter of the market place was a more profitable location than the market place center. Some argued that by being on the outer edge, they were able to gain access to tourists before any of the traders on the inside of the market place. Others, however, felt that tourists often used the traders on the perimeter to gauge the price of objects, and then moved toward the inside of the market place to make their actual purchases.

22 The rental fee for a stall varies depending on its size and structure. There is a significant initial purchase price for a stall – an entire covered stall may cost up to

400,000 CFA ($1,300). The monthly fee can range from 800 CFA ($2.50) to 6,000 CFA ($20) depending on the quality of the stall. The high price of stalls discourages traders from getting started in the market-place business. This was made clear by the abandonment of a stall in the Plateau market place, which has remained unoccupied since the death of its owner in 1985. Since the onset of the *crise économique*, traders say that nobody has been able to afford the cost of this stall.

23 Because the city never took a census of the makeshift wooden stalls that were erected in the 1970s after the division of the market place among ethnic coalitions, there is no schedule of stall fees which applies to these newly constructed structures. Traders report that the *contrôleurs* exploit this ambiguity in the system by negotiating a fee with the individual stallholder and then pocketing this "unofficial" stall right.

24 While the Wolofs tend to control the Plateau market place, the Hausa have historically controlled the Treichville storehouses.

25 Although this situation is in some ways parallel to the landlord–broker relationships described by Abner Cohen (1969) for Hausa kola and cattle trade in Nigeria, it nonetheless differs considerably. Rather than rely on the host's network of buyers, the visiting traders have their own circuits of distribution within Abidjan. Also, unlike the kola and cattle diaspora, the art traders are not housed by the storehouse owners; they often stay in inexpensive hotels near the storehouses.

26 The Hotel Ivoire first opened in 1963 with the construction of a twelve-floor building; an additional thirty-floor tower was built in 1969; an extension to the original building was added in 1972. The hotel now has a total of 750 rooms (Nedelec 1974: 76–78).

27 The couple who originally ran the Rose d'Ivoire died in the late 1970s. According to several sources, the couple had been poisoned by villagers who accused them of selling a mask which had been stolen from their village (Susan Vogel, personal communication).

28 Since it is not advertised anywhere, access to this area of the gallery is available only to those who are serious enough about collecting African art to inquire with one of the salesmen or with the gallery owner.

29 The gallery is actually run by the owner's wife. He says he opened the room at her request in order to give her "something to do."

30 It is interesting to note the variation in the pattern of price disclosure. The mid-level of the market is the only place where prices are fixed and visibly marked. At neither the low level of the trade (in the open-air market places) nor at the highest level of the trade (the up-scale gallery) are prices visible. Recently, the Department of Consumer Affairs invoked New York City's so-called Truth-in-Pricing law, ordering art galleries to post their selling prices. Gallery owners protested that this intrusion of monetary tags went against the desired atmosphere of an art gallery (Gast 1988).

31 For a discussion of Anoh Acou's work see Jules-Rosette 1990.

32 For different perspectives on the classification of Ivoirian arts, see Anquetil 1977; Etienne-Nugue and Laget 1985.

33 Sometimes objects are brought to the city with specific economic targets in mind. One day, for instance, I saw a young man bring to the market an ivory trumpet which

he said was given to him by his father in order to buy his mother a wax-print outfit. See Hopkins (1973: 60) on the concept of "target marketeers" and the role of seasonal economic cycles.

34 Although one would think that students would research the value of the objects they sell, market information is so tightly restricted that it is impossible to gauge the value of an object in the abstract. Prices are not posted in the market place, and they are not revealed until serious verbal bargaining begins. Traders will never offer an initial purchase price. They wait for uninformed sellers to state how much they want for particular objects (see Chapter 2).

35 *Nyama-nyama* is a Dioula-derived term meaning literally something that is petty or trite. The word, however, has been appropriated by traders speaking many different languages, and is used to refer to any sort of insignificant and inexpensive souvenir or export craft.

36 One particular type of object which has appeared on the market in the past several years should be classified as a sort of transitional piece between commercial fine art and souvenir. These are Baule face masks and Asante fertility figures which are inlaid with small multicolored glass beads. The beads are arranged in such a way as to form a geometric pattern on the surface of the figure. They are made by Senegalese traders, who say the style was originated in Dakar. Traders refer to these objects as "Kenya" (*masques Kenya* and *poupées Kenya*). The name, according to traders, refers to the pan-African qualities which the objects represent. They are not generally marketed as representations of Baule or Asante art – they are simply marketed as "things from Africa."

37 A few itinerant traders have "instamatic" cameras with which they can photograph objects in the villages, and then circulate the pictures among urban traders in order to find prospective buyers – before even purchasing the object(s) from the village owner.

38 In an article on the art market in Liberia, Blake Robinson notes a similar pattern among itinerant merchants. "Since they are strangers," he writes, "the merchants hire guides and interpreters from among the local population. These tend more often than not to be youngsters who have rather a diminished sense of awe for traditional objects, whether cult or not, and who have a ready appreciation for things the money economy has brought such as lanterns, cutlasses, and clothes" (1975: 75). A Dioula trader told me about the time he spent collecting objects in Mali. While he was walking through a village, two young boys came up to him and asked him if he was interested in buying "two fetish statues covered with blood and stuff." The boys asked for money, but he only agreed to pay them after they had taken him to the place where the statues were located. They brought him to an abandoned shrine. "The pieces were good," he said, "so I paid them 2000 [CFA] a piece, which in the village is like a million francs." The boys told him that if he wanted more, they knew where their father had stored other objects. "No," the trader recounts having told them, "your father needs those other things, I'll just take these" (10/25/87).

39 In some cases, a camera is used to photograph the object so that the carver can sell the original and still have a model from which to create a copy.

40 Often called *courtiers* in francophone countries (Staatz 1979: 102).

41 Cohen does note the difference, however, between cattle which are transported by foot and those transported by train. When the cattle which travel by foot arrive at the

market in Ibadan, they are thin and weak from their exposure to the disease-carrying tse-tse fly in the forest belt. The life expectancy of cattle transported by foot is only two weeks after their arrival in the city; those transported by train can live for up to two and a half months. Since the "train cattle" are much healthier than the "foot cattle," they are more valued by city butchers and they are therefore sold for a higher market price (1965: 9).

2 THE DIVISION OF LABOR AND THE MANAGEMENT OF CAPITAL

1 The hierarchy of market personnel is no different in structure than that for many other commodity markets in West Africa. In her description of the marketwomen of Bobo-Dioulasso, Ellie Bosch notes that most fruits and vegetables in the market go through the hands of three to five different traders before they reach the consumer. Each trader in the network occupies a different socio-economic position in the produce economy – ranging from wholesalers who might buy forty to a hundred baskets of tomatoes at a time, to retailers who would only buy ten baskets from the wholesaler, to the young girls who sell the tomatoes from metal trays which are balanced on the top of their head (1985: 71).

2 Before her involvement in the African art trade, Bembe Aminata was a receptionist at the Air Ivoire ticket office in Man. After losing her job, she became involved in the art trade through a cousin who knew one of the artists in the doll *atelier*.

3 At the time that I was conducting research in Côte d'Ivoire, none of these dolls were being sold in the Treichville market place.

4 For a discussion and history of the Kumase workshop, see Ross and Reichert (1983).

5 When men transport art objects they must pay fairly large amounts of money in custom bribes. Since the value of artworks are so hard to specify, the duty on works of art are always unclear. Following a widely publicized auction in Paris, where a Baule mask sold for 125 million FF (Anon. 1987c: 33), traders say that the price of bribes went way up. "Now everybody thinks that whatever you transport is worth a million francs [CFA]" (10/20/87). Traders are sometimes thrown in jail until they agree to pay a sufficient fee. More often, their merchandise is held (*bloquer*) until they come up with the required sum of money. Traders must often travel to Abidjan, from wherever their goods are being held, in order to solicit money from regular clients and/or prospective buyers.

6 In the past few years, the market for African art in the United States, like the market for other arts, has suffered as a result of the American economic recession.

7 Because the women are not professional art traders, they do not have their own network of American clients. As a result, they must sell to the African art traders (many of whom stay in the same hotel in Manhattan's upper East Side).

8 In the United States, an atmosphere of neglect and confusion is often created on purpose by auctioneers. "The best setting for a liquidation auction," writes Charles Smith, "is one that conveys a sense of disorder, fostering the hope and belief that there may be treasures here that no one has yet discovered" (1989: 113).

9 Objects which are piled up in art storehouses are sometimes referred to by traders as *gool-gool* (a Wolof-derived term referring to something insignificant or to a hodgepodge).

10 In the West, for instance, it is reported that dealers keep paintings for years, even decades, waiting for a propitious time to put them on the market (Bates 1979: 167). Objects are either kept in closed storage facilities, in the dealer's private collection, or they are placed in the gallery, but with a price so high that it is unlikely to sell (however, if the object did sell, the profit would be equal to the dealer's most favorable anticipated long-term gain).

11 In his ethnography of Hausa traders in the region of Tibiri, Niger, Gerd Spittler reports that most traders are unable to store goods in order to wait for possible price increases. The only exception, which he notes as being rather unusual, is a wealthy sugar trader who is able to store his merchandise long enough to anticipate seasonal price fluctuations. "Alhaji K. always puts many tons of sugar into storage until the month of Ramadan," writes Spittler, "when he sells at higher prices. Most traders, however, lack the capital reserves required to pursue this strategy. Traders usually sell their goods during their weekly cycles, using the proceeds to buy new supplies" (1977: 374).

12 This second service, of course, is more necessary in the food and produce sections of the market than it is in the sections where art is sold.

13 As Lewis described the marketwomen of Treichville: "Thus the market's predominant tone is one of intense competitiveness: each woman has struggled, maneuvered, and even bribed her way onto 'her' space, and her keenest desire is to maximize returns for the money and energy she has expended" (1976: 139).

14 Cf. the economy of Chinese wholesale merchants in Indonesia and the operation of their warehouse-style stores as described by Geertz (1963: 52).

15 Some members of the American expatriate community in Abidjan refer to these traders collectively as the "mask men."

16 Time has very little value to many of the art traders. Because they have such a limited network of distribution, it is almost always to their advantage to wait for as long as it takes to see a client. Time can also be used as a status-signaling device by the more prosperous traders in the market. One might hear, for instance, a trader telling another trader whom he considers to be his social inferior, "Hurry up! I'm in a rush."

17 Since the objects travel through the hands of different traders it is conceivable that an object which had been in the possession of trader X would eventually end up with trader Y. If the collector was shown an object by trader X, turned it down, but then bought the very same object, on another occasion, from trader Y, then trader X would probably feel a certain resentment either toward the collector or toward the other trader. By not showing his collection to African traders, the collector avoids such conflicts or potential confrontations.

18 The reason he would probably not buy the objects from the stock of a European collector relates to the "discovery" element in the art trade which is explored in Chapter 6.

19 The term is probably derived from the Wolof verb *porah*, meaning to enter quickly and exit through the side, as in the action of a thief passing quickly through a house (Kobé and Abiven 1923).

20 At various points in the fieldwork, I observed different numbers of internal market peddlers, ranging from one to five at any given time.

21 The role of the internal market peddler is in some ways reminiscent of the

ambulant auctioneers (*dellala*) described by Geertz for the Moroccan bazaar (1979: 186).

22 The question of status is an important feature which helps to distinguish different levels in the trading hierarchy. Traders can often be sensitive to the sort of treatment they receive by Europeans. Outtara Youssouf, for example, refused to sell art to a certain American collector whom he said treated him with disrespect and with no concern for his proper status. "I am not a little boy selling trinkets [*nyama-nyama*] on the street," he said. Although he was still willing to conduct business with the collector, he would not go there himself. "I sometimes sit in the truck outside, and send the young people in to deal with him" (1/10/88). In her study of woodcarvers in Benin City, Paula Ben-Amos also notes the relationship between status and marketing. "The manner in which a carver sells his work has important social implications. Hawking is of very low status, particularly in comparison to working on commission . . . [and] to whom a carver sells [Hausa trader versus European customer] is also important and reflects on his status" (1971: 171–72).

23 Pronounced "täp", but derived from such English language expressions as "top quality," "top grade," "top of the line," or "top drawer."

24 Robinson also notes the presence of truckers, cattle merchants, and other wayfaring workers who transport art from rural areas to urban centers (1975: 76). However, he does not discuss any cases in which this practice of transporting art objects becomes a kind of "second" career.

25 Ibrahim was so familiar with such a large corpus of Lobi art, that when he showed his collection to prospective buyers he would group pieces according to the different artists which he could identify as having carved various pieces.

26 See Chapter 3 for Muslim attitudes toward selling African art.

27 Whether Tijjani had indeed made an error in his first major purchase is not entirely clear. Surely, it is not difficult to imagine that his uncle would have wanted to hinder his nephew's prospects at commercial independence, in order to keep him as an apprentice or hired hand.

28 On the sociology of trust see, for example, Bailey, ed. (1971) and Gambetta, ed. (1988). For a specific African case-study of trust among urban migrants, see Hart (1988).

29 The word *lèk* is derived from the Wolof verb "to eat" (*lèka*). Thus, the phrase, *donne moi mon lèk*, literally means "give me the wherewithal to eat" (cf. Kobé and Abiven 1923). Although it is a Wolof expression, the term *lèk* has been incorporated into the lingua franca vocabulary of all African art traders – i.e., regardless of their ethnic or linguistic background. Even when speaking French, the traders insert the word *lèk* into their conversation (cf. Turcotte 1981). Although the Hausa have their own word to describe a commission on sales, *la'ada*, in the market place Hausa art traders always use the Wolof word *lèk* (cf. Hill 1966: 365).

30 For the anthropologist trying to study the organization of the market, however, the system is at first more confusing than anything else. It seems almost impossible to figure out who owns what.

31 The word *ràngu* is a Wolof term which literally means to "carry something slung over the shoulder." When a trader takes an object on credit from another trader in order to try to sell it to a prospective buyer in the market place or to one of his clients (see below), he says that he is taking the object on *ràngu* (*je prends la pièce en*

ràngu). The metaphor of carrying something in a sling is thus transposed to the market context where objects are taken or transported from one trader's network of clients to another's. The *ràngu* system parallels what Geertz identifies in the Moroccan market economy as the *qirad* or what is called, in the Western tradition, the *commenda* – both of which refer to short-term credit on merchandise (1979: 133–36).

32 The problem of establishing client relationships is made even more difficult by the fact that many of the collectors and expatriates are only in Côte d'Ivoire for a limited period of time.

33 This last reason is the most commonly discussed problem with the *ràngu* system. Traders say that if they lend an object to a particular individual who is known for his lack of creditworthiness, he will sell the item and then *boufer l'argent* (literally, "eat the money"). He may avoid coming back to the market place, or he may return and say that he took a price below the one which was set for *ràngu*. In such cases, the object-owner's only recourse is to spread a bad reputation so that others will avoid dealing with this particular reseller. If the person does not have his own stock but relies for his livelihood on the profits from *ràngu* credit, such a bad reputation could indeed ruin his trading career. For the most part, then, the *ràngu* system is kept in balance by corporate liability among market-place traders (cf. Moore 1978: 111–26). Noteworthy parallels exist here with the organization of credit in the Jewish diamond and pearl trades.

34 It is interesting to note, however, that among Hausa merchants themselves there *is* an unwritten code which allows for the possibility that goods be returned if they are found to be of inferior quality. Reporting on Hausa trade in Kano, Hugh Clapperton wrote in 1823: "The market is regulated with the greatest fairness and the regulations are strictly and impartially enforced. If a *tobe* or *turkadee* [man's robe], purchased here, is carried to Bornou or any other distant place, without being opened, and is there discovered to be of inferior quality, it is immediately sent back as a matter of course . . . [and] by the laws of Kano, [the seller] is forthwith obliged to refund the purchase money" (quoted in Colvin 1971: 108).

3 AN ECONOMY OF WORDS: BARGAINING AND THE SOCIAL PRODUCTION OF VALUE

1 In addition, the bargaining process shares features with the purchasing of art in the West. As Constance Bates notes for the pricing of art in US galleries: "Usually the asking price is higher than the gallery owner expects to receive and bargaining is expected in order to reach a more agreeable price. Prices may be different for different customers, depending on whether the owner thinks the customer can bring in additional business" (Bates 1979: 16).

2 The term "extractive bargaining" is intended to signal the contrast between a mode of extraction and a mode of production (see Bunker 1985: 22–31).

3 In English, the wholesale/retail contrast has two meanings: (1) gross vs. retail selling and (2) intra-trade sale vs. sale to the general public. In this context, the terms wholesale bargaining and retail bargaining refer to the second usage of the wholesale/retail contrast.

4 Anne Chapman (1980) has argued that in most contexts the notion of barter ought

to include the exchange of both money and goods – i.e., that the "two types of exchange fade into one another."

5 In its original context, an African art object has no true monetary worth. Although it may have been commissioned in exchange for cash or goods, its intrinsic value is determined largely by such factors as ritual efficacy and social prestige. In the market place, however, the object becomes a true commodity. Its value is determined directly through the rational calculation of its potential resale price.

6 It is reported, for instance, that vast quantities of art have been placed on the international market during the most intense periods of Sahelian drought. People either migrate with art objects which they sell along the way, or itinerant merchants travel to their communities buying from those who need to make up lost revenues from a failed agricultural harvest (cf. Forde and Amin 1978). The surfacing of objects in Côte d'Ivoire also seems to coincide with state-sponsored urban renewal projects. During the razing of Korhogo (when many traditional forms of architecture were replaced with concrete buildings), for instance, vast quantities of Senufo masks were appearing on the Abidjan market (Susan Vogel, personal communication).

7 It is interesting to note that the justification used by the trader is an extension of the reasoning used by Western collectors and academics. As Price summarizes, "The most commonly cited justification for field collecting, even at some ethical cost, is that the documentation and preservation of Primitive Art constitutes a contribution to human knowledge. In this view, Westerners have a moral obligation to protect the artifacts of Primitive cultures regardless of the original owners' assessment of the scientific importance of the rescue operation" (1989: 75).

8 It should be pointed out that at this level of the trade *both* sellers and buyers have little sense of what an object is worth in the international market. Although urban traders (i.e., market-place stallholders and storehouse owners) have some idea of the price range of African objects in the West, they purposely keep their suppliers uninformed of these values.

9 Some of the urban middlemen may have traveled to Europe or America on selling trips. A few traders of my acquaintance held active subscriptions to the sale catalogues of the Guy Loudmer and Hôtel Drouot auction houses in Paris (cf. Himmelheber 1966: 192).

10 This point parallels Charles Smith's distinction between, what he calls "wholesale auctions" and all other auctions. "In [wholesale] situations," he writes, "auctions are more a means of allocating than of pricing goods. The issue is who will be 'allowed' to purchase the item, or, as may also be the case, who is expected to absorb the item. Uncertainty in these cases bears on the respective rights and responsibilities of the players more than on the value of the items" (1989: 42).

11 In some instances, I have witnessed traders trying to outbid one another for the purchase of a lot of commissioned goods. A carver brought to the market place a group of objects which a stallholder had ordered. Rather than hold to the agreed "contract," however, the trader tried to get more money by offering the objects to other stallholders. The objects were ultimately purchased by the stallholder who had initially ordered the objects, however, because of competition from other buyers he was forced to pay more than he had originally agreed or anticipated.

12 As Ravenhill noted in 1980, "In the past two or three years . . . untold thousands of pieces of Lobi art – wooden statues and bronze amulets and pendants – have

appeared on the market and while, currently, the supply is drying up, the price of Lobi art has increased five to ten-fold over two years. There is even talk, in informed circles, that the Lobi are the 'next Dogon' for the international African art scene" (1980a: 19).

13 Professional art dealers have told me that when buying from African middlemen they purposely try to show no emotion in their demeanor – lest the seller perceive that they are keenly interested in obtaining a certain object.

14 Since both the supply of spoons and the market for spoons are *limited*, the trader cannot seek higher profits by increasing the number of transactions. Rather he can only increase his earnings by augmenting the profit margin of each individual transaction (cf. Fanselow 1990: 257).

15 In a slightly more understated tone, Sol Tax made the same point in his study of the Guatemalan market economy. "The general market custom," he wrote, "is for the seller to name a price higher than he expects to receive, and to reduce it if necessary after an interval of haggling. A travel-book notion that his method is pursued because the people enjoy it is exaggerated" (1953: 137).

16 In an essay on bargaining in the Middle East, Fuad Khuri notes the same sort of interaction between seller and buyer: "The seller uses . . . key associations to link the commodity to the buyer's background, in an effort to show him that other people of his status do consume the same goods" (1968: 701).

17 It is ironic that in satisfying the foreign buyer's quest for an "authentic" cultural experience in the African market place, the traders have devised what is in fact the most *inauthentic* form of bargaining (see Chapter 4 on the permutations of authenticity).

18 Some tourists will try to augment their market knowledge by first pricing objects in the hotel giftshop or gallery, where prices are affixed to individual items. Then, they will go to the market place and try to buy what they want for anything *less* than the price posted in the stores.

19 In his study of the "bazaar economy" in Morocco, Geertz divided the bargaining process into three distinct phases: (1) initial bidding; (2) movement toward a settlement region, and, if that region is in fact entered; (3) settlement itself (1979: 227). The tempo or inner rhythm of bargaining is as indicative of the outcome as the monetary figures which are exchanged. Rapid movement between seller's initial price and buyer's initial bid indicates greater likelihood of sale. A slow or hesitant pace, on the other hand, usually indicates greater distance between buyer and seller and thus there is an increased likelihood of not striking a deal.

20 Expatriates who regularly visit the market place say that they are even afraid to look at particular objects for fear that the trader will pick it up and try to sell it to them. They walk through the market place, they say, without looking directly at anyone.

21 If an initial asking price is much too high, a Hausa trader buying from a member of his own ethnic group says in Hausa *àlbàrkà* (which, in this context, is a sarcastic rendering of the phrase "thank you very much"), after which he would either wait for a lower asking price or walk away in disgust.

22 By which it is understood that the trader probably means either another member of his ethnic group or simply a friend.

23 Some traders would insist that the "disclosure" of purchase price by other traders is *invariably* untrue, except in rare instances where a trader has a reputation for always

speaking the truth. One of the only traders that I knew who possessed such a reputation was Yusufu Bakano, a Hausa trader at the Treichville market place. Without exception, traders told me that if Bakano quoted the price which he had paid he would be telling the truth.

24 This technique finds its parallel in auctions where prices can sometimes be inflated artificially by inventing bids, a practice referred to as "taking bids off the wall" (C. Smith 1989: 150).

25 The technique draws upon one of the fundamental principles of buyer self-legitimation which is found among participants at an auction sale. "If one buys through an auction," writes Smith, "one can always rationalize a high price by observing that others were willing to pay almost as much" (1989: 90). Although in this bargaining strategy the implication is that someone was willing to pay even *more*, the pattern of manipulation of buyer consciousness is roughly the same in both cases.

26 Since buyer and seller are often equally aware that these price fixing mechanisms are purely invented, it is difficult to locate specifically the level at which this dialogue is effective. I would argue, however, that the language of bargaining is not simply an empty or meaningless mechanism for structuring an economic transaction but is in fact a very powerful device for persuading the buyer to purchase an object. Karl Polanyi hinted at this aspect of bargaining when he wrote, "[B]arter is the behavior of persons who exchange goods on the assumption that each makes the most of it. Higgling and haggling are the essence here, *since there is no other way each person can make sure he is gaining as much as possible from the bargain*" (1977: 42, emphasis added).

27 According to Rees, there are two types of market information sought by those who sell. He labels one type of information as "intensive" and the other as "extensive." He writes: "The search for information in any market has both an extensive and an intensive margin. A buyer can search at the extensive margin by getting a quotation from one more seller. He can search at the intensive margin by getting additional information concerning an offer already received. Where the goods and services sold are highly standardized, the extensive margin is the more important; when there is great variation in quality, the intensive margin moves to the forefront" (Rees 1971: 110). The two categories of market information searches identified by Rees parallel the distinction between the two market situations found in the Côte d'Ivoire art trade: (1) the contemporary fine art and souvenir trades, and (2) the traditional/antique art market. Since the former consists of the sale of similar type objects, an extensive market search would yield the greatest amount of market information and the best possible market price. And, since the latter consists of the sale of more-or-less unique items, "comparison shopping" or extensive searching would not be possible. In this type of market situation, one would need to get as much information about the particular object being sold – i.e., the greatest amount of intensive market information. In her ethnography of Jamaican middlemen (higglers), Margaret Katzin describes the search for information which is carried out by rural producers who attempt to assess urban resale prices before selling to itinerant middlemen: "Farmers congregate at the truck stops on Fridays and Saturdays where the principal topic of conversation is the prices of locally grown goods in town markets. Also, some country people always go to the market every week and return with

information about prices. Thus, the growers often know the prices received by the higglers for their goods before the higglers return" (Katzin 1960: 309–10).

28 When referring to the "faking" process, traders usually use the French verb *trafiquer* which implies the conduct of a dishonest activity.

29 Some people might have a better instinct about whether or not a car is good, but inevitably problems will only arise after the car has been purchased and driven for a certain amount of time.

4 THE POLITICAL ECONOMY OF ETHNICITY IN A PLURAL MARKET

1 The neighborhood is similar to the Hausa quarter in Ibadan described by Abner Cohen (1969). It is sometimes also referred to by the Hausa term *zango* (or *zongo*), meaning a camp or quarter which has been settled by diaspora Muslim traders (see Schildkrout 1978: 67).

2 The Wolof presence in Korhogo is far smaller than that of the Hausa. Unlike Abidjan, where the Wolofs dominated the art trade for many years, in Korhogo the Hausa have controlled the art market from very early on.

3 The Kulebele carvers were the subject of an excellent ethnographic study undertaken during the early 1970s by anthropologist Dolores Richter (1980).

4 The practice of trading in several classes of objects represents a diversification aimed at spreading economic risk. If errors are made in the marketing of one type of good, losses can be recouped in other areas. If there is a sudden drop in the demand for a certain type of goods, profits can still be made in other sectors.

5 Most traders who sell *antiquités* also sell contemporary trade pieces. Although the margin of profit in the tourist market is much smaller than it is in the antique/collectible market, it provides a more regular form of income which can be reinvested in *antiquités* and other higher value objects.

6 The honorific Alhadji (or, in Arabic, El Hadj) is accorded to Muslims who have made the pilgrimage to Mecca. Because the pilgrimage is so costly, the title Alhadji is as much a symbol of economic success as it is of Islamic piety (cf. Grégoire 1992: 2). Traders who possess this title are generally referred to in conversation simply as "Alhadji." Distinctions are understood by the context of the conversation, or individuals are specified by "nicknames" (*lakàbi*) which draw on a person's idiosyncratic qualities: e.g., "*grand* Alhadji" refers to Alhadji Inusa who is very tall, "Alhadji Peugeot" refers to Alhadji Haruna who drives a Peugeot pickup truck, etc. Since many of the traders in Aoussabougou bear the honorific title of Alhadji, it would be too confusing to follow this practice in my writing. For the sake of brevity and clarity, therefore, I refer to Alhadji Usuman simply by his family name.

7 Hausa traders avoid direct reference to commerce when they first arrive in a trade context or when they first meet another trader. According to them, it is considered extremely impolite to simply walk up to a person and directly state your business. An exchange of extensive greetings and non-trade-oriented conversation are expected to precede all discussions of a commercial nature. The length of this dialogue will vary depending on the relationship between the two traders. The most striking instance of this etiquette practice was observed when a trader of my acquaintance, Ahmed Arachi, gave a number of art objects on credit to a trader who

was traveling to France. After spending a month in France, the trader returned to Abidjan. He had been in the market place for over a week but had not yet told Arachi whether or not he had in fact been able to sell the pieces. Arachi was anxious to know the outcome of the sale, but was unable to learn anything. When I asked him how his creditor had made out with the goods he had taken to France, he told me that he had not yet been informed, and that it would be totally unacceptable and improper for him to ask.

8 Richter notes that in order to purchase the best quality objects from the Kulebele carvers, Hausa traders compete among themselves to curry the favor of master carvers in Koko quarter. She writes: "Dealers encourage, present gifts to and make social calls on the carvers they deal with, as well as extend invitations to them to share food. One dealer begins his rounds every morning at six, calling on all carvers who carve for him, as well as those he hopes will find time to carve for him in the future . . . By creating a multiplex bond with his carvers, he can expect rapid completion of his orders and favored treatment from those he commissions. 'I'll finish Aladijen's work first; he's always very nice to me and gives me lots of presents,' was the comment made by a master carver who was being pressured by Aladijen and another trader to complete each one's work first" (1980: 70–71).

9 Again, I would postulate that my presence may have altered the interaction between the traders and the Senufo carver. Coulibaly asked several times if I was an art dealer, and whether the masks were being purchased on my behalf. The traders were insistent that I was just an observer (writing a book on African art), and that the objects were strictly of interest to Madu. Their insistence on this point was not merely an issue of role clarification. If Coulibaly had indeed thought that I was a dealer buying the masks through two Hausa agents, the asking price would surely have been much higher.

10 Just as my presence may have been instrumental in initiating the contact between Tanko and Madu, it also became a factor in settling the dispute between the Senufo carvers and the Hausa traders. Because of my forced involvement in the situation as an invented American dealer, the Senufo men grew nervous at the prospect of calling in the police.

11 Grappling with the controversy surrounding the relationship between segmentation and unilineal descent is beyond the scope of this book (see Karp and Maynard 1983; Herzfeld 1987: 156–85), however, it might be useful to point out that among art traders in Côte d'Ivoire those living and working in the region of Korhogo expressed a far more explicit sense of ethnic cohesiveness than those working in the sprawling urban landscape of Abidjan. That is to say, in the more rural sectors of the country, ethnic affiliation and notions of segmentation were more important to group cohesiveness and commercial viability than they were in Abidjan, where the bond of friendship was often stronger than the ties of blood (cf. Hart 1988).

12 Some of these migrant ethnic groups are thought to possess a certain skill and shrewdness in business which is attributable to a putative psychological "genius" for trade. Cohen points out, however, that inherent personality has little to do with the success of these commercial ethnic groups. "[M]uch of this 'genius'," writes Cohen, "turns out to be associated, not with a basic personality trait, but with a highly developed economic-political organization which has been evolved over a long period of time" (1969: 9).

13 Although these ethnic groups are currently involved in the art trade, their original migration to the regions of West Africa in which they now work dates to a much earlier time and to previous commercial pursuits. The Hausa, for example, are reported to have spread south from their homeland into areas populated by non-Hausa-speaking peoples as a result of the ivory trade which began in the sixteenth century (M. G. Smith 1962; Johnson 1978; Hiskett 1984: 78). The Dioula are thought to have begun their southward migration into the Upper Niger region and, eventually, the Akan territories as a result of their participation in the gold trade which began during the late fourteenth and early fifteenth centuries (Perinbam 1980: 458). The Wolof spread east from Senegambia during the fifteenth and sixteenth centuries as a result of their involvement in the gum and salt markets, as well as in the Atlantic slave trade (Curtin 1975).

14 The negative attitude of Muslim Africans toward non-Muslim Africans was clearly expressed by Marcel Griaule when he wrote in *Conversations with Ogotemmêli*, "The Dogon, in short, were thought to present one of the best examples of primitive savagery, and this view has been shared by some Muslim Africans, who are no better equipped intellectually than Europeans to understand those of their brothers who cling to the traditions of their ancestors" (1948: 1). This attitude is felt not only in certain aspects of the Muslim relationship with non-Muslim communities, but applies also to the relationship between those who associated themselves with the colonial regime and those who resisted it. Myron Echenberg's description of the attitude of the *tirailleurs sénégalais* toward other Africans is apposite in this context: "The Wolof soldier was spoiled by his long association with France and had become 'a terrible snob' toward other Africans whom he regarded as 'savages'" (1991: 14).

15 The hadith is a collection of thoughts and sayings ascribed directly to Muhammad which was amassed and compiled by Muslim scholars after his death.

16 In his 1975 essay on the art dealers of Man in western Côte d'Ivoire, Himmelheber makes a very curious assertion for which I find no evidence in support. He writes: "The Hausa and Senegalese people are still Muslims who display no figural art even though they could perhaps earn more money from this" (1975: 19, translated from the German). In my experience, the Muslim Hausa and Wolof sell as much figural art as anyone else in the market.

17 Although the claim to Baule ethnicity is most prominent in the tourist market, some Wolof merchants falsify their ethnic identity in the *antiquités* market as well. A prominent Abidjan-based European collector of African art refused to deal with the Senegalese, claiming that they were generally dishonest and that their prices were too high. On any given afternoon, however, one could see scores of Wolof merchants lined up in front of the collector's home, waiting to show him their latest shipment of goods. They simply told him they were Dioula, Mossi, or whatever ethnic group they could think of. "We tell him whatever we want," said Barane Diop, "he doesn't know any different" (3/29/88).

18 A parallel to this practice can be found in the "authentication" of certain types of auctions in the United States. "The cowboy hat and boots affected not only by cattle auctioneers but many Midwestern and Western general auctioneers," writes Smith, "serves to authenticate their auctions by introducing a historical and traditional element" (1989: 115).

19 The social link between clothing and cultural identity was first made explicit during the 1930s in the writings of Petr Bogatyrev (1937). The semiotic link was best spelled out by Roland Barthes (1967).

20 The Dan are located in southwestern Côte d'Ivoire and northeastern Liberia; the Lobi are in northeastern Côte d'Ivoire, southeastern Burkina Faso, and northwestern Ghana.

21 For the arts of Côte d'Ivoire in particular, I am thinking here of the publications by Bohumil Holas (e.g., 1960; 1966).

22 Côte d'Ivoire has had only modest success in its promotion of tourism: 109,000 tourists visited Côte d'Ivoire in 1975, 220,000 in 1979, and 210,000 in 1980. Tourism accounted for only 2.4 percent of export receipts in 1976, yet it was still ranked as the fifth most important "product" of Côte d'Ivoire, after coffee, cocoa, wood, and refined petroleum. The Ministry of Tourism was created in 1975 (Mundt 1987: 129). In the late 1980s, tourism dropped as a result of growing economic trouble. The decline accelerated during the Gulf War, and reached stagnation when the international media began to report on the student manifestations at the Université Nationale in Abidjan.

23 The process is also reflected in the use of "traditional" symbols on West African bank notes (Francs CFA) used jointly by nations of former Afrique Occidentale Française (see S. Vogel, ed. 1991: 233).

24 It is interesting to compare Houphouët-Boigny's argument against the power of myth and the invention of a "traditional" African past with one of Marx's assertions in *The Eighteenth Brumaire of Louis Bonaparte*: "The social revolution of the nineteenth century cannot draw its poetry from the past but only from the future. It must strip off all superstition in regard to the past, else the tradition of all the dead generations weighs like a nightmare on the brain of the living" (quoted in Harvey 1989: 109).

25 The link between an African festival and the Carnival in Rio was first made by the government of Senegal in 1974 when they tried (without success) to launch a series of "ethnic" dances which "would become as famous as the Carnival of Rio or of Nice" (Copans 1978: 119).

26 In addition to the masked dancers, the new festival of masks also featured the election of "Miss Festimask," who was to be elected in a contest called "Awoulaba" according to standards of beauty set by the First Lady of Côte d'Ivoire, Thérèse Houphouët-Boigny (Lakpé 1987). Out of a group of twenty female contestants, the judges selected the most slender and light-colored woman. Strong local criticism was levelled against the contest for having selected someone with the most Western traits. The Miss Festimask contest may have been inspired by the Miss Sénégal contest which has been held from time to time in Dakar since 1974. As in Côte d'Ivoire, the Miss Sénégal competition received some harsh criticism from the local press. An article in the Dakar newspaper *Le Soleil* declared: "We have reached the epitome of the ridiculous when Senegal, confronted with the innumerable problems of a major drought . . . finds nothing better to do than to organize a national election of Miss Sénégal" (quoted in Copans 1978: 119, my translation from the French).

27 The town of Yamoussoukro, which is located about 250 kilometers north of Abidjan, is the birthplace of Félix Houphouët-Boigny and since 1983 the official

political and administrative capital of Côte d'Ivoire. Yamoussoukro houses the massive "Maison du Parti" (the administrative home of the central government or PDCI-RDA). It is also the site of the Houphouët-Boigny Foundation, the President's residence/retreat which is surrounded by a moat of sacred crocodiles, the Hôtel Président (which is the most luxurious hotel in Côte d'Ivoire), and the controversial and highly publicized basilica which Houphouët-Boigny is constructing at the cost of over $100 million (Bentsi-Enchill 1988: 1428).

28 Côte d'Ivoire is often held up by economists as one of Africa's few post-colonial success stories. From independence in 1960 through the late 1970s, Côte d'Ivoire enjoyed tremendous export earnings from coffee, cocoa, and lumber. This rapid growth of the agricultural sector, which resulted in a quadrupling of the gross national product from 1960 to 1978, is sometimes glossed as the Ivoirian "miracle" (see Shepard 1978).

29 Nelson Graburn has clearly pointed out the extent to which ethnicity can be manipulated by the state, especially in various parts of the developing world. "As countries industrialize and modernize their physical structures," he writes, "they become, at least to tourist eyes, more and more similar, yet the tourists demand that they be significantly different from the homeland, for without any differences why would one bother to travel . . . [As a result of this sort of pressure] the vast majority of modern nations emphasize their 'specialness' by playing up their distinctive ethnic minorities and their material production" (1984: 413; cf. Wood 1984).

5 THE QUEST FOR AUTHENTICITY AND THE INVENTION OF AFRICAN ART

1 Perhaps it is not surprising that the small amount which has been published on the subject of authenticity has largely been written by dealers, collectors, and others who have a direct stake in the definition of authenticity because of their economic involvement in the African art trade (see, for example, Baker 1973; Kamer 1974; Allison, *et al.* 1976; Silberman 1976; Provost 1980).

2 In some instances, the Western desire for age in African art has been communicated beyond the world of traders to the art producers themselves. Walter Van Beek reports that among the Dogon of Mali the demands of the Western art trade have triggered a series of quick and specific responses. "The Dogon are very much aware of the value that outsiders attach to . . . old things, so [when selling to outsiders] they identify their *dege* [statues] as ancestors, or they give them the look of age . . . Whereas the production of statues for traditional use has declined, that of statues for sale has become a thriving industry. The Dogon smear these new carvings with kitchen soot, bury them for a few weeks near a termite mound, soak them in millet gruel, and dry them in hot ashes, all to give them the highly valued patina of long ritual use" (1988: 64).

3 Although these objects still circulate among Westerners, they are out of *his* commercial realm. The fact that something becomes authentic because it is unattainable finds parallels in the way signs of authenticity and aesthetic merit are conveyed through subtle cues in the structuring and arrangement of space in Western art museums (see Greenblatt 1990).

4 The creation of authenticity through temporalization could fruitfully be compared

to Johannes Fabian's (1983) notion of allochronism – the creation of anthropology's epistemological domain through the distancing of subjects in time.

5 It is arguable that this view of authenticity has parallels to a religious doctrine which grounds its authority in an earlier period of revelation (cf. Tambiah 1985: 25–29).

6 An interesting parallel can also be found in the rapid rise in demand for art by the insane. Because of more responsive medical treatments for mental disorders, the production of art by the insane, which once was a thriving branch of psychoanalytic industry, has been suffering a steady decline. Limited by the efficacy of powerful psychiatric drugs, the supply of art by the insane now constitutes a finite resource whose monetary value is swiftly increasing (Kuspit 1990).

7 The techniques of object alteration are explored further in Chapter 6.

8 The idea that there exists a direct relationship between ethnic identity and aesthetic style grows out of a Durkheimian assumption that African art is the creation of a collective mind instead of an individual artist. This assumption was carried to its logical conclusion by William Fagg (1965) who was one of the first to develop systematically a "one tribe, one style" model for the classification and analysis of African art. The model assumes that every ethnic group constitutes a closed system, and that the art of one ethnic group is incomprehensible to the members of another. For an insightful critique of the "one tribe, one style" paradigm in the study of African art, see Kasfir 1984.

9 In the language of economics, one would say that the category authentic African art constitutes a "depleting non-self-renewing resource."

10 Wooden slingshots are used by young Baule boys to shoot birds and bats. Some of the slingshots are decorated with carvings of animals, humans, or masks. Many, however, are crudely hewn as they are usually created in haste by the boys themselves.

11 Slingshots were, however, known to some collectors before the publication of *Potomo Waka*. Two examples were published from the collection of André Blandin in Anonymous (1977) and one from the collection of William Kohler in Cole and Ross (1977). Slingshots were purchased from local market places in Côte d'Ivoire during the 1970s by anthropologists and art historians working in the region (Philip Ravenhill, personal communication).

12 This passage from *Potomo Waka* is strikingly similar to a point which Bourdieu and Derbel raise in *The Love of Art*: "The same people who will no doubt be amazed that so much trouble has been taken to express a few obvious truths will be annoyed at not recognizing in these truisms the flavour, at once obvious and inexpressible, of their experience of works of art. What is the point, they will say, of knowing where and when Van Gogh was born, of knowing the ups and downs of his life and the periods of his work? When all is said and done, what counts for true art lovers is the pleasure they feel in seeing a Van Gogh painting" (1990: 108).

13 In a sample of about 500 slingshots seen over the course of one year in the stock of several traders, over half were painted with bright enamel colors.

14 Potassium permanganate is used extensively by traders both to stain wood and darken brass. It is purchased in pharmacies in the form of a dark-purple crystalline compound which is then mixed with water and applied with a toothbrush. The stain imitates the effect of natural indigenous dyes used by African carvers. It is interesting to note that potassium permanganate is also widely used by European

and American antique dealers to finish and restore antique furniture. "The most reliable method of toning down the colour of wood is to use potassium permanganate, which becomes wedded to the fibres by a process of chemical decomposition" (Arnau 1961: 188).

15 Objects are often damaged or broken during their transportation along the trade routes which move objects from rural to urban areas. Traders are often skilled at repairing cracks and fractures with scraps of wood and a mixture of sawdust and commercial white glue. In the Plateau market place in Abidjan, two carvers made their living by repairing broken objects for art traders. Because the quality of their work was vastly superior to anything a trader could produce, a trader would employ their services if he needed to repair something major on an important piece – e.g., a broken carved horn on the side of a mask; a broken leg on a chair; a damaged foot on a statue, etc.

16 Although there are now at least four African art framing galleries in Abidjan, Césaréo's gallery is the only one to have developed out of a conventional African art gallery. The others started out as poster and print framing galleries and expanded their operation to include the framing of three-dimensional African art objects. While Césaréo relies entirely on the proceeds from framing African art, the other galleries continue to earn revenues from traditional picture framing.

17 The French term *cadre* is suggestive of both a frame and a context.

18 The re-circulation of beads spans a broad trajectory through the entire economic world system. It has been reported, for instance, that beads from the African trade which once flooded the US market in the 1970s are now being resold to accommodate demand in Asia, the Philippines, Mexico, and Guatemala (Liu 1975: 1). In the American Southwest, Northwest Coast, and Alaska beads from the African trade are being marketed to tourists as authentic beads from the Native American trade (*ibid.*: 1).

19 In an essay on bargaining in the Middle East, Fuad Khuri describes a similar pattern of marketing among carpet dealers in Lebanon. "The seller uses . . . key associations to link the commodity to the buyer's background, in an effort to show him that other people of his status do consume the same goods by citing specific incidents – a carpet trader told an American customer who wanted to buy a Bukhara carpet: 'I deal mostly with Americans and with the people of the American University of Beirut. They like my carpets and my prices. See the post cards they send me, all are pleased with my dealings'" (1968: 701).

6 CULTURAL BROKERAGE AND THE MEDIATION OF KNOWLEDGE

1 In his introduction to *The Social Life of Things*, Arjun Appadurai signals the distance between producers and consumers as a primary device which sustains the power and livelihood of the middleman. "[T]he translation of external demands to local producers," he writes, "is the province of the trader and his agents, who provide logistical and price bridges between worlds of knowledge that may have minimal direct contact" (1986: 42). In an essay on the social organization of the Maine lobster market in the US, James Acheson provides a specific example of the distancing mechanism – noting a similar institutional barrier such as the one

which exists between suppliers and consumers in the African art trade. "From the point of view of the fisherman," writes Acheson, "the whole marketing process is shrouded in fog, mystery and rumor. He has very little idea where his lobsters actually go, or how many hands they pass through on their way to the consumer" (1985: 105).

2 Like Hermes – who inhabited both Olympus and the underworld – traders in ancient Greece were referred to as "professional boundary crossers," who, because of their profession, "bridged the abyss between different social worlds" (Agnew 1986: 24).

3 There is an assumption which grows at least in part out of Barth's *Models of Social Organization* (1966) that entrepreneurs or brokers function as integrators of culture. In his critique of Barth's essay, *Second Thoughts About Barth's Models*, Robert Paine points out the paradox of the broker's role (1974: 24–28).

4 Because village-level traders have little contact with the ultimate consumers of African art, they tend to buy anything that they feel may be resold. And, indeed, some of the items they bring back from villages are not African at all. Among the more unusual items which I have seen brought to the market have been European military helmets and gas masks, European-made furniture hardware (some of which may have been used as goldweights, see Garrard 1983), a pair of brass factory-made Buddha figures, and an Oriental wooden statue of a Chinese peasant which was completely encrusted with sacrificial blood and feathers (collected by a Hausa trader among the Lobi in southern Burkina Faso).

5 For definitions and concepts of authenticity and fakery see Chapter 5.

6 There is a long tradition in travel writing involving an author's difficult search for authentic cultural objects. Compare, for example, the language of Lehuard's description of art collecting in West Africa in the 1970s to Sir Martin Conway's account of his exploits in North Africa at the turn of the century. "I had spent two or three evenings in the *dark* native houses of Luxor," wrote Conway, "finding nothing but the ordinary poor rubbish that came to the surface everywhere in Egypt. At last I was taken . . . into an *inner* room within the courtyard of a *specially secluded* house, and there, to my astonishment, they showed me a few quite extraordinary treasures . . . Women, *veiling their faces*, brought things in from the *background*, one by one" (1914: 77; emphasis added). For a self-critical deconstruction of this genre of art-travel-discovery literature, see Greenwald 1990.

7 In a recent sociological study of auctions in the United States, Charles Smith points out that certain auction sales are carefully staged to foster the impression that the sellers do not know what it is they are selling. "The best setting for a liquidation auction," he says, "is one that conveys a sense of disorder, fostering the hope and belief that there may be treasures here that no one has yet discovered" (1989: 113). And further on, he also notes that the mix of front- and backstage areas (in Goffman's sense of the term) makes it more likely that "a given participant may believe that he sees something that no one else does, which, in turn may induce him to act on his own judgment" (*ibid.*: 115).

8 On the psychology of discovery, Desmond Coke notes the following in his book entitled *Confessions of an Incurable Collector*: "The mania for 'finds', which I have already condemned, leads to a kind of Hope that almost seeks to be deceived. Indeed its victim does actually deceive himself. 'It wouldn't be worth their while to fake *this*', he exclaims hopefully at one pole: and at the other, 'They couldn't possibly

fake *that*'. He forgets, in his fool's paradise, that even a hair-pin would be worth faking if one sold enough . . . He forgets everything. He merely wants to have a 'find'. He scarcely ever fails" (1928: 164).

9 Field observations confirm the view that art buyers attempt to misguide dealers in order to get a better buy. On this subject, Desmond Coke's observations are illuminating: "I know that, according to tradition, there is an eternal enmity between buyer and dealer: where, otherwise, that find, that bargain? The more simple-minded of collectors go, indeed, to every sort of artifice as help in this long-standing warfare. They solemnly believe in disparaging an object before buying it. Perhaps, even, they will put it down and take up another. They will certainly not enter, bluntly, and ask how much the silhouette is: they will register ignorance, astonishment, disappointment, doubt, conviction, in their proper order" (1928: 91).

10 Although inexperience among traders is often genuine (rather than staged), this practice of "situational ignorance" has often led people to believe that *all* traders are naive about the objects they sell. In his essay on the art market in Man, Côte d'Ivoire, Hans Himmelheber wrote, "[Traders] have no clue as to artistic quality or as to the merit of the workmanship of a piece. The meaning of a form within the culture of its origin is unknown to the dealers and they are not interested in it if one wants to explain it to them, which almost none of the European or American art dealers do anyway. Even their lack of knowledge as entrepreneurs is astonishing. They set out the same objects fifty-fold, one next to another, so that the customer sees at once that everything is mass-produced" (1975: 20).

11 This aspect of presentation is a characteristic feature of any competitive market place. In her analysis of the international art market in the West, Bates remarks that the "intensity of competition is affected by the ability (or lack of ability) among the various sellers in an industry to engage in product differentiation. Whether or not the physical products are actually different, the sellers will try to establish in consumers' minds that they are" (1979: 134).

12 One African art collector underscores the value of collected information when she writes, "To collect in the field . . . is not just to accumulate objects. To collect, more significantly, offers an important opportunity to record specific data about an artist and the context in which the arts are used. To collect in this manner is to observe firsthand: *from the boisterous encounters in the market place to the private courtyard of an African dealer*" (Werner 1982: 12, emphasis added).

13 A striking parallel to the phenomenon of "pedigree" in the art market can be found in the context of pharmaceuticals: "Typically a drug is valued not just because it works, but because it comes from a certain source . . . It is part of the meaning of medicine that it was recommended by an authoritative doctor, or manufactured in Switzerland, that it was expensive or that it was advertised and packaged in a particular way" (Whyte and Van der Geest 1988: 4).

14 As in many art/antique markets, the "discovery" element plays a crucial role both in the assessment of value by consumers and in their perception of authenticity. If a collector believes that s/he is the first to whom a trader has shown a particular object, then there is a greater chance that s/he will be interested in buying the item. The more often an object is shown among a particular circuit of buyers, the less it is worth in the eyes of the local collectors. In this sense, one could argue that although

art objects are not "perishable" in the same way as are fruits or vegetables, art traders share some of the risks of a perishable-commodity market. They must be careful to whom they show their stock, and they must cautiously control the number of merchants who take their goods around on credit (see Chapter 2). Traditional/antique objects have only a limited "life" within the parameters of a local art economy.

15 When traders sell art door-to-door (see Chapter 2), they will sometimes say that there is an old man from the village who is sitting outside; that the object really belongs to him, but that he is too shy or embarrassed to come inside. If the buyer agrees to purchase the object, the trader says, he will give the old man the money afterwards. Whether or not the statement is true (sometimes it is and other times it is not), the message which is being communicated to the buyer is that the art is "fresh" from its "native context" and that it has not yet been "spoiled" by its circulation through market networks and the defiling touch of commoditizing hands.

16 The tourist's intense quest for cultural knowledge relating to the objects which they purchase for their private collection, agrees with Susan Stewart's hypothesis that taxonomy and knowledge are what differentiate collection from fetishism – a proper relation with objects versus a "savage" or deviant relation (idolatry or erotic fixation). In her study *On Longing*, Stewart writes: "The boundary between collection and fetishism is mediated by classification and display in tension with accumulation and secrecy" (1984: 163). This point is also picked up by Clifford who goes on to say, "If the passion is for Egyptian figurines, the collector will be expected to label them, to know their dynasty (it is not enough that they simply exude power or mystery), *to tell 'interesting' things about them*, to distinguish copies from originals' (1988: 219, emphasis added).

17 The trade in market lore can be traced as far back as the trade in Near Eastern artifacts in pre-Carolingian times. Arnau reports that: "[the traders] contented themselves with simple deception, forcing up the market price of works of art by 'treating' them slightly *and surrounding them with some mythical tale regarding their origin*. It was not enough for a chalice to exhibit good workmanship; what was more important was that it should be wrapped in a cloak of legend. The cup must either have been used by Jesus Christ at the wedding of Cana or by Pontius Pilate at some banquet – whichever story filled the bill" (1961: 26, emphasis added).

18 The explanations are meant to confirm the buyer's stereotype of Africa and African art. As one writer put it: "The sculptures and masks are thrilling to look at and exciting to touch. *The stories behind the pieces of tribal rituals can inflame the imagination*" (Bordogna 1981: 36, emphasis added).

19 Susan Vogel reports that in the 1970s many traders described Akan brass boxes (*kuduo*) as spittoons (*crachoirs*) for the king – enticing the buyer with mixed metaphors of pollution and royalty (personal communication 1987). In point of fact, these vessels are actually used to store valuables (such as gold dust), and are often buried with the deceased (R. Silverman 1983: 21).

20 Definitions of "fakery" have focussed on the physical aspect of object alteration. Although this is the most obvious aspect of the post-production manipulation, I would argue that it is only one aspect of a larger process which also includes object placement and verbal elaboration. For some widely held views on African art

fakery see "Fakes, Fakers, and Fakery: Authenticity in African Art." Special issue of *African Arts* 9 (3), 1976.

21 Traders often face a double-edged problem with regard to the mounting of unbalanced objects. On the one hand, for the reasons elucidated above, traders cannot sell objects with wooden mounts. On the other hand, however, buyers are sometimes reluctant to purchase objects which do not stand up on their own. More often than not this reluctance is actually a manipulation of the bargaining process, i.e., if the buyer feels that the price of the object is too high s/he will complain that the figure does not even stand up, and, therefore, it ought to be sold for less money. Traders have responded to this pattern in bargaining strategies, by repairing the breakages and erosions at a figure's feet or base. If a buyer questions the ability of a figure to stand on its own, the trader will place the figure on the ground and demonstrate that it does indeed stand upright (*pas de problème, sa tiens*). Many collectors tell traders that they ought not alter the natural erosion of an object – i.e., that their repairs mask the true age of a figure by covering up the eroded wood. At the same time, however, many of these same collectors complain to the traders, during the bargaining process, if a figure does not stand up on its own.

22 The sea bean (*Entada phaseoloides*) is found in coastal West Africa and is used traditionally for medicinal purposes. When rubbed, the bean exudes a greasy fluid which, if applied to wood, imitates quickly the effects of prolonged human contact. My thanks to Robert Gustafson for his help in identifying the botanical name of the bean.

23 Summarizing a point by Bakhtin on inter-textual utterances (and, in particular, relating to a homology which Bakhtin draws between a parody and a forgery), Gary Morson writes: "[S]uccessful forgers do not imitate an original as *they* perceive it, but rather the way such an original is likely to appear *to their intended victims*, perhaps subtly exaggerating what the latter are likely to regard as the original's most characteristic features and hence as the marks of its authenticity" (1989: 66, original emphasis).

24 The nakedness of the statue also draws on the collector's emphasis on the sexual nature of "primitive" art. For a discussion of this particular facet of Western taste, see Price 1989: 47.

25 Traders are faced with a double standard which prevails among many collectors/dealers. On the one hand, traders know that Western buyers are concerned with the age of an object and its overt signs of indigenous wear. On the other hand, however, traders know that when bargaining for individual items, many Western buyers will try to lower the asking price by pointing out that a figure is eroded, that it does not stand up on its own anymore, or simply that it is "ugly" because it is missing a limb or some other feature. Traders are faced with the decision of either not repairing objects (and capitalizing on their outward signs of use) or repairing objects so as to avoid potential losses in the bargaining process (cf. Wilsmore 1986). For the most part, traders have responded by repairing breaks and missing parts. Two artists in the Plateau market place earned their livelihood by restoring objects for art traders.

26 I once showed a group of traders a photograph from a Sotheby's auction catalogue (New York, May 20, 1987). The image is of an eroded Dogon face mask, described in the catalogue as "Highly weathered, the left side of the face and lower section

eroded away" (1987: 38). The mask was estimated in the catalogue at $5,000–7,000. One of the traders in the market place, a Mossi from Burkina Faso, said: "White people are too clever [*malin*]. If we tried to sell this [mask] in the market place, they would say that it was worthless, that it was ruined, that it was too ugly. So they would try to buy it for as little money as possible. Then they put it in this book here and they say, 'now see how nice it is; you must pay us a lot of money'. No, it's too much" (5/19/88).

27 Shifting demand, diminishing supply, and the associated transformation of "gift" into "commodity" is, of course, not unique to the African art market. Commenting on the vicissitudes of commercial fishing in New England, for instance, a Boston fisherman clearly echoes the remarks made by the Wolof trader from Bouaké. "Who can say what fish is worth. It's not like we have any control over what we bring in. I've had trips when everything went right. All we had to do was drop the nets and the nicest-looking fish you ever saw jumped in. Other times we go out, work our butts off, and still come back with *stuff we used to give away*" (quoted in C. Smith 1989: 26, emphasis added).

28 I do not believe that it is merely a coincidence that the market for African colonial figures should be expanding at the same moment as the market for American folk art (see C. Vogel 1988). Many colonial figures and folk sculptures, I would argue, bear a number of formal aesthetic similarities (rigid posture, elaborate detail of garments, sharply angled pant legs, etc.). Furthermore, both have comparable surface wear and paint erosion, and both are often carved on their own stands (see examples of American folk art in Miele 1988).

29 In 1974, Daniel Crowley reported on similar "treatment" of painted Guro masks. "[Although] the enameled masks have actually been used in Guro rituals, and hence are 'traditional art', [this] does little to improve their popularity with foreigners, so the wise traders often cover the shiny polychrome with dull paint before putting them on the market" (1974: 57).

30 In many instances those who use or produce African art objects have far less understanding of Western taste than the traders who sell directly to European and American buyers. While the traders busily *add* dirt to new objects to satisfy the Western demand for old, "authentic" African art, I have been told about several instances in which villagers have been known to *remove* dirt from old pieces in order to make them more "presentable" for sale to outsiders. One American collector in Abidjan gives the following example: "Years ago I had the experience of buying a mask [directly] from the dancer. Although I knew it wasn't terribly old, I knew for sure that it was used; I knew that it was authentic in that sense, and it was a beautiful piece. It had a couple of places on it that showed its wear and use – it was broken in one place and worn down in another, but for a collector that's appealing. In any case, they told me to come back and pick it up in a few hours. When I went back to get it, I found they had shoe-polished it and had repaired the place where it was broken on the chin. Of course I almost broke out in tears and asked why did you do that? They said 'we couldn't possibly have given you that old dirty mask'" (6/19/91).

In short, while one participant in the trade paints objects in brilliant colors, those closer to the Western market remove the freshly applied paint; and while one participant adds dirt and fake sacrificial materials to new objects, those further from the Western buyers remove the real layers of encrustation that testify to an object's authenticity.

31 Because there is no sharp distinction between the marketing of old and new objects (i.e. both types of objects are sold in the same places and often by the same merchants), one must combine under the same analytic canopy the market for antique colonial statues and the one for contemporary figures manufactured for external trade.

32 For a perceptive discussion of the mediating role of the art trader in the West, see Hauser 1982: 506–17.

33 Lack of control over the forces of production is characteristic of the cross-cultural diaspora in many other commodities. Writing about the Hausa in Nigeria, Cohen points out that the Hausa have no control over production of cattle or kola, yet they dominate the trade in both (1969: 21). This lack of direct control over the forces of consumption and, in particular, the capacity for "taste-making" stands in sharp contrast to the function and role of art dealers in the West, where *they* are expected to "discover" new artists, set trends, and create interest and demand (see Taylor and Brooke 1969; Gee 1981; De Coppet and Jones 1984; Moulin 1987; Montias 1988).

34 This point deviates slightly from Cohen and Comaroff's assertion that in certain cases the broker can actually create a demand for the relationship between patron and client. "Thus we suggest," write Cohen and Comaroff, "that the broker role is not called into being simply by the desire of two parties to transact valued items: rather, it creates or may create the demand for those values in the 'transacting' parties" (1976: 89). In other markets, Hausa traders have a long history of being able to develop local tastes for their goods. When they entered new trading territories, for instance, Hausa merchants would offer gifts (e.g. cloth, spices, etc.) to local chiefs. These gifts not only gained the support of the local authorities, but it also created a demand for the goods among other members of the local population. "Even things like *kwalli* (antimony)," writes Adamu, "which appears to have captured the northern markets from the second half of the eighteenth century, was advertised by the wives of resident Hausa traders using it skillfully in their make-up" (1978: 83).

CONCLUSION: AFRICAN ART AND THE DISCOURSES OF VALUE

1 For a neo-classical economic critique of this perspective, see Grampp (1989).

2 Although it is true that a large portion of objects exchanged in the international African art trade today are expressly manufactured for the tourist and export markets – and therefore not immediately part of the spiritual world to which I refer – the general argument that I propose is still valid in so far as almost all the objects in the trade are either modelled by the African producers on objects associated with indigenous categories of meaning and beliefs, or assumed (often incorrectly) by the Western consumers to have once belonged to a religious or ritual world. A similar situation is analyzed by Stanley Tambiah for the "vulgar materialization" of Thai amulets (1984: 335–47).

3 I would follow Baudrillard's broad definition of use value as "the expression of a whole metaphysic: that of utility. It registers itself as a kind of *moral law* at the heart

of the object – and it is inscribed there as the finality of the 'need' of the subject" (1988: 67).

4 To say that African art objects in the West have lost their utility or use value, is not to say that collectors ignore the spiritual aspects of the art. Indeed, it is often these "magical" elements that draw the Western collector to African art. As prominent African art collector Paul Tishman once put it: "[N]o matter in what part of Africa the material originated and no matter how diverse the tribal or regional styles, through most of it runs the same spirituality, the same invocation of magic, the same attempt to express in image-form the unknown forces with which the African had to deal in his daily life . . . Although my interest in collecting African art was at first not fully shared by my wife, it was not long before the purity and force of the material worked their spell on her" (Sieber and Rubin 1968: [i]).

5 For another perspective on the "failed metaphor," see Raymond Williams's distinction between "art" and "non-art" (1982: 124–25; see also Thompson 1979: 118–20).

6 Although it is true that the separation of the categories useful and aesthetic is rather peculiar to the West, I would caution against overgeneralizing the label "the West." Just as anthropologists have become increasingly sensitive to the multiple perspectives and nuanced differences within the specific cultural groups they study, it is important to keep in mind that the same heterogeneity exists within what we designate as the West. In its reference to a specific geographic and cultural domain, the label "the West" is thus more a term of convenience – an oversimplification for the sake of brevity – than a useful anthropological or sociological term. My thanks to Ivan Karp for insisting on the subtleties of this important issue.

7 The Victorian distinction between magic and religion is apposite to the present contrast between use value and aesthetic value. According to writers like Edward Tylor and James Frazer, "magic" assumes the possibility of private advancement through personal access to the spiritual world, while "religion," by contrast, refers to a more dispassionate and, therefore, more socially acceptable form of behavior, which works for the good of society not the individual. The putative vulgarities attributed to the realm of magic translate faithfully to the negative qualities imputed to the use value of art, while the language that is used to characterize religion finds its equivalent in the creed of immaculate aesthetics. (For a discussion of the religious qualities of the art museum, see Bourdieu 1968: 610–12.)

8 My reference to an object's "life" is derived largely from Igor Kopytoff's excellent essay on "The Cultural Biography of Things" (1986).

9 Herbert Baker, a noted African art dealer/collector once remarked, "As an 'explorer' seeking treasures in second-hand stores, attics and basements, I found that I could afford African and Oceanic 'curios' that looked like copies [sic] of a Picasso, Braque, Modigliani or Vlaminck" (1969: 4).

10 When the profit motive is introduced into a discussion of African art collecting, it is done with a "light-hearted realist's" attitude. "With an indulgent chuckle," writes Price, "adherents of [art as economic investment] hover somewhere between a gentle 'tsk tsk' and a recognition that 'boys will be boys'" (1989: 77). Many of the editorials in the front of *African Arts* magazine are "clearly designed to assuage, through humor, any malaise that subscribers might feel about approaching their shared passion with an eye toward material gain" (*ibid.*: 77).

11 Heinrich Heine's thoughts on the emergence of the commodity as the new fetish of the bourgeois world is also germane here (see Betz 1971: 30–36).
12 Similar examples of this logic are to be found in Price 1989: 100–7.

References

Anonymous n.d.(a) *Artisanat d'art de Côte d'Ivoire*. Abidjan: Office National du Tourisme.

n.d.(b) *Places of Interest In and Around Abidjan*. Abidjan: US Embassy.

1977 *African Art: A Selection from Two Private Collections*. Private printing.

1982 African Art. *Collector-Investor* (Chicago), May, pp. 14–17.

1983 Festival de masques à Man. *Fraternité Matin*, February 7, p. 11.

1984 African Influence. *The Washington Post*. August 12, p. 4.

1987a *Exploring West Africa* (travel itinerary). San Francisco: Gulliver's Travel.

1987b Communiqué from the Ministry of Information to the Ministry of Tourism. Archives of the Ministry of Tourism, Abidjan.

1987c Prix record pour un masque baoulé. *Jeune Afrique* 1384 (July 15): 33.

1988 *African Art: The Segy Gallery*. Bolton, Mass.: Skinner Auctioneers and Appraisers.

Acheson, James 1985 Social Organization of the Maine Lobster Market, pp. 105–32. In *Markets and Marketing*. Monographs in Economic Anthropology, No. 4, edited by Stuart Plattner. Lanham, Maryland: University Press of America.

Adams, Monni 1989 African Visual Arts From an Art Historical Perspective. *African Studies Review* 32 (2): 55–103.

Adamu, Mahdi 1978 *The Hausa Factor in West African History*. Zaria and Ibadan: Ahmadu Bello University Press and Oxford University Press of Nigeria.

Agnew, Jean-Christophe 1986 *Worlds Apart: The Market and the Theater in Anglo-American Thought, 1550–1750*. Cambridge: Cambridge University Press.

Akerlof, George A. 1970 The Market for "Lemons": Quality Uncertainty and the Market Mechanism. *The Quarterly Journal of Economics* 84 (3): 488–500.

Alexander, Jennifer, and Paul Alexander 1987 Striking a Bargain in Javanese Markets. *Man* 22 (1): 42–68.

Allison, Philip, *et al.* 1976 Fakes, Fakers, and Fakery: Authenticity in African Art. *African Arts* 9 (3): 21–31, 48–74, 92.

Alperton, Matty 1981 Decorating with Primitive Art. *Primitive Art Newsletter* 4 (3): 1–3.

Alsop, Joseph 1982 *The Rare Art Traditions: The History of Art Collecting and Its Linked Phenomena*. Princeton: Princeton University Press.

Amselle, Jean-Loup 1977 *Les Négociants de la savane: Histoire et organisation sociale des Kooroko (Mali)*. Paris: Éditions Anthropos.

Anderson, Benedict 1983 *Imagined Communities: Reflections on the Origin and Spread of Nationalism*. London and New York: Verso.

Angoulvant, Gabriel Louis 1911 *Guide du Commerce et de la Colonization à la Côte d'Ivoire*. Paris: Office Colonial.

Anquetil, Jacques 1977 *L'Artisanat créateur en Côte d'Ivoire*. Paris: Agence de Coopération Culturelle et Technique.

Appadurai, Arjun 1986 Introduction: Commodities and the Politics of Value, pp. 3–63. In *The Social Life of Things: Commodities in Cultural Perspective*, edited by Arjun Appadurai. Cambridge: Cambridge University Press.

Arnau, Frank 1961 *Three Thousand Years of Deception in Art and Antiques*. Translated by J. Maxwell Brownjohn. London: Jonathan Cape.

Babb, Florence E. 1989 *Between Field and Cooking Pot: The Political Economy of Marketwomen in Peru*. Austin: University of Texas Press.

Bailey, F. G. 1969 *Stratagems and Spoils: A Social Anthropology of Politics*. New York: Schocken Books.

Bailey, F. G. (ed.) 1971 *Gifts and Poison*. Oxford: Blackwell.

Baillet, Emile 1957 Le Rôle de la marine de commerce dans l'implantation de la France en A.O.F. *Revue Maritime* 135: 832–40.

Baker, Herbert 1969 *The Herbert Baker Collection*. New York: Museum of Primitive Art.

1973 On Appraising African and Oceanic "Primitive" Art. *Valuation* (American Society of Appraisers) December: 1–7.

Balandier, Georges 1967 *Ambiguous Africa*. Translated by Helen Weaver. New York: Pantheon Books. First published in French in 1957.

Barth, Fredrik 1956 Ecologic Relationships of Ethnic Groups in Swat, North Pakistan. *American Anthropologist* 58: 1079–89.

1966 *Models of Social Organization*. Royal Anthropological Institute Occasional Paper No. 23. London: Royal Anthropological Institute of Great Britain and Ireland.

Barth, Fredrik (ed.) 1969 *Ethnic Groups and Boundaries: The Social Organization of Culture Difference*. Boston: Little, Brown and Company.

Barthes, Roland 1982 *Mythologies*. Translated by Annette Lavers. New York: Hill and Wang. First published in French in 1957.

1984 *The Fashion System*. Translated by Matthew Ward and Richard Howard. New York: Hill and Wang. First published in French in 1967.

Bascom, William 1976 Changing African Art, pp. 303–19. In *Ethnic and Tourist Arts*, edited by Nelson H. H. Graburn. Berkeley: University of California Press.

Bates, Constance 1979 An Economic Analysis of the International Art Market. Ph.D. Dissertation. Graduate School of Business, Indiana University.

Bateson, Gregory 1951 Conventions of Communication: Where Validity Depends upon Belief, pp. 212–27. In *Communication: The Social Matrix of Psychiatry*, edited by Jurgen Reusch and Gregory Bateson. New York: W. W. Norton.

Baudrillard, Jean 1972 *Pour une critique de l'économie politique du signe*. Paris: Gallimard.

1988 *Selected Writings*. Edited and translated by Mark Poster. Stanford: Stanford University Press.

Ben-Amos, Paula 1971 Social Change in the Organization of Wood Carving in Benin City, Nigeria. Ph.D. dissertation, Department of Anthropology, Indiana University.

1976 "A la Recherche du Temps Perdu": On Being an Ebony-Carver in Benin, pp. 320-33. In *Ethnic and Tourist Arts: Cultural Expressions from the Fourth World*, edited by Nelson H. H. Graburn. Berkeley: University of California Press.

1977 Pidgin Languages and Tourist Arts. *Studies in the Anthropology of Visual Communication* 4 (2): 128–39.

1989 African Visual Arts from a Social Perspective. *African Studies Review* 32 (2): 1–54.

Benjamin, Walter 1969 The Work of Art in the Age of Mechanical Reproduction, pp. 217–51. In *Illuminations*. Translated by Harry Zohn. New York: Schocken Books. First published in German in 1955.

Bentley, G. Carter 1987 Ethnicity and Practice. *Comparative Studies in Society and History* 29 (1): 24-55.

Bentsi-Enchill, Nii K. 1988 Liberalism at a Price: An Assessment of the Troubled Ivorian Economy. *West Africa* 3704: 1428.

Berger, John 1972 *Ways of Seeing*. London: Penguin.

Betz, Albrecht 1971 *Ästhetik und Politik: Heinrich Heine's Prosa*. Munich: Carl Hanser.

Blier, Suzanne Preston 1988–89 Art Systems and Semiotics: The Question of Art, Craft, and Colonial Taxonomies in Africa. *The American Journal of Semiotics* 6 (1): 7–18.

1990 African Art Studies at the Crossroads: An American Perspective, pp. 91–110. In *African Art Studies: The State of the Discipline*. Washington, DC: National Museum of African Art.

Blumenthal, Susan 1974 The World's Best Traveled Art. *Africa Report* 19 (1): 4–10.

Bogatyrev, Petr 1937 *The Functions of Folk Costume in Moravian Slovakia*. The Hague: Mouton (1971).

Bonacich, Edna 1973 A Theory of Middleman Minorities. *American Sociological Review* 38 (5): 583–94.

Boorstin, Daniel J. 1962 *The Image: A Guide to Pseudoevents in America*. New York: Harper & Row.

Bordogna, Charles 1981 The Lure of African Art. *Collector-Investor* (Chicago), September, p. 36.

Bosch, Ellie 1985 *Les Femmes du marché de Bobo: La vie et le travail des commerçantes dans la ville de Bobo-Dioulasso au Burkina-Faso*. Leiden: Centre de Recherche et de Documentation Femmes et Développement.

Bouabré, Paul 1987a Les Petits fabricants de masques: Le commerce des "bannis". *Fraternité Matin* (Abidjan), March 12, p. 11.

1987b Festimask 1987: Le masque doit servir à la paix. *Fraternité Matin* (Abidjan), July 16, p. 8.

Boulding, K. E. 1971 The Economics of Knowledge and the Knowledge of Economics, pp. 21–36. In *Economics of Information and Knowledge*, edited by D. M. Lamberton. London: Penguin Books.

Bourdieu, Pierre 1968 Outline of a Sociological Theory of Art Perception. *International Social Science Journal* 20 (4): 589–612.

1977 *Outline of a Theory of Practice*. Translated by Richard Nice. Cambridge: Cambridge University Press. First published in French in 1972.

1980 The Production of Belief: Contribution to an Economy of Symbolic Goods. Translated by Richard Nice. *Media, Culture and Society* 2: 261–93.

1984 *Distinction: A Social Critique of the Judgement of Taste*. Translated by Richard Nice. Cambridge, Mass.: Harvard University Press. First published in French in 1979.

Bourdieu, Pierre, and Alain Derbel, with Dominique Schnapper 1990 *The Love of Art: European Art Museums and their Public*. Translated by Caroline Beattie and Nick Merriman. Stanford: Stanford University Press. First published in French in 1969.

Bouret, Jean 1970 Une amitié esthétique au début du siècle, Apollinaire et Paul Guillaume (1911–1918), d'après une correspondance inédite. *Gazette des Beaux-Arts* 76: 373–99.

Boutillier, Jean-Louis, Michèle Fiéloux, and Jean-Louis Ormières 1978 Le Tourisme en Afrique de l'ouest, pp. 5–83. In *Le Tourisme en Afrique de l'ouest*, edited by Jean-Louis Boutillier, Jean Copans, Michèle Fiéloux, Suzanne Lallemand, and Jean-Louis Ormières. Paris: François Maspero.

Braudel, Fernand 1973 *Capitalism and Material Life, 1400–1800*. Translated by Miriam Kochan. New York: Harper & Row. First published in French in 1967.

1982 *The Wheels of Commerce*. Translated by Sian Reynolds. New York: Harper.

Bravmann, René A. 1973 *Open Frontiers: The Mobility of Art in Black Africa*. Seattle: University of Washington Press.

1974 *Islam and Tribal Art in West Africa*. Cambridge: Cambridge University Press.

Briggs, Jean 1971 Strategies of Perception: The Management of Ethnic Identity, pp. 55–73. In *Patrons and Brokers in the East Arctic*, edited by Robert Paine. Newfoundland Social and Economic Papers, no. 2. St. John's: Memorial University.

Brooke, James 1988 Ivory Coast Gambles to Prop up Cocoa Prices. *The New York Times*, November 21, section D, p. 10.

Brown, Patricia Leigh 1990 Lauren's Wink at the Wild Side. *The New York Times*, February 8, section C, pp. 1, 3.

Bunker, Stephen G. 1985 *Underdeveloping the Amazon: Extraction, Unequal Exchange, and the Failure of the Modern State*. Urbana: University of Illinois Press.

Bunn, James H. 1980 The Aesthetics of British Mercantilism. *New Literary History* 11 (2): 303-21.

Carpenter, Edmund 1975 Collecting Northwest Coast Art, pp. 9–49. In *Indian Art of the Northwest Coast*, edited by Bill Holm and Bill Reid. Seattle: University of Washington Press.

Cassady, Ralph Jr. 1968 Negotiated Price-Making in Mexican Traditional Markets: A Conceptual Analysis. *America Indigena* 28 (1): 51–79.

Challenor, Herschelle Sullivan 1979 Strangers as Colonial Intermediaries: The Dahomeyans in Francophone Africa, pp. 67–103. In *Strangers in African Societies*, edited by William A. Shack and Elliott P. Skinner. Berkeley: University of California Press.

Chapman, Anne 1980 Barter as a Universal Mode of Exchange. *L'Homme* 20 (3): 33–83.

Chauvin, Françoise 1987 Les Gentils petits colons. *Paris Match*, May 8, pp. 84–85.
Clifford, James 1985 Histories of the Tribal and the Modern. *Art in America* 73 (4): 164–77, 215. Reprinted in James Clifford, *The Predicament of Culture*, pp. 189–214. Cambridge, Mass.: Harvard University Press.
1987 Of Other Peoples: Beyond the "Salvage Paradigm", pp. 121–30. In *Discussions in Contemporary Culture*, no. 1, edited by Hal Foster. Seattle: Bay Press.
1988 On Collecting Art and Culture. In *The Predicament of Culture: Twentieth-Century Ethnography, Literature, and Art*, pp. 215–51. Cambridge, Mass.: Harvard University Press.
Cohen, Abner 1965 The Social Organization of Credit in a West African Cattle Market. *Africa* 35: 8–20.
1969 *Custom and Politics in Urban Africa: A Study of Hausa Migrants in Yoruba Towns*. Berkeley: University of California Press.
1971 Cultural Strategies in the Organization of Trading Diasporas, pp. 266–84. In *The Development of Indigenous Trade and Markets in West Africa*, edited by Claude Meillassoux. Oxford: Oxford University Press.
Cohen, A. P., and J. L. Comaroff 1976 The Management of Meaning: On the Phenomenology of Political Transactions, pp. 87–107. In *Transaction and Meaning: Directions in the Anthropology of Exchange and Symbolic Behavior*, edited by Bruce Kapferer. Philadelphia: Institute for the Study of Human Issues.
Cohen, William B. 1980 *The French Encounter with Africans: White Response to Black, 1530–1880*. Bloomington: Indiana University Press.
Coke, Desmond 1928 *Confessions of an Incurable Collector*. London: Chapman and Hall.
Cole, Herbert M. 1975 Artistic and Communicative Values of Beads in Kenya and Ghana. *The Bead Journal* 1 (3): 29–37.
Cole, Herbert M., and Doran H. Ross 1977 *The Arts of Ghana*. Los Angeles: Museum of Cultural History, University of California.
Colvin, Lucie G. 1971 The Commerce of Hausaland, 1780–1833, pp. 101–36. In *Aspects of West African Islam*, edited by Daniel F. McCall and Norman R. Bennett. Boston University Papers on Africa, no. 5. Boston: African Studies Center.
Comaroff, John L. 1987 Of Totemism and Ethnicity. *Ethos* 52 (4): 301–23.
Conway, Sir Martin 1914 *The Sport of Collecting*. London: Fisher Unwin.
Copans, Jean 1978 Idéologies et idéologues du tourisme au Sénégal: Fabrications et contenus d'une image de marque, pp. 108–40. In *Le Tourisme en Afrique de l'ouest*, edited by Jean-Louis Boutillier, Jean Copans, Michèle Fiéloux, Suzanne Lallemand, and Jean-Louis Ormières. Paris: François Maspero.
Coquery-Vidrovitch, Catherine, and Paul E. Lovejoy (eds.) 1985 *The Workers of African Trade*. Beverly Hills: Sage Publications.
Crowley, Daniel J. 1970 The Contemporary-Traditional Art Market in Africa. *African Arts* 4 (1): 43–49.
1974 The West African Art Market Revisited. *African Arts* 7 (4): 54–59.
1979 The Art Market in Cameroon and the Central African Empire. *African Arts* 12 (3): 74–75.
Culler, Jonathan 1981 Semiotics of Tourism. *American Journal of Semiotics* 1 (1–2): 127–40.

Curtin, Philip D. 1975 *Economic Change in Precolonial Africa: Senegambia in the Era of the Slave Trade*. Madison: University of Wisconsin Press.

1984 *Cross-Cultural Trade in World History*. Cambridge: Cambridge University Press.

Danto, Arthur 1981 *Transfiguration of the Commonplace*. Cambridge, Mass.: Harvard University Press.

Dark, Philip J. C. 1967 The Study of Ethno-Aesthetics: The Visual Arts, pp. 131–48. In *Essays on the Verbal and Visual Arts: Proceedings of the 1966 Annual Spring Meeting of the American Ethnological Society*, edited by June Helm. Seattle: University of Washington Press.

Davis, Whitney 1989 Review of *Object and Intellect: Interpretations of Meaning in African Art*, special issue of *Art Journal* 47 (2), 1988, edited by Henry John Drewal. In *African Arts* 22 (4): 24–32, 85.

De Coppet, Laura, and Alan Jones 1984 *The Art Dealers*. New York: Clarkson N. Potter.

Delcourt, Jean Paul, and Giovanni Franco Scanzi 1987 *Potomo Waka*. Milan: Editions Lediberg.

Den Tuinder, Bastiaan A. 1978 *Ivory Coast: The Challenge of Success*. Baltimore: The Johns Hopkins University for the World Bank.

Djidji, Ambroise 1983 Réflexion sur le festival des masques. *Fraternité Matin*. February 22, p. 10.

Donne, J. B. 1978 African Art and Paris Studios, 1905–20, pp. 105–20. In *Art in Society*, edited by Michael Greenhalgh and Vincent Megaw. New York: St. Martin's Press.

Dozon, Jean-Pierre 1985 Les Bété: Une création coloniale, pp. 49–85. In *Au coeur de l'ethnie: Ethnies, tribalisme et état en Afrique*, edited by Jean-Loup Amselle and Elikia M'Bokolo. Paris: Editions La Decouverte.

Drewal, Henry John 1990 African Art Studies Today, pp. 29–62. In *African Art Studies: The State of the Discipline*. Washington, DC: National Museum of African Art.

Dubin, Lois Sherr 1981 *The History of Beads*. New York: Harry N. Abrams.

Durkheim, Emile 1933 *The Division of Labor in Society*. Translated by George Simpson. New York: The Free Press. First published in French in 1893.

Echenberg, Myron 1991 *Colonial Conscripts: The* Tirailleurs Sénégalais *in French West Africa, 1857–1960*. Portsmouth, NH: Heinemann Educational Books.

Elkan, Walter 1958 The East African Trade in Woodcarving. *Africa* 33 (4): 314–23.

Errington, Shelly 1989 Fragile Traditions and Contested Meanings. *Public Culture* 1 (2): 49–59.

forthcoming Artifacts into Art. In *Collections and Culture: Museums and the Development of American Life and Thought*, edited by Michael Lacey and Sally Kohlstedt.

Etienne-Nugue, Jocelyne, and Elisabeth Laget 1985 *Artisanats traditionnels: Côte d'Ivoire*. Dakar: Institut Culturel Africain.

Evans-Pritchard, E. E. 1940 *The Nuer*. Oxford: Clarendon Press.

Fabian, Johannes 1983 *Time and the Other: How Anthropology Makes its Object*. New York: Columbia University Press.

Fagg, William B. 1965 *Tribes and Forms in African Art*. New York: Tudor.

Fanselow, Frank S. 1990 The Bazaar Economy, Or How Bizarre is the Bazaar Really? *Man* 25 (2): 250–65.

Faris, James C. 1988 "ART/artifact": On the Museum and Anthropology. *Current Anthropology* 29 (5): 775–79.

Fischer, Eberhard, and Hans Himmelheber 1984 *The Arts of the Dan in West Africa.* Translated by Anne Bundle. Zurich: Rietberg Museum. First published in German in 1976.

Forde, Daryl, and Samir Amin 1978 *Drought and Migration in the Sahel.* New York: Oxford University Press.

Foster, Brian L. 1974 Ethnicity and Commerce. *American Ethnologist* 1 (3): 437–48.

Francis, Peter Jr. 1979 *The Story of Venetian Beads.* The World of Beads Monograph Series, vol. 1. New York: Lapis Route Books.

Frank, Barbara E. 1987 Open Borders: Style and Ethnic Identity. *African Arts* 20 (4): 48–55.

Galerie des Arts 1985, vol. 227.

Gambetta, Diego, ed. 1988 *Trust: Making and Breaking Cooperative Relations.* New York and Oxford: Basil Blackwell.

Garnham, Nicholas, and Raymond Williams 1980 Pierre Bourdieu and the Sociology of Culture. *Media, Culture and Society* 2 (3): 209–23.

Garrard, Timothy F. 1983 Akan Pseudo-Weights of European Origin, pp. 70–81. In *Akan Transformations: Problems in Ghanaian Art History*, edited by Doran H. Ross and Timothy F. Garrard. Los Angeles: Museum of Cultural History.

Gast, Dwight V. 1988 Pricing New York Galleries. *Art in America* (July): 86–87.

Gaudio, Attilio, and Patrick Van Roekeghem 1984 Le Charme surranné des "sondja" ou statuette "colon", pp. 197–99. In *Étonnante Côte d'Ivoire.* Paris: Éditions Karthala.

Gee, Malcolm 1981 *Dealers, Critics, and Collectors of Modern Painting: Aspects of the Parisian Art Market Between 1910 and 1930.* New York: Garland Publishing.

Geertz, Clifford 1963 *Peddlers and Princes: Social Development and Economic Change in Two Indonesian Towns.* Chicago: University of Chicago Press.

1979 Suq: The Bazaar Economy in Sefrou, pp. 123–313. In *Meaning and Order in Moroccan Society*, edited by Clifford Geertz, Hildred Geertz, and Lawrence Rosen. Cambridge: Cambridge University Press.

Gellner, Ernest 1983 *Nations and Nationalism.* Ithaca: Cornell University Press.

Gerbrands, Adrian A. 1990 The History of African Art Studies, pp. 11-28. In *African Art Studies: The State of the Discipline.* Washington, DC: National Museum of African Art, Smithsonian Institution.

Gnangnan, Desiré 1987 Festivale de masque 1987. *Fraternité Matin*, May 4, p. 10.

Gnobo, J. Z. 1976 Le Rôle des femmes dans le commerce pré-colonial à Daloa. *Godo-Godo* (Abidjan) 2: 79–105.

Goffman, Erving 1959 *The Presentation of Self in Everyday Life.* Garden City, NY: Anchor Books.

Goldsmith, Oliver 1891 *The Citizen of the World.* London: J. M. Dent and Company.

Gorer, Geoffrey 1935 *Africa Dances.* New York: Alfred A. Knopf.

Graburn, Nelson H. H. 1976 Introduction: Arts of the Fourth World, pp. 1–32. In *Ethnic and Tourist Arts: Cultural Expressions from the Fourth World*, edited by Nelson H. H. Graburn. Berkeley: University of California Press.

1984 The Evolution of Tourist Arts. *Annals of Tourism Research* 11: 393–419.
Graburn, Nelson H. H. (ed.) 1976 *Ethnic and Tourist Arts: Cultural Expressions from the Fourth World*. Berkeley: University of California Press.
Grampp, William D. 1989 *Pricing the Priceless: Art, Artists, and Economists*. New York: Basic Books.
Greenblatt, Stephen 1990 Resonance and Wonder. *Bulletin of the American Academy of Arts and Sciences* 43 (4): 11–34.
Greenwald, Jeff 1990 *Shopping for Buddhas*. San Francisco: Harper & Row.
Grégoire, Emmanuel 1992 *The Alhazai of Maradi: Traditional Hausa Merchants in a Changing Sahelian City*. Translated by Benjamin H. Hardy. Boulder and London: Lynne Rienner Publishers. First published in French in 1986.
Griaule, Marcel 1965 *Conversations with Ogotemmêli*. London: Oxford University Press for the International African Institute. First published in French in 1948.
Griggs, Anthony 1974 Dealing in African Art. *Race Relations Reporter* 5: 17–21.
Guido (Abidjan) 1987 Ces drôles de colons. No. 231: 42–46.
1988a Mary-Laure Grange Césaréo: A la recherche d'une esthétique. No. 254: 64–67.
1988b Statues colons. No. 255: 74–75.
Guillaume, Paul, and Thomas Munro 1926 *Primitive Negro Sculpture*. New York: Harcourt, Brace & Co.
Handler, Richard 1986 Authenticity. *Anthropology Today* 2 (1): 2-4.
1988 *Nationalism and the Politics of Culture in Quebec*. Madison: University of Wisconsin Press.
Haraway, Donna 1984/85 Teddy Bear Patriarchy: Taxidermy in the Garden of Eden, New York City, 1908–1936. *Social Text* 11: 19–64. Reprinted in Donna Haraway, *Primate Vision: Gender, Race, and Nature in the World of Modern Science*, pp. 26–58. New York: Routledge, 1989.
Hart, Keith 1988 Kinship, Contract, and Trust: The Economic Organization of Migrants in an African City Slum, pp. 176–93. In *Trust: Making and Breaking Cooperative Relations*, edited by Diego Gambetta. New York and Oxford: Basil Blackwell.
Harvey, David 1989 *The Condition of Postmodernity*. Oxford: Basil Blackwell.
Hauser, Arnold 1982 *The Sociology of Art*. Translated by Kenneth J. Northcott. Chicago: University of Chicago Press. First published in German in 1974.
Hersey, Irwin 1979 The Times They Are A-Changing. *Primitive Art Newsletter* 2 (12): 1–2, 5–6.
1982 The African Art Market. Lecture delivered at the National Museum of African Art, Washington, DC, April 21. Notes compiled by Janet L. Stanley.
Herskovits, Melville J. 1930 The Culture Areas of Africa. *Africa* 3: 59–77.
Herzfeld, Michael 1987 *Anthropology Through the Looking-Glass: Critical Ethnography in the Margins of Europe*. Cambridge: Cambridge University Press.
Hill, Polly 1966 Landlords and Brokers: A West African Trading System. *Cahiers d'Etudes Africaines* 6: 349–66.
Himmelheber, Hans 1966 The Present Status of Sculptural Art Among the Tribes of the Ivory Coast, pp. 192–97. In *Essays on the Verbal and Visual Arts: Proceedings of the 1966 Annual Spring Meeting of the American Ethnological Society*, edited by June Helm. Seattle and London: University of Washington Press.

1975 Afrikanische Kunsthandler an der Elfenbeinkuste. *Zeitschrift für Ethnologie* 100 (1): 16–26.

Hiskett, Mervyn 1984 *The Development of Islam in West Africa.* London and New York: Longman.

Holas, Bohumil 1960 *Cultures Matérielles de la Côte d'Ivoire.* Paris: Presses Universitaires de France.

1966 *Arts traditionnels de la Côte d'Ivoire.* Paris: Presses Universitaires de France.

Hopkins, A. G. 1973 *An Economic History of West Africa.* New York: Columbia University Press.

Imperato, Pascal J. 1976 The Whims of Termites. *African Arts* 9 (3): 72–73.

Jahn, Jens (ed.) 1983 *Colon: Das schwarze Bild vom weiben Mann.* Munich: Rogner and Bernhard.

Jamin, Jean 1982 Objets trouvés des paradis perdues: A propos de la Mission Dakar-Djibouti, pp. 69–100. In *Collections passion*, edited by Jacques Hainard and Roland Kaehr. Neuchâtel: Musée d'Ethnographie.

1985 Les Objets ethnographiques sont-ils des choses perdues? pp. 51–74. In *Temps perdu, temps retrouvé: Voir les choses du passé au présent*, edited by Jacques Hainard and Roland Kaehr. Neuchâtel: Musée d'Ethnographie.

Johnson, Marion 1978 By Ship or By Camel: The Struggle for the Cameroons Ivory Trade in the Nineteenth Century. *Journal of African History* 19 (4): 539–49.

Jules-Rosette, Bennetta 1984 *The Messages of Tourist Art: An African Semiotic System in Comparative Perspective.* New York: Plenum Press.

1990 Simulations of Postmodernity: Images of Technology in African Tourist and Popular Art. *Visual Anthropology Review* 6 (1): 29–37.

Kaeppler, Adrienne L. 1978 *"Artificial Curiosities": An Exposition of Native Manufactures Collected on the Three Pacific Voyages of Captain James Cook, R.N.* Honolulu: Bishop Museum Press.

Kamer, Henri 1974 De l'authenticité des sculptures africaines/The Authenticity of African Sculptures. *Arts d'Afrique Noire* 12: 17–40.

Karp, Ivan, and Kent Maynard 1983 Reading *The Nuer. Current Anthropology* 24: 481–503.

Kasfir, Sidney Littlefield 1984 One Tribe, One Style: Paradigms in the Historiography of African Art. *History in Africa* 11: 163–93.

1992 African Art and Authenticity: A Text With a Shadow. *African Arts* 25 (3): 40–53, 96–97.

Katzin, Margaret Fisher 1960 The Business of Higglering in Jamaica. *Social and Economic Studies* 9 (3): 297-331.

Khuri, Fuad I. 1968 The Etiquette of Bargaining in the Middle East. *American Anthropologist* 70: 698–706.

Kobé, A., and O. Abiven 1923 *Dictionnaire Volof-Français.* Dakar: Mission Catholique.

Kopytoff, Igor 1986 The Cultural Biography of Things, pp. 64–91. In *The Social Life of Things: Commodities in Cultural Perspective*, edited by Arjun Appadurai. Cambridge: Cambridge University Press.

Kuper, Hilda 1973 Costume and Identity. *Comparative Studies in Society and History* 15 (3): 348–67.

Kuspit, Donald B. 1990 The Psychoanalyst-Collector's Relation to Art Objects. Lecture delivered at the Arthur M. Sackler Museum, Harvard University. May 24.

Lakpé, Raphael 1987 Interview avec Pierre Poulou (Commissaire Général du Festimask). *I.D.: Le Magazine de Côte d'Ivoire*, no. 857, July 12.

Laude, Jean 1971 *The Arts of Black Africa.* Translated by Jean Decock. Berkeley: University of California Press. First published in French in 1966.

Launay, Robert 1982 *Traders without Trade: Responses to Change in Two Dyula Communities.* Cambridge: Cambridge University Press.

Le Pape, Marc 1985 De l'espace et des races à Abidjan, entre 1903 et 1934. *Cahiers d'Études Africaines* 25 (3): 295–307.

Leach, Edmund R. 1954 *Political Systems of Highland Burma.* Boston: Beacon Press.

Lehuard, Raoul 1976 Sacred and Profane. *African Arts* 9 (3): 73–74.

1977 Un Voyage en Côte d'Ivoire. *Arts d'Afrique Noire* 23: 26–33.

1979 Un Voyage en Haute-Volta. *Arts d'Afrique Noire* 31: 11–19.

Lemann, Nicholas 1987 Fake Masks. *The Atlantic* 260 (5): 24-38.

Leuzinger, Elsy 1960 *Africa: The Art of the Negro People.* New York: McGraw-Hill.

Levine, Donald N. 1979 Simmel at a Distance: On the History and Systematics of the Sociology of the Stranger, pp. 21–36. In *Strangers in African Societies*, edited by William A. Shack and Eliott P. Skinner. Berkeley: University of California Press.

Lewis, Barbara 1971 The Dioula in the Ivory Coast, pp. 273–307. In *Papers on the Manding*, edited by Carleton T. Hodge. Bloomington: Indiana University Press.

1976 The Limitations of Group Action Among Entrepreneurs: The Market Women of Abidjan, Ivory Coast, pp. 135–56. In *Women in Africa: Studies in Social and Economic Change*, edited by Nancy J. Hafkin and Edna G. Bay. Stanford: Stanford University Press.

Lips, Julius E. 1937 *The Savage Hits Back.* Reissued 1966. New Hyde Park, NY: University Books.

Liu, Robert K. 1975 Editorial. *The Bead Journal* 2 (2): 1.

Lovejoy, Paul E. 1980 *Caravans of Kola: The Hausa Kola Trade, 1700–1900.* Zaria and Ibadan: Ahmadu Bello University Press and Oxford University Press.

Lowenthal, David 1985 *The Past is a Foreign Country.* Cambridge: Cambridge University Press.

MacCannell, Dean 1976 *The Tourist: A New Theory of the Leisure Class.* New York: Schocken Books.

1984 Reconstructed Ethnicity: Tourism and Cultural Identity in Third World Communities. *Annals of Tourism Research* 11: 375–91.

MacClancy, Jeremy 1988 A Natural Curiosity: The British Market in Primitive Art. *RES: Anthropology and Aesthetics* 15: 164–76.

McLeod, M. D. 1976 Limitations of the Genuine. *African Arts* 9 (3): 31, 48-51.

Mailfert, André 1968 *Au pays des antiquaires: confidences d'un "maquilleur" professionel.* Paris: Flammarion.

Malinowski, Bronislaw 1927 The Life of Culture, pp. 26–46. In *Culture: The Diffusion Controversy*, edited by Grafton Elliot Smith. New York: W. W. Norton.

Malinowski, Bronislaw and Julio de la Fuente 1985 *Malinowski in Mexico: The Economics of a Mexican Market System.* Edited by Susan Drucker-Brown. London: Routledge & Kegan Paul. First published in 1957.

Manning, Patrick 1985 Primitive Art and Modern Times. *Radical History Review* 33: 165–81.

1988 *Francophone Sub-Saharan Africa, 1880–1985*. Cambridge: Cambridge University Press.

Marcus, George E., and Michael M. J. Fischer 1986 *Anthropology as Cultural Critique*. Chicago: University of Chicago Press.

Marx, Karl 1906 *Capital*. Translated by Ernest Untermann. New York: The Modern Library. First published in German in 1867.

Maurer, Evan 1981 Caveat Ethnos: Unmasking Frauds in Ethnographic Art. *National Arts Guide* 3 (2): 22–25.

Maybury-Lewis, David (ed.) 1984 *The Prospects for Plural Society*. 1982 Proceedings of the American Ethnological Society. Washington, DC: American Ethnological Society.

Mazrui, Ali 1970 The Robes of Rebellion: Sex, Dress and Politics in Africa. *Encounter* 34 (2): 19–30. Reprinted in *The Body Reader: Social Aspects of the Human Body*, pp. 196–217, edited by Ted Polhemus. New York: Pantheon Books, 1978.

Melikian, Souren 1987 African Sculpture From the Colonial Era. *International Herald Tribune*, October 31, p. 22.

Meyer, Piet 1981 *Kunst und Religion der Lobi*. Zürich: Museum Rietberg.

Miele, Frank J. 1988 *Source and Inspiration*. New York: Hirschl and Adler Folk.

Mintz, Sidney W. 1961 *Pratik*: Haitian Personal Economic Relationships. *Proceedings of the 1961 Annual Spring Meeting of the American Ethnological Society*, edited by Viola E. Garfield. Seattle: University of Washington Press.

1964a The Employment of Capital by Market Women in Haiti, pp. 256–86. In *Capital, Saving and Credit in Peasant Societies*, edited by Raymond Firth and B. S. Yamey. Chicago: Aldine.

1964b *Peasant Market Places and Economic Development in Latin America*. The Graduate Center for Latin American Studies, Vanderbilt University, Occasional Paper No. 4.

1971 Men, Women, and Trade. *Comparative Studies in Society and History* 13 (3): 247–69.

1985 *Sweetness and Power: The Place of Sugar in Modern History*. New York: Viking.

Miracle, Marvin P., and Bruce Fetter 1970 Backward-Sloping Labor-Supply Functions and African Economic Behavior. *Economic Development and Cultural Change* 18 (2): 240–51.

Monson, Terry D., and Garry G. Pursell 1979 The Use of DRC's to Evaluate Indigenization Programs: The Case of the Ivory Coast. *Journal of Development Economics* 6 (1): 119–39.

Montias, John Michael 1988 Art Dealers in the Seventeenth-Century Netherlands. *Simiolus* 18 (4): 244–53.

Moore, Sally Falk 1978 *Law as Process: An Anthropological Approach*. London: Routledge & Kegan Paul.

1989 The Production of Cultural Pluralism as a Process. *Public Culture* 1 (2): 26–48.

Moore, Sally Falk, and Barbara G. Myerhoff 1977 "Introduction," pp. 3–23. In *Secular Ritual*, edited by Sally Falk Moore and Barbara G. Myerhoff. Assen, The Netherlands: Van Gorcum.

Morson, Gary Saul 1989 Parody, History, and Metaparody, pp. 63–86. In *Rethinking Bakhtin*, edited by Gary Saul Morson and Caryl Emerson. Evanston: Northwestern University Press.

Moulin, Raymonde 1987 *The French Art Market: A Sociological View*. Translated by Arthur Goldhammer. New Brunswick: Rutgers University Press. First published in French in 1967.

Mudimbe, V. Y. 1988 *The Invention of Africa*. Bloomington: Indiana University Press.

Mundt, Robert J. 1987 *Historical Dictionary of the Ivory Coast (Côte d'Ivoire)*. African Historical Dictionaries, no. 41. Metuchen, NJ and London: Scarecrow Press.

Nash, Dennison 1977 Tourism as a Form of Imperialism, pp. 33–47. In *Hosts and Guests: The Anthropology of Tourism*, edited by Valene L. Smith. Philadelphia: University of Pennsylvania Press.

Nedelec, Michel 1974 *Equipements touristiques et recreatifs dans la région d'Abidjan*. Abidjan: Centre de Recherches Architecturales et Urbaines.

Northern, Tamara 1986 *Expressions of Cameroon Art: The Franklin Collection*. Beverly Hills, CA: Rembrandt Press.

Paine, Robert 1971 A Theory of Patronage and Brokerage, pp. 3–28. *Patrons and Brokers in the East Arctic*, edited by Robert Paine. Newfoundland Social and Economic Papers, no. 2. St. John's: Memorial University.

1974 *Second Thoughts About Barth's Models*. London: Royal Anthropological Institute of Great Britain and Ireland.

1976 Two Modes of Exchange and Mediation, pp. 63–86. In *Transaction and Meaning: Directions in the Anthropology of Exchange and Symbolic Behavior*, edited by Bruce Kapferer. Philadelphia: Institute for the Study of Human Issues.

Paudrat, Jean-Louis 1984 The Arrival of Tribal Objects in the West from Africa, pp. 125–75. Translated by John Shepley. In *"Primitivism" in 20th-Century Art: Affinity of the Tribal and the Modern*, edited by William Rubin. New York: The Museum of Modern Art.

Paulme, Denise 1962 *African Sculpture*. Translated by Michael Ross. London: Elek Books. First published in French in 1956.

Perinbam, B. Marie 1977 *Homo Africanus*: *Antiquus* or *Oeconomicus*? Some Interpretations of African Economic History. *Comparative Studies in Society and History* 19 (2): 156–78.

1980 The Julas in Western Sudanese History: Long-Distance Traders and Developers of Resources, pp. 455–75. In *West African Culture Dynamics: Archaeological and Historical Perspectives*, edited by B. K. Swartz, Jr. and Raymond E. Dumett. The Hague: Mouton Publishers.

Philmon, Thierry O. 1982 Conference du Ministre Duon Sadia. *Fraternité Matin*. December 2, pp. 13-16.

Picard, John and Ruth 1986 *Tabular Beads from the West African Trade*. Carmel: Picard African Imports.

Pitt-Rivers, A. H. 1875 Principles of Classification. *Journal of the Anthropological Institute* 4: 293–308.

Plattner, Stuart 1985 Equilibrating Market Relationships, pp. 133–52. In *Markets and Marketing*, edited by Stuart Plattner. Monographs in Economic Anthropology, no. 4. Lanham, Maryland: University Press of America.

Polanyi, Karl 1977 *The Livelihood of Man*. Edited by H. W. Pearson. New York: Academic Press.

Price, Sally 1989 *Primitive Art in Civilized Places*. Chicago: University of Chicago Press.

Provost, Carl K. 1980 The Valuation of Traditional Art: Special Problems in Connoisseurship. *Valuation* (American Society of Appraisers), November: 136–49.

Ranger, Terence 1983 The Invention of Tradition in Colonial Africa, pp. 211–62. In *The Invention of Tradition*, edited by Eric Hobsbawm and Terence Ranger. Cambridge: Cambridge University Press.

Ravenhill, Philip L. 1980a Art. Ivory Coast Supplement to the *Financial Times* (London), December 9, p. 19.

1980b *Baule Statuary Art: Meaning and Modernization*. Working Papers in the Traditional Arts, vol. 5. Philadelphia: Institute for the Study of Human Issues.

1988 The Passive Object and the Tribal Paradigm: Colonial Museography in French West Africa. Paper presented at the Workshop on African Material Culture in Bellagio, Italy.

Rees, A. 1971 Information Networks in Labour Markets, pp. 109–18. In *Economics of Information and Knowledge*, edited by D. M. Lamberton. London: Penguin Books.

Reif, Rita 1988 It May Look Like a Hoe, but It's Really Money. *The New York Times*, October 30, section H, p. 37.

1990 Auctions. *The New York Times*, February 9, p. 32.

Rémy, Mylène 1976 *La Côte d'Ivoire aujourd'hui*. Paris: Éditions J.A.

Revel, Jean-François 1966 Paul Guillaume par lui-même. *L'Oeil* 135: 35–40, 70.

Rheims, Maurice 1961 *The Strange Life of Objects: Thirty Five Centuries of Art Collecting and Collectors*. Translated by David Pryce-Jones. New York: Atheneum. First published in French in 1959.

Richter, Dolores 1980 *Art, Economics and Change: The Kulebele of Northern Ivory Coast*. La Jolla: Psych/Graphic Publishers.

Robinson, Blake W. 1975 The African Art Trade in Monrovia. *Liberian Studies Journal* 6 (1): 73–79.

Ross, Doran H., and Raphael X. Reichert 1983 Modern Antiquities: A Study of a Kumase Workshop, pp. 82–91. In *Akan Transformations: Problems in Ghanaian Art History*, edited by Doran H. Ross and Timothy F. Garrard. Los Angeles: Museum of Cultural History.

Roy, Joseph 1987 Marché de l'art: Le blanc et le noir. *L'Express*, November 13, p. 57.

Rubin, William 1984 Modernist Primitivism: An Introduction, pp. 1–79. In *"Primitivism" in 20th-Century Art: Affinity of the Tribal and the Modern*, 2 vols. New York: The Museum of Modern Art.

Sagoff, Mark 1981 On the Aesthetic and Economic Value of Art. *British Journal of Aesthetics* 21 (4): 318–29.

Schildkrout, Enid 1978 *People of the Zongo: The Transformation of Ethnic Identities in Ghana*. Cambridge: Cambridge University Press.

Schoffel, Alain 1989 Notes on the Art Fakes Which Have Recently Appeared in the Northern Philippines. *Art Tribal: Bulletin of the Musée Barbier-Mueller* 1: 11–23.

Segy, Ladislas 1958 *African Sculpture*. New York: Dover Publications.

Shepard, John H. 1978 Ivory Coast and the Ivorian Miracle. *Crisis* 85: 240–42.
Shils, Edward 1981 *Tradition*. Chicago: University of Chicago Press.
Sieber, Roy 1976 Forgeries without Forgers. *African Arts* 9 (3): 22–24.
Sieber, Roy, and Arnold Rubin 1968 *Sculpture of Black Africa: The Paul Tishman Collection*. Los Angeles: Los Angeles County Museum of Art.
Silberman, James M. 1976 Preparing the Appraisal for Traditional or Primitive Art Properties. *Monographs* (American Society of Appraisers) 7 (3): 58–64.
Silver, Harry R. 1976 The Mind's Eye: Art and Aesthetics in an African Craft Community. Ph.D. dissertation, Department of Anthropology, Stanford University.
1979a Ethnoart. *Annual Review of Anthropology* 8: 267–307.
1979b Beauty and the "I" of the Beholder: Identity, Aesthetics, and Social Change Among the Ashanti. *Journal of Anthropological Research* 35 (2): 191–207.
1981a Carving Up the Profits: Apprenticeship and Structural Flexibility in a Contemporary African Craft Market. *American Ethnologist* 8: 41–52.
1981b Calculating Risks: The Socioeconomic Foundations of Aesthetic Innovation in an Ashanti Carving Community. *Ethnology* 20 (2): 101–14.
Silverman, Raymond A. 1983 Akan *Kuduo*: Form and Function, pp. 10–29. In *Akan Transformations: Problems in Ghanaian Art History*, edited by Doran H. Ross and Timothy F. Garrard. Los Angeles: Museum of Cultural History.
Silverman, Sydel 1965 Patronage and Community-Nation Relationships in Central Italy. *Ethnology* 4: 172–89.
Simmel, Georg 1950 *The Sociology of Georg Simmel*. Translated and edited by Kurt H. Wolff. New York: The Free Press.
1978 *The Philosophy of Money*. Translated and edited by Tom Bottomore and David Frisby. Boston: Routledge & Kegan Paul. First published in German in 1907.
Simpson, Donald 1975 *Dark Companions: The African Contribution to the European Exploration of East Africa*. New York: Barnes & Noble.
Skinner, Elliott P. 1975 Competition Within Ethnic Systems in Africa, pp. 131–57. In *Ethnicity and Resource Competition in Plural Societies*, edited by Leo A. Despres. The Hague: Mouton.
Smith, Barbara Herrnstein 1988 *Contingencies of Value: Alternative Perspectives for Critical Theory*. Cambridge, Mass.: Harvard University Press.
Smith, Charles W. 1989 *Auctions: The Social Construction of Value*. New York: Free Press.
Smith, M. G. 1962 Exchange and Marketing Among the Hausa, pp. 299–334. In *Markets in Africa*, edited by Paul Bohannan and George Dalton. Evanston: Northwestern University Press.
Sotheby's 1987 *Important Tribal Art*. May 20. New York: Sotheby's.
Spittler, Gerd 1977 Traders in Rural Hausaland. *Bulletin de l'I.F.A.N.*, series 2, 39 (2): 362–85.
Spooner, Brian 1986 Weavers and Dealers: The Authenticity of an Oriental Carpet, pp. 195–235. In *The Social Life of Things: Commodities in Cultural Perspective*, edited by Arjun Appadurai. Cambridge: Cambridge University Press.
Staatz, John 1979 *The Economics of Cattle and Meat Marketing in Ivory Coast*. Livestock Production and Marketing in the Entente States of West Africa, Monograph no. 2. Ann Arbor: Center for Research on Economic Development, University of Michigan.

Stanley, Janet L. 1987 The African Art Market: An Essay and Bibliography. *Africana Journal* 14 (2/3): 157–70.

Steiner, Christopher B. 1985 Another Image of Africa: Toward an Ethnohistory of European Cloth Marketed in West Africa, 1873–1960. *Ethnohistory* 32 (2): 91–110.

1986a Of Drums and Dancers: Convention and Reality in Portrayals of Non-Western Peoples in European Accounts of Discovery and Exploration. *The Harvard Review* 1 (1): 104–29.

1986b Interpreting African Masks: The Harley Collection at the Peabody Museum. *Symbols* (Fall): 6–10.

1990 Body Personal and Body Politic: Adornment and Leadership in Cross-Cultural Perspective. *Anthropos* 85 (4–6): 431–45.

Stewart, Susan 1984 *On Longing: Narratives of the Miniature, the Gigantic, the Souvenir, the Collection.* Baltimore: The Johns Hopkins University Press.

Swan, Jon 1989 Guaranteed Genuine. *Connoisseur.* April: 96–101.

Tambiah, Stanley Jeyaraja 1984 *The Buddhist Saints of the Forest and the Cult of Amulets.* Cambridge: Cambridge University Press.

1985 The Magical Power of Words, pp. 17–59. In *Culture, Thought, and Social Action: An Anthropological Perspective.* Cambridge, Mass.: Harvard University Press.

Tax, Sol 1953 *Penny Capitalism: A Guatemalan Indian Economy.* Institute of Social Anthropology Publication, no. 16. Washington, DC: Smithsonian Institution.

Taylor, John Russell, and Brian Brooke 1969 *The Art Dealers.* New York: Charles Scribner's Sons.

Thompson, E. P. 1966 *The Making of the English Working Class.* New York: Vintage Books.

Thompson, Michael 1979 *Rubbish Theory: The Creation and Destruction of Value.* Oxford: Oxford University Press.

Torgovnick, Marianna 1990 *Gone Primitive: Savage Intellects, Modern Lives.* Chicago: University of Chicago Press.

Touré, Abdou 1985 Les Femmes et le noms des pagnes, pp. 127–38. In *Les petits métiers à Abidjan.* Paris: Karthala.

Turcotte, Denis 1981 *La Politique linguistique en Afrique francophone: Une étude comparative de la Côte d'Ivoire et de Madagascar.* Québec: Presses de l'Université Lawal.

Uchendu, Victor C. 1967 Some Principles of Haggling in Peasant Markets. *Economic Development and Cultural Change* 16 (1): 37–50.

Ulin, Robert C. 1984 *Understanding Cultures: Perspectives in Anthropology and Social Theory.* Austin: University of Texas Press.

Van Beek, Walter E. A. 1988 Functions of Sculpture in Dogon Religion. *African Arts* 21 (4): 58–65, 91.

Van Binsbergen, Wim M. J. 1981 The Unit of Study and the Interpretation of Ethnicity. *Journal of Southern African Studies* 8 (1): 51–81.

Vansina, Jan 1984 *Art History in Africa: An Introduction to Method.* London: Longman.

Vasari, Giorgio 1927 *Lives of the Painters, Sculptors and Architects.* 4 vols. London: Dent. First published in Italian in 1551.

Visonà, Monica Blackmun 1987 The Limitations of Labels. *African Arts* 20 (4): 38.

Vogel, Carol 1988 Nothing Folksy About Folk Art Market. *The New York Times*, January 14, section C, p. 12.

Vogel, Susan 1981a Baule Female Figure, p. 73. In *For Spirits and Kings: African Art from the Paul and Ruth Tishman Collection*, edited by Susan Vogel. New York: The Metropolitan Museum of Art.

1981b Collecting African Art at the Metropolitan Museum of Art. *Quaderni Poro* 3: 75–83.

Vogel, Susan (ed.) 1988 *The Art of Collecting African Art*. New York: The Center for African Art.

1991 *Africa Explores: 20th Century African Art*. New York: The Center for African Art.

Wallerstein, Immanuel 1960 Ethnicity and National Integration in West Africa. *Cahiers d'Études Africaines* 3: 129–39.

1974 *The Modern World-System, I: Capitalist Agriculture and the Origins of European World-Economy in the Sixteenth Century*. New York: Academic Press.

Wallis, Brian 1991 Selling Nations. *Art in America* 79 (9): 84–91.

Ward, Michael 1988 Introduction. In *Symbols of Wealth: Abstractions in African Metalwork*, text by Peter Westerdijk. New York: Michael Ward Gallery.

Werewere-Liking 1987 *Statues colons*. Paris: Les Nouvelles Editions Africaines.

Werner, Ellen C. H. 1982 Field Collecting in West Africa. *Triptych* (San Francisco), March–April: 12.

Whyte, Susan Reynolds, and Sjaak Van der Geest 1988 Medicines in Context: An Introduction, pp. 3–11. In *The Context of Medicines in Developing Countries: Studies in Pharmaceutical Anthropology*. Dordrecht: Kluwer Academic Publishers.

Willett, Frank 1971 *African Art: An Introduction*. New York: Praeger.

Williams, Raymond 1982 *The Sociology of Culture*. New York: Schocken Books.

Wilsmore, S. J. 1986 Authenticity and Restoration. *British Journal of Aesthetics* 26 (3): 228–38.

Wolf, Eric R. 1956 Aspects of Group Relations in a Complex Society: Mexico. *American Anthropologist* 58: 1065–78.

1982 *Europe and the People Without History*. Berkeley: University of California Press.

Wolff, Norma H. 1985 Adugbologe's Children: Continuity and Change in a Yoruba Woodcarving Industry. Ph.D. dissertation, Department of Anthropology, Indiana University.

Wood, Robert E. 1984 Ethnic Tourism, the State, and Cultural Change in Southeast Asia. *Annals of Tourism Research* 11 (3): 353–74.

Yudelman, Montague 1964 *Africans on the Land*. Cambridge, Mass.: Harvard University Press.

Index